Lecture Notes in Mathematics

2044

Editors:
J.-M. Morel, Cachan
B. Teissier, Paris

For further volumes:
http://www.springer.com/series/304

Kendall Atkinson • Weimin Han

Spherical Harmonics and Approximations on the Unit Sphere: An Introduction

 Springer

Kendall Atkinson
University of Iowa
Department of Mathematics
and Department of Computer Science
Iowa City, IA 52242
USA

Weimin Han
University of Iowa
Department of Mathematics
Iowa City, IA 52242
USA

ISBN 978-3-642-25982-1   e-ISBN 978-3-642-25983-8
DOI 10.1007/978-3-642-25983-8
Springer Heidelberg Dordrecht London New York

Lecture Notes in Mathematics ISSN print edition: 0075-8434
                    ISSN electronic edition: 1617-9692

Library of Congress Control Number: 2012931330

Mathematics Subject Classification (2010): 41A30, 65N30, 65R20

Springer is part of Springer Science+Business Media (www.springer.com)

Preface

Spherical harmonics have been studied extensively and applied to solving a wide range of problems in the sciences and engineering. Interest in approximations and numerical methods for problems over spheres has grown steadily. These notes provide an introduction to the theory of spherical harmonics in an arbitrary dimension as well as a summarizing account of classical and recent results on some aspects of approximation by spherical polynomials and numerical integration over the sphere. The notes are intended for graduate students in the mathematical sciences and researchers who are interested in solving problems involving partial differential and integral equations on the sphere, especially on the unit sphere $\mathbb{S}^2$ in $\mathbb{R}^3$. We also discuss briefly some related work for approximation on the unit disk in $\mathbb{R}^2$, with those results being generalizable to the unit ball in more dimensions. The subject of theoretical approximation of functions on $\mathbb{S}^d$, $d > 2$, using spherical polynomials has been an active area of research over the past several decades. We summarize some of the major results, giving some insight into them; however, these notes are not intended to be a complete development of the theory of approximation of functions on $\mathbb{S}^d$ by spherical polynomials.

There are a number of other approaches to the approximation of functions on the sphere. These include spline functions on the sphere, wavelets, and meshless discretization methods using radial basis functions. For a general survey of approximation methods on the sphere, see Fasshauer and Schumaker [46]; and for a more complete development, see Freeden et al. [47]. For more recent books devoted to radial basis function methods, see Buhmann [24], Fasshauer [45], and Wendland [118]. For a recent survey of numerical integration over $\mathbb{S}^2$, see Hesse et al. [63].

During the preparation of the book, we received helpful suggestions from numerous colleagues and friends. We particularly thank Feng Dai (University of Alberta), Mahadevan Ganesh (Colorado School of Mines), Olaf Hansen

(California State University, San Marcos), and Yuan Xu (University of Oregon). We also thank the anonymous reviewers for their comments that have helped improve the final manuscript. This work was partially supported by a grant from the Simons Foundation (# 207052 to Weimin Han).

Contents

Chapter 1
Preliminaries

The study of spherical harmonics has a long history, over 200 years by now. Classical spherical harmonics on the unit sphere of three dimensional Euclidean space can be viewed as extensions of trigonometric functions on the unit circle. Originally introduced for the study of gravitational theory in the works of Laplace and Legendre in the 1780s, spherical harmonics have been studied extensively and applied to solving a wide range of problems in the natural sciences and engineering, including geosciences, neutron transport theory, astronomy, heat transfer theory, optics, atmospheric physics, oceanic physics, quantum mechanics and other areas [29, 30, 38, 43, 49, 82, 86, 116, 122]. As an example, in a number of disciplines of science and engineering, an equation of central importance is the radiative transfer equation or transport equation. The steady-state monoenergetic version of the radiative transfer equation is

$$\boldsymbol{\omega} \cdot \nabla u(\boldsymbol{x}, \boldsymbol{\omega}) + \mu_t(\boldsymbol{x}) \, u(\boldsymbol{x}, \boldsymbol{\omega}) = \mu_s(\boldsymbol{x}) \, (Su)(\boldsymbol{x}, \boldsymbol{\omega}) + f(\boldsymbol{x}, \boldsymbol{\omega}). \qquad (1.1)$$

Here $\boldsymbol{x} \in \mathbb{R}^3$ is a spatial variable, $\boldsymbol{\omega} \in \mathbb{S}^2$ is a direction variable, $\mathbb{S}^2$ being the unit sphere in $\mathbb{R}^3$, $\mu_t = \mu_a + \mu_s$. In optics, μ_a is the absorption coefficient, μ_s is the scattering coefficient, f is a light source function. The symbol S on the right side of (1.1) is an integral operator defined by

$$(Su)(\boldsymbol{x}, \boldsymbol{\omega}) = \int_{\mathbb{S}^2} \eta(\boldsymbol{x}, \boldsymbol{\omega} \cdot \hat{\boldsymbol{\omega}}) \, u(\boldsymbol{x}, \hat{\boldsymbol{\omega}}) \, d\sigma(\hat{\boldsymbol{\omega}}),$$

η being a nonnegative normalized phase function:

$$\int_{\mathbb{S}^2} \eta(\boldsymbol{x}, \boldsymbol{\omega} \cdot \hat{\boldsymbol{\omega}}) \, d\sigma(\hat{\boldsymbol{\omega}}) = 1 \quad \forall \, \boldsymbol{x}, \ \forall \, \boldsymbol{\omega} \in \mathbb{S}^2.$$

In the literature, the function η is usually assumed to be independent of $\boldsymbol{x}$, and then we write $\eta(\boldsymbol{\omega} \cdot \hat{\boldsymbol{\omega}})$ instead of $\eta(\boldsymbol{x}, \boldsymbol{\omega} \cdot \hat{\boldsymbol{\omega}})$. One well-known example is

K. Atkinson and W. Han, *Spherical Harmonics and Approximations on the Unit Sphere: An Introduction*, Lecture Notes in Mathematics 2044, DOI 10.1007/978-3-642-25983-8_1, © Springer-Verlag Berlin Heidelberg 2012

the Henyey–Greenstein phase function (cf. [62])

$$\eta(t) = \frac{1 - r^2}{4\pi(1 + r^2 - 2rt)^{3/2}}, \quad t \in [-1, 1], \tag{1.2}$$

where the parameter $r \in (-1, 1)$ is the anisotropy factor of the scattering medium. Note that $r = 0$ for isotropic scattering, $r > 0$ for forward scattering, and $r < 0$ for backward scattering.

Such issues as function approximation and numerical integration over the unit sphere arise naturally in numerically solving the radiative transfer equation (1.1). Approximation of a solution of the equation through a linear combination of spherical harmonics of an order up to N leads to the P_N methods in the literature [74]. Since the radiative transfer equation is a high dimensional problem with five independent variables, numerical methods that allow easy parallel implementation are attractive. In this regard, discontinuous Galerkin methods appear to be a good choice for the discretization of the radiative transfer equation. Some discontinuous Galerkin methods for the radiative transfer equation are studied in [57]. In recent years, inverse problems with the radiative transfer equation as the forward model have found applications in optical molecular imaging [6, 7, 19, 56].

This book intends to present the theory of the spherical harmonics on the unit sphere of a general d-dimensional Euclidean space and provides a summarizing account of function approximation and numerical integration over the unit sphere $\mathbb{S}^2$ and the related problem of approximation over the unit disk in $\mathbb{R}^2$, including recent research results. Several excellent books are available on spherical harmonics, e.g., [47, 78, 84, 85]. The presentation of the theory of the spherical harmonics is given here in a manner similar to [85], yet more easily accessible to a reader with only a basic knowledge of analysis. This is done in Chaps. 2 and 3. Function approximation over the unit sphere in $\mathbb{R}^3$ and the unit disk in $\mathbb{R}^2$ are discussed in Chap. 4. Numerical integration over the unit sphere in $\mathbb{R}^3$ and the unit disk in $\mathbb{R}^2$ is the topic of Chap. 5. The book ends with Chap. 6 on examples of spectral methods over the unit sphere and the unit disk. A boundary integral equation in $\mathbb{R}^3$ is converted to one over the unit sphere in $\mathbb{R}^3$, and a boundary value problem for a partial differential equation in $\mathbb{R}^2$ is converted to one over the unit disk in $\mathbb{R}^2$. Spectral numerical methods are then proposed and analyzed for the solution of these transformed problems. This chapter of applications ends with a spectral method for solving the Laplace–Beltrami equation over $\mathbb{S}^d$.

This preliminary chapter presents notation that will be used regularly throughout the book, as well as a brief introduction to the Γ-function that will be needed later in the study of spherical harmonics. In the final section of the chapter, we introduce some basic results related to the sphere.

1.1   Notations

We adopt the symbol ":=" for equality by definition. We will use the following sets of numbers:

$\mathbb{N}$: the set of positive integers
$\mathbb{N}_0$: the set of non-negative integers
$\mathbb{Z}$: the set of integers
$\mathbb{R}$: the set of real numbers
$\mathbb{R}_+$: the set of positive numbers
$\mathbb{C}$: the set of complex numbers

For $x \in \mathbb{R}$, $[x]$ denotes the smallest integer that is larger than or equal to x. For $m, n \in \mathbb{N}_0$, $m \geq n$, the binomial coefficient

$$\binom{m}{n} := \frac{m!}{n!\,(m-n)!}.$$

Here $m!$ is the m factorial,

$$m! := 1 \cdot 2 \cdots m \quad \text{for } m \in \mathbb{N}, \quad 0! := 1.$$

We also recall the notation of double factorial for later use,

$$m!! = \begin{cases} m\,(m-2)\,(m-4)\cdots 2, & m \text{ even}, \\ m\,(m-2)\,(m-4)\cdots 1, & m \text{ odd}, \end{cases} \qquad 0!! = 1.$$

Throughout this work, except in Chap. 5, we use $d \in \mathbb{N}$ to represent the dimension of a geometric set. The set

$$\mathbb{R}^d := \left\{ \boldsymbol{x} = (x_1, \cdots, x_d)^T : x_j \in \mathbb{R}, \ 1 \leq j \leq d \right\}$$

is the d-dimensional Euclidean space with the inner product and norm

$$(\boldsymbol{x}, \boldsymbol{y}) := \sum_{j=1}^{d} x_j y_j, \quad |\boldsymbol{x}| := (\boldsymbol{x}, \boldsymbol{x})^{1/2} \quad \text{for } \boldsymbol{x}, \boldsymbol{y} \in \mathbb{R}^d.$$

In $\mathbb{R}^d$, we use the canonical basis

$$\boldsymbol{e}_1 = (1, 0, \cdots, 0)^T, \quad \cdots, \quad \boldsymbol{e}_d = (0, \cdots, 0, 1)^T$$

and write

$$\boldsymbol{x} = \sum_{j=1}^{d} x_j \boldsymbol{e}_j$$

for $\boldsymbol{x} \in \mathbb{R}^d$. Sometimes it is helpful to show the dimension explicitly and then we will write $\boldsymbol{x}_{(d)}$ instead of $\boldsymbol{x}$. Thus,

$$\boldsymbol{x}_{(d)} = \boldsymbol{x}_{(d-1)} + x_d \boldsymbol{e}_d.$$

Here, $\boldsymbol{x}_{(d-1)} = (x_1, \ldots, x_{d-1}, 0)^T \in \mathbb{R}^d$. For convenience, we will use the symbol $\boldsymbol{x}_{(d-1)}$ to also mean the $(d-1)$-dimensional vector $(x_1, \ldots, x_{d-1})^T$. This will not cause confusion from the context. We will frequently use the unit sphere in $\mathbb{R}^d$,

$$\mathbb{S}^{d-1} := \left\{ \boldsymbol{\xi} \in \mathbb{R}^d : |\boldsymbol{\xi}| = 1 \right\}.$$

Usually we simply say the sphere to mean the unit sphere. For any $\boldsymbol{0} \neq \boldsymbol{x} \in \mathbb{R}^d$, we have $\boldsymbol{x} = |\boldsymbol{x}| \, \boldsymbol{\xi}$ with $\boldsymbol{\xi} \in \mathbb{S}^{d-1}$.

The (Euclidean) distance between two points $\boldsymbol{\xi}, \boldsymbol{\eta} \in \mathbb{S}^{d-1}$ is

$$|\boldsymbol{\xi} - \boldsymbol{\eta}| = \sqrt{2 \, (1 - \boldsymbol{\xi} \cdot \boldsymbol{\eta})}.$$

The geodesic distance between $\boldsymbol{\xi}$ and $\boldsymbol{\eta}$ on $\mathbb{S}^{d-1}$ is the angle between the two vectors:

$$\theta(\boldsymbol{\xi}, \boldsymbol{\eta}) := \arccos(\boldsymbol{\xi} \cdot \boldsymbol{\eta}) \in [0, \pi].$$

It is also the arc-length of the shortest path connecting $\boldsymbol{\xi}$ and $\boldsymbol{\eta}$. From the elementary inequalities

$$\frac{2}{\pi} t \leq \sin t \leq t, \quad t \in [0, \pi/2],$$

we can deduce the following equivalence relation between the two definitions of distance:

$$\frac{2}{\pi} \theta(\boldsymbol{\xi}, \boldsymbol{\eta}) \leq |\boldsymbol{\xi} - \boldsymbol{\eta}| \leq \theta(\boldsymbol{\xi}, \boldsymbol{\eta}), \quad \boldsymbol{\xi}, \boldsymbol{\eta} \in \mathbb{S}^{d-1}.$$

It is convenient to use the multi-index notation. A multi-index with d components is $\boldsymbol{\alpha} = (\alpha_1, \cdots, \alpha_d)$, $\alpha_1, \cdots, \alpha_d \in \mathbb{N}_0$. When we need to indicate explicitly the dependence on the dimension, we write $\boldsymbol{\alpha}_{(d)}$ instead of $\boldsymbol{\alpha}$. The length of $\boldsymbol{\alpha}$ is $|\boldsymbol{\alpha}| = \sum_{j=1}^d \alpha_j$. We write $\boldsymbol{\alpha}!$ to mean $\alpha_1! \cdots \alpha_d!$. With $\boldsymbol{x} = (x_1, \cdots, x_d)^T$ we define

$$\boldsymbol{x}^{\boldsymbol{\alpha}} := x_1^{\alpha_1} \cdots x_d^{\alpha_d}.$$

Similarly, with the gradient operator $\boldsymbol{\nabla} = (\partial_{x_1}, \cdots, \partial_{x_d})^T$, we define

$$\boldsymbol{\nabla}^{\boldsymbol{\alpha}} := \frac{\partial^{|\boldsymbol{\alpha}|}}{\partial x_1^{\alpha_1} \cdots \partial x_d^{\alpha_d}}.$$

Note in passing that the Laplacian operator

$$\Delta = \boldsymbol{\nabla} \cdot \boldsymbol{\nabla} = \sum_{j=1}^{d} \left(\frac{\partial}{\partial x_j} \right)^2 .$$

When it is necessary, we write $\Delta_{(d)}$ and $\boldsymbol{\nabla}_{(d)}$ to indicate the spatial dimension d explicitly.

1.2 The Γ-Function

The Γ-function will be frequently used throughout this book.

Definition 1.1.

$$\Gamma(x) := \int_0^\infty t^{x-1} e^{-t} dt, \quad x \in \mathbb{R}_+. \tag{1.3}$$

The following formulas can be verified [5, Chap. 1].

$$\int_0^\infty t^{x-1} e^{-at^b} dt = b^{-1} a^{-x/b} \Gamma(x/b), \quad x, a, b \in \mathbb{R}_+, \tag{1.4}$$

$$\int_0^1 |\ln t|^{x-1} dt = \Gamma(x), \quad x \in \mathbb{R}_+, \tag{1.5}$$

$$\Gamma(x+1) = x \Gamma(x), \quad x \in \mathbb{R}_+, \tag{1.6}$$

$$\Gamma^{(k)}(x) = \int_0^\infty (\ln t)^k t^{x-1} e^{-t} dt, \quad k \in \mathbb{N}_0, \ x \in \mathbb{R}_+. \tag{1.7}$$

Obviously, $\Gamma(1) = 1$. Hence, from the formula (1.6) we deduce that

$$\Gamma(n+1) = n!, \quad n \in \mathbb{N}_0. \tag{1.8}$$

In other words, the Γ-function extends the factorial operator from positive integers to positive numbers. We also have the value

$$\Gamma(1/2) = \sqrt{\pi}. \tag{1.9}$$

Using (1.6), we have

$$\Gamma\left(n + \frac{1}{2}\right) = \left(n - \frac{1}{2}\right) \left(n - \frac{3}{2}\right) \cdots \frac{1}{2} \Gamma\left(\frac{1}{2}\right)$$

$$= \frac{1 \cdot 3 \cdot 5 \cdots (2n-1)}{2^n} \Gamma\left(\frac{1}{2}\right),$$

which can be expressed as, with the help of (1.9),

$$\Gamma\left(n + \frac{1}{2}\right) = \frac{(2n-1)!!}{2^n}\sqrt{\pi} = \frac{(2n)!}{2^{2n}n!}\sqrt{\pi}. \tag{1.10}$$

Stirling's formula provides the asymptotic order of the Γ-function when its argument tends to ∞:

$$\lim_{x\to\infty} \frac{\Gamma(x)}{\sqrt{2\pi}\, x^{x-1/2}e^{-x}} = 1. \tag{1.11}$$

Choosing $x = n + 1 \in \mathbb{N}$ in (1.11), we deduce that

$$\lim_{n\to\infty} \frac{n!}{\sqrt{2\pi n}\, n^n e^{-n}} = 1.$$

We write

$$n! \sim \sqrt{2\pi n}\left(\frac{n}{e}\right)^n \quad \text{for } n \text{ sufficiently large.}$$

Pochhammer's symbol is defined as follows. Let $x \in \mathbb{R}$. Then,

$$(x)_0 := 1, \quad (x)_n := x\,(x+1)\,(x+2)\cdots(x+n-1), \; n \in \mathbb{N}.$$

It is handy to note

$$(x)_n = \frac{\Gamma(x+n)}{\Gamma(x)} \quad \text{for } x \in \mathbb{R}_+. \tag{1.12}$$

A closely related function is the beta-function,

$$B(x,y) := \int_0^1 t^{x-1}(1-t)^{y-1}dt, \quad x,y \in \mathbb{R}_+. \tag{1.13}$$

We have the following relation:

$$B(x,y) = \frac{\Gamma(x)\,\Gamma(y)}{\Gamma(x+y)}, \quad x,y \in \mathbb{R}_+. \tag{1.14}$$

1.3   Basic Results Related to the Sphere

We use dV^d for the d-dimensional volume element and dS^{d-1} for the $(d-1)$-dimensional surface element over the unit sphere $\mathbb{S}^{d-1}$. Over the surface of a general domain, we use $d\sigma$ for the surface element.

For $d \geq 3$, we write, for $\boldsymbol{\xi} = \boldsymbol{\xi}_{(d)} \in \mathbb{S}^{d-1}$,

$$\boldsymbol{\xi}_{(d)} = t\,\boldsymbol{e}_d + \sqrt{1 - t^2}\,\boldsymbol{\xi}_{(d-1)}, \quad t \in [-1, 1], \ \boldsymbol{\xi}_{(d-1)} \in \mathbb{S}^{d-1}. \tag{1.15}$$

Here and below, similar to $\boldsymbol{x}_{(d-1)}$, depending on the context, we use $\boldsymbol{\xi}_{(d-1)}$ to represent both a d-dimensional vector $(\xi_1, \ldots, \xi_{d-1}, 0)^T \in \mathbb{S}^{d-1}$ and a $(d-1)$-dimensional vector $(\xi_1, \ldots, \xi_{d-1})^T \in \mathbb{S}^{d-2}$. Then it can be shown that [85, Sect. 1] for $d \geq 3$,

$$dS^{d-1}(t\,\boldsymbol{e}_d + \sqrt{1 - t^2}\,\boldsymbol{\xi}_{(d-1)}) = (1 - t^2)^{\frac{d-3}{2}}\,dt\,dS^{d-2}(\boldsymbol{\xi}_{(d-1)}), \tag{1.16}$$

or simply,

$$dS^{d-1} = (1 - t^2)^{\frac{d-3}{2}}\,dt\,dS^{d-2}. \tag{1.17}$$

As an example, let $d = 3$. For a generic point in $\mathbb{S}^2$, in spherical coordinates,

$$\boldsymbol{\xi}_{(3)} = \begin{pmatrix} \cos\phi\,\sin\theta \\ \sin\phi\,\sin\theta \\ \cos\theta \end{pmatrix}, \quad 0 \leq \phi < 2\pi, \ 0 \leq \theta \leq \pi.$$

Denote

$$t = \cos\theta \in [-1, 1], \quad \boldsymbol{\xi}_{(2)} = \begin{pmatrix} \cos\phi \\ \sin\phi \\ 0 \end{pmatrix} \in \mathbb{S}^2.$$

Then,

$$\boldsymbol{\xi}_{(3)} = t\,\boldsymbol{e}_3 + \sqrt{1 - t^2}\,\boldsymbol{\xi}_{(2)}.$$

Furthermore, $dS^1 = d\phi$ and (1.17) takes the form $dS^2 = dt\,d\phi$.

The formula (1.17) will be applied frequently later in the book. In particular, we can use this formula to compute the surface area of the unit sphere

$$|\mathbb{S}^{d-1}| := \int_{\mathbb{S}^{d-1}} dS^{d-1}.$$

We have

$$|\mathbb{S}^{d-1}| = \int_{-1}^{1} (1 - t^2)^{\frac{d-3}{2}}\,dt \int_{\mathbb{S}^{d-2}} dS^{d-2} = |\mathbb{S}^{d-2}| \int_{-1}^{1} (1 - t^2)^{\frac{d-3}{2}}\,dt.$$

To compute the integral, we use the change of variable $s = t^2$. Then,

$$\int_{-1}^{1} (1 - t^2)^{\frac{d-3}{2}}\,dt = \int_{0}^{1} s^{-\frac{1}{2}}(1 - s)^{\frac{d-3}{2}}\,ds = B\left(\frac{1}{2}, \frac{d-1}{2}\right)$$

$$= \frac{\Gamma(\frac{1}{2})\,\Gamma(\frac{d-1}{2})}{\Gamma(\frac{d}{2})} = \frac{\sqrt{\pi}\,\Gamma(\frac{d-1}{2})}{\Gamma(\frac{d}{2})}.$$

Thus, we have the recursive relation

$$|\mathbb{S}^{d-1}| = \frac{\sqrt{\pi}\,\Gamma(\frac{d-1}{2})}{\Gamma(\frac{d}{2})}\,|\mathbb{S}^{d-2}| \text{ for } d \geq 3, \quad |\mathbb{S}^1| = 2\,\pi. \tag{1.18}$$

We derive from (1.18) the following formula

$$|\mathbb{S}^{d-1}| = \frac{2\,\pi^{\frac{d}{2}}}{\Gamma(\frac{d}{2})}. \tag{1.19}$$

This formula is proved for $d \geq 2$; for $d = 2$, $\mathbb{S}^1$ is the unit circle and $|\mathbb{S}^1| = 2\pi$ is the circumference of the unit circle. We will use the formula (1.19) also for $d = 1$ by defining $|\mathbb{S}^0| := 2$. Note that the formula (1.19) can also be stated as follows:

$$|\mathbb{S}^{2k-1}| = \frac{2\,\pi^k}{(k-1)!}, \qquad |\mathbb{S}^{2k}| = \frac{2^{k+1}\pi^k}{(2k-1)!!}, \qquad k \in \mathbb{N}.$$

We record some useful relations in computing integrals with change of variables. Their proofs can be found in [85, Sect. 1].

If $A \in \mathbb{R}^{d \times d}$ is orthogonal, then $dS^{d-1}(A\boldsymbol{\xi}) = dS^{d-1}(\boldsymbol{\xi})$, $dV^d(A\boldsymbol{x}) = dV^d(\boldsymbol{x})$.

With polar coordinates $\boldsymbol{x}_{(d)} = r\boldsymbol{\xi}_{(d)}$, $r = |\boldsymbol{x}_{(d)}|$ and $\boldsymbol{\xi}_{(d)} = \boldsymbol{\xi} \in \mathbb{S}^{d-1}$, we have $dV^d(r\boldsymbol{\xi}) = r^{d-1}dr\,dS^{d-1}(\boldsymbol{\xi})$.

We denote by $C(\mathbb{S}^{d-1})$ the space of complex-valued or real-valued continuous functions on $\mathbb{S}^{d-1}$. This is a Banach space with its canonical norm

$$\|f\|_\infty := \sup\{|f(\boldsymbol{\xi})| : \boldsymbol{\xi} \in \mathbb{S}^{d-1}\}.$$

We denote by $L^2(\mathbb{S}^{d-1})$ the space of complex-valued or real-valued squared integrable functions on $\mathbb{S}^{d-1}$. This is a Hilbert space with the canonical inner product

$$(f, g) := \int_{\mathbb{S}^{d-1}} f\,\overline{g}\,dS^{d-1}$$

and its induced norm

$$\|f\|_2 := (f, f)^{1/2}.$$

We will also use $(f, g)_{L^2(\mathbb{S}^{d-1})}$ or $(f, g)_{\mathbb{S}^{d-1}}$ for the $L^2(\mathbb{S}^{d-1})$ inner product, and use $\|\cdot\|_{L^2(\mathbb{S}^{d-1})}$ for the $L^2(\mathbb{S}^{d-1})$ norm.

We will consider the space $C(\mathbb{S}^{d-1})$ with the $L^2(\mathbb{S}^{d-1})$ inner product and norm. Note that then $C(\mathbb{S}^{d-1})$ is not complete. The closure of $C(\mathbb{S}^{d-1})$ with respect to the $\|\cdot\|_2$ norm is $L^2(\mathbb{S}^{d-1})$. In other words, given an $f \in L^2(\mathbb{S}^{d-1})$, there exists a sequence $\{f_n\} \subset C(\mathbb{S}^{d-1})$ such that

$$\|f_n - f\|_2 \to 0 \quad \text{as } n \to \infty.$$

For $f \in C(\mathbb{S}^{d-1})$, define its modulus of continuity

$$\omega(f; \delta) := \sup\{|f(\boldsymbol{\xi}) - f(\boldsymbol{\eta})| : \boldsymbol{\xi}, \boldsymbol{\eta} \in \mathbb{S}^{d-1}, |\boldsymbol{\xi} - \boldsymbol{\eta}| \le \delta\} \qquad (1.20)$$

for $\delta \in (0, 1)$. Then

$$\omega(f; \delta) \to 0 \quad \text{as } \delta \to 0.$$

For $\delta \in (0, 1)$, define the set

$$\Omega_\delta = \left\{ \boldsymbol{x} \in \mathbb{R}^d : |\boldsymbol{x}| \in [1 - \delta, 1 + \delta] \right\}. \qquad (1.21)$$

In studying a function f defined on $\mathbb{S}^{d-1}$, a useful trick is to consider its following extension

$$f^*(\boldsymbol{x}) := f\left(\frac{\boldsymbol{x}}{|\boldsymbol{x}|}\right) = f(\boldsymbol{\xi}) \text{ for } \boldsymbol{x} \in \Omega_\delta. \qquad (1.22)$$

For $k \in \mathbb{N}$, we say f is k-times continuously differentiable on $\mathbb{S}^{d-1}$ whenever f^* is k-times continuously differentiable in Ω_δ. Obviously, this definition of differentiability of f on $\mathbb{S}^{d-1}$ does not depend on the choice of $\delta \in (0, 1)$. For $k \in \mathbb{N}_0$, we define $C^k(\mathbb{S}^{d-1})$ to be the space of complex-valued or real-valued functions on $\mathbb{S}^{d-1}$ that are k-times continuously differentiable; $C^0(\mathbb{S}^{d-1}) \equiv C(\mathbb{S}^{d-1})$. We define the norm

$$\|f\|_{C^k(\mathbb{S}^{d-1})} := \|f^*\|_{C^k(\Omega_\delta)}. \qquad (1.23)$$

The right side of (1.23) does not depend on the choice of $\delta \in (0, 1)$. We will use the simplified notation $\|f\|_\infty$ for $\|f\|_{C(\mathbb{S}^{d-1})}$. Under the norm (1.23), $C^k(\mathbb{S}^{d-1})$ is a Banach space.

Chapter 2
Spherical Harmonics

This chapter presents a theory of spherical harmonics from the viewpoint of invariant linear function spaces on the sphere. It is shown that the system of spherical harmonics is the only system of invariant function spaces that is both complete and closed, and cannot be reduced further. In this chapter, the dimension $d \geq 2$. Spherical harmonics are introduced in Sect. 2.1 as the restriction to the unit sphere of harmonic homogeneous polynomials. Two very important properties of the spherical harmonics are the addition theorem and the Funk–Hecke formula, and these are discussed in Sects. 2.2 and 2.5, respectively. A projection operator into spherical harmonic function subspaces is introduced in Sect. 2.3; this operator is useful in proving various properties of the spherical harmonics. Since several polynomial spaces are used, it is convenient to include a discussion on relations of these spaces and this is done in Sect. 2.4. Legendre polynomials play an essential role in the study of the spherical harmonics. Representation formulas for Legendre polynomials are given in Sect. 2.6, whereas numerous properties of the polynomials are discussed in Sect. 2.7. Completeness of the spherical harmonics in $C(\mathbb{S}^{d-1})$ and $L^2(\mathbb{S}^{d-1})$ is the topic of Sect. 2.8, and this refers to the property that linear combinations of the spherical harmonics are dense in $C(\mathbb{S}^{d-1})$ and in $L^2(\mathbb{S}^{d-1})$. As an extension of the Legendre polynomials, the Gegenbauer polynomials are introduced in Sect. 2.9. The last two sections of the chapter, Sects. 2.10 and 2.11, are devoted to a discussion of the associated Legendre functions and their role in generating orthonormal bases for spherical harmonic function spaces.

2.1 Spherical Harmonics Through Primitive Spaces

We start with more notation. We use $\mathbb{O}^d$ for the set of all real orthogonal matrices of order d. Recall that $A \in \mathbb{R}^{d \times d}$ is orthogonal if $A^T A = I$, or alternatively, $AA^T = I$, $I = I_d$ being the identity matrix of order d.

K. Atkinson and W. Han, *Spherical Harmonics and Approximations on the Unit Sphere: An Introduction*, Lecture Notes in Mathematics 2044, DOI 10.1007/978-3-642-25983-8_2, © Springer-Verlag Berlin Heidelberg 2012

The product of two orthogonal matrices is again orthogonal. In algebra terminology, $\mathbb{O}^d$ is a group; but in this book, we will avoid using this term. It is easy to see that $\det(A) = \pm 1$ for any $A \in \mathbb{O}^d$. The subset of those matrices in $\mathbb{O}^d$ with the determinant equal to 1 is denoted as $\mathbb{SO}^d$. For any non-zero vector $\boldsymbol{\eta} \in \mathbb{R}^d$,

$$\mathbb{O}^d(\boldsymbol{\eta}) := \left\{ A \in \mathbb{O}^d : A\boldsymbol{\eta} = \boldsymbol{\eta} \right\}$$

is the subset of orthogonal matrices that leave the one-dimensional subspace $\mathrm{span}\{\boldsymbol{\eta}\} := \{\alpha \boldsymbol{\eta} : \alpha \in \mathbb{R}\}$ unchanged.

For a function $f : \mathbb{R}^d \to \mathbb{C}$ and a matrix $A \in \mathbb{R}^{d \times d}$, we define $f_A : \mathbb{R}^d \to \mathbb{C}$ by the formula

$$f_A(\boldsymbol{x}) = f(A\boldsymbol{x}) \quad \forall \boldsymbol{x} \in \mathbb{R}^d.$$

We will use this definition mainly for $A \in \mathbb{O}^d$ and for study of symmetry properties of functions.

Proposition 2.1. *If $f_A = f$ for any $A \in \mathbb{O}^d$, then $f(\boldsymbol{x})$ depends on $\boldsymbol{x}$ through $|\boldsymbol{x}|$, so that f is constant on a sphere of an arbitrary radius.*

Proof. For any two vectors $\boldsymbol{x}, \boldsymbol{y} \in \mathbb{R}^d$ with $|\boldsymbol{x}| = |\boldsymbol{y}|$, we can find a matrix $A \in \mathbb{O}^d$ such that $A\boldsymbol{x} = \boldsymbol{y}$. Thus, $f(\boldsymbol{x}) = f_A(\boldsymbol{x}) = f(\boldsymbol{y})$ and the proof is completed. $\qquad\square$

Consider the subset $\mathbb{O}^d(\boldsymbol{e}_d)$. It is easy to show that any $A \in \mathbb{O}^d(\boldsymbol{e}_d)$ is of the form

$$A = \begin{pmatrix} A_1 & \mathbf{0} \\ \mathbf{0}^T & 1 \end{pmatrix}, \quad A_1 \in \mathbb{O}^{d-1}. \tag{2.1}$$

Similar to Proposition 2.1, if $f_A = f$ for any $A \in \mathbb{O}^d(\boldsymbol{e}_d)$, then $f(\boldsymbol{x})$ depends on $\boldsymbol{x}$ through $|\boldsymbol{x}_{(d-1)}|$ and x_d.

We will introduce spherical harmonic spaces of different orders as primitive subspaces of $C(\mathbb{S}^{d-1})$. Consider a general subspace $\mathbb{V}$ of functions defined in $\mathbb{R}^d$ or over a subset of $\mathbb{R}^d$.

Definition 2.2. $\mathbb{V}$ is said to be *invariant* if $f \in \mathbb{V}$ and $A \in \mathbb{O}^d$ imply $f_A \in \mathbb{V}$.

Assume $\mathbb{V}$ is an invariant subspace of an inner product function space with the inner product $(\cdot, \cdot)$. Then $\mathbb{V}$ is said to be *reducible* if $\mathbb{V} = \mathbb{V}_1 + \mathbb{V}_2$ with $\mathbb{V}_1 \neq \emptyset$, $\mathbb{V}_2 \neq \emptyset$, both invariant, and $\mathbb{V}_1 \perp \mathbb{V}_2$. $\mathbb{V}$ is *irreducible* if it is not reducible. $\mathbb{V}$ is said to be *primitive* if it is both invariant and irreducible.

We note that $\mathbb{V}_1 \perp \mathbb{V}_2$ refers to the property that $(f, g) = 0 \; \forall f \in \mathbb{V}_1$, $\forall g \in \mathbb{V}_2$.

Definition 2.3. Given $f : \mathbb{R}^d \to \mathbb{C}$, define $\mathrm{span}\left\{ f_A : A \in \mathbb{O}^d \right\}$, the space of functions constructed through f and $\mathbb{O}^d$, to be the space of all the convergent combinations of the form $\sum_{j \geq 1} c_j f_{A_j}$ with $A_j \in \mathbb{O}^d$ and $c_j \in \mathbb{C}$.

For the above definition, it is easy to see span $\left\{ f_A : A \in \mathbb{O}^d \right\}$ is a function subspace. Moreover, if $\mathbb{V}$ is a finite dimensional primitive space, then

$$\mathbb{V} = \text{span} \left\{ f_A : A \in \mathbb{O}^d \right\} \quad \forall 0 \neq f \in \mathbb{V}.$$

2.1.1   Spaces of Homogeneous Polynomials

We start with $\mathbb{H}_n^d$, the space of all homogeneous polynomials of degree n in d dimensions. The space $\mathbb{H}_n^d$ consists of all the functions of the form

$$\sum_{|\alpha|=n} a_\alpha x^\alpha, \ a_\alpha \in \mathbb{C}.$$

As some concrete examples,

$$\mathbb{H}_2^2 = \left\{ a_1 x_1^2 + a_2 x_1 x_2 + a_3 x_2^2 : a_j \in \mathbb{C} \right\},$$
$$\mathbb{H}_2^3 = \left\{ a_1 x_1^2 + a_2 x_1 x_2 + a_3 x_1 x_3 + a_4 x_2^2 + a_5 x_2 x_3 + a_6 x_3^2 : a_j \in \mathbb{C} \right\},$$
$$\mathbb{H}_3^2 = \left\{ a_1 x_1^3 + a_2 x_1^2 x_2 + a_3 x_1 x_2^2 + a_4 x_2^3 : a_j \in \mathbb{C} \right\}.$$

It is easy to see that $\mathbb{H}_n^d$ is a finite dimensional invariant space. To determine the dimension $\dim \mathbb{H}_n^d$, we need to count the number of monomials of degree n: x^α with $\alpha_i \geq 0$ and $\alpha_1 + \cdots + \alpha_d = n$. We consider a set of $n + d - 1$ numbers: $1, 2, \ldots, n + d - 1$. Let us remove from the set $d - 1$ numbers, say $\beta_1 < \cdots < \beta_{d-1}$. Denote $\beta_0 = 0$ and $\beta_d = n + d$. Then define

$$\alpha_i = \beta_i - \beta_{i-1} - 1, \quad 1 \leq i \leq d,$$

i.e., define α_i to be the number of integers between β_{i-1} and β_i, exclusive. Note that $\sum_{i=1}^n \alpha_i = d$. This establishes a one-to-one correspondence between the set of non-negative integers $\alpha_1, \ldots, \alpha_d$ with a sum n and the set of $d - 1$ distinct positive integers $\beta_1 < \cdots < \beta_{d-1}$ between 1 and $n + d - 1$. Since the number of ways of selecting $d - 1$ different numbers from a set of $n + d - 1$ numbers is

$$\binom{n + d - 1}{d - 1},$$

we have

$$\dim \mathbb{H}_n^d = \binom{n + d - 1}{d - 1} = \binom{n + d - 1}{n}. \tag{2.2}$$

In particular, for $d = 2$ and 3, we have

$$\dim \mathbb{H}_n^2 = n + 1, \qquad \dim \mathbb{H}_n^3 = \frac{1}{2}(n+1)(n+2). \qquad (2.3)$$

We give in passing a compact formula for the generating function of the sequence $\{\dim \mathbb{H}_n^d\}_{n \geq 0}$,

$$\sum_{n=0}^{\infty} \left(\dim \mathbb{H}_n^d \right) z^n.$$

Recall the Taylor expansion (e.g., deduced from [9, (1.1.7)])

$$(1+x)^s = \sum_{n=0}^{\infty} \binom{s}{n} x^n, \quad |x| < 1, \quad \binom{s}{n} := \frac{s(s-1)\cdots(s-n+1)}{n!}.$$

Replacing x by $(-x)$ and choosing $s = -d$, we obtain

$$(1-x)^{-d} = \sum_{n=0}^{\infty} \binom{n+d-1}{n} x^n, \quad |x| < 1. \qquad (2.4)$$

Thus,

$$\sum_{n=0}^{\infty} \left(\dim \mathbb{H}_n^d \right) z^n = \frac{1}{(1-z)^d}, \quad |z| < 1. \qquad (2.5)$$

For $n \geq 2$,

$$|\cdot|^2 \mathbb{H}_{n-2}^d := \left\{ |\boldsymbol{x}|^2 H_{n-2}(\boldsymbol{x}) : H_{n-2} \in \mathbb{H}_{n-2}^d \right\}$$

is a proper invariant subspace of $\mathbb{H}_n^d$. Hence $\mathbb{H}_n^d|_{\mathbb{S}^{d-1}}$, the restriction of $\mathbb{H}_n^d$ to $\mathbb{S}^{d-1}$, is reducible. Let us identify the subspace of $\mathbb{H}_n^d$ that does not contain the factor $|\boldsymbol{x}|^2$.

Any $H_n \in \mathbb{H}_n^d$ can be written in the form

$$H_n(\boldsymbol{x}) = \sum_{|\boldsymbol{\alpha}|=n} a_{\boldsymbol{\alpha}} \boldsymbol{x}^{\boldsymbol{\alpha}}, \ a_{\boldsymbol{\alpha}} \in \mathbb{C}.$$

For this polynomial H_n, define

$$H_n(\boldsymbol{\nabla}) = \sum_{|\boldsymbol{\alpha}|=n} a_{\boldsymbol{\alpha}} \boldsymbol{\nabla}^{\boldsymbol{\alpha}}.$$

Given any two polynomials in $\mathbb{H}_n^d$,

$$H_{n,1}(\boldsymbol{x}) = \sum_{|\boldsymbol{\alpha}|=n} a_{\boldsymbol{\alpha},1} \boldsymbol{x}^{\boldsymbol{\alpha}}, \qquad H_{n,2}(\boldsymbol{x}) = \sum_{|\boldsymbol{\alpha}|=n} a_{\boldsymbol{\alpha},2} \boldsymbol{x}^{\boldsymbol{\alpha}},$$

it is straightforward to show

$$H_{n,1}(\nabla)\overline{H_{n,2}(\boldsymbol{x})} = \sum_{|\boldsymbol{\alpha}|=n} \boldsymbol{\alpha}! a_{\boldsymbol{\alpha},1}\overline{a_{\boldsymbol{\alpha},2}} = \overline{H_{n,2}(\nabla)\overline{H_{n,1}(\boldsymbol{x})}}.$$

Thus,

$$(H_{n,1}, H_{n,2})_{\mathbb{H}_n^d} := H_{n,1}(\nabla)\overline{H_{n,2}(\boldsymbol{x})} \tag{2.6}$$

defines an inner product in the subspace $\mathbb{H}_n^d$.

Recall that a function f is *harmonic* if $\Delta f(\boldsymbol{x}) = 0$. Being harmonic is an invariant property for functions.

Lemma 2.4. *If* $\Delta f = 0$*, then* $\Delta f_A = 0 \ \forall A \in \mathbb{O}^d$.

Proof. Denote $\boldsymbol{y} = A\boldsymbol{x}$. Then $\nabla_{\boldsymbol{x}} = A\nabla_{\boldsymbol{y}}$. Since $A \in \mathbb{O}^d$, we have

$$\Delta_{\boldsymbol{x}} = \nabla_{\boldsymbol{x}} \cdot \nabla_{\boldsymbol{x}} = \nabla_{\boldsymbol{y}} \cdot \nabla_{\boldsymbol{y}} = \Delta_{\boldsymbol{y}}.$$

So the stated property holds.   □

We now introduce an important subspace of $\mathbb{H}_n^d$.

Definition 2.5. The space of the homogeneous harmonics of degree n in d dimensions, $\mathbb{Y}_n(\mathbb{R}^d)$, consists of all homogeneous polynomials of degree n in $\mathbb{R}^d$ that are also harmonic.

We comment that non-trivial functions in $\mathbb{Y}_n(\mathbb{R}^d)$ do not contain the factor $|\boldsymbol{x}|^2$. This is shown as follows. Suppose $Y_n(\boldsymbol{x}) = |\boldsymbol{x}|^2 Y_{n-2}(\boldsymbol{x})$ is harmonic, where $Y_{n-2}(\boldsymbol{x})$ is a homogeneous polynomial of degree $(n-2)$. Then

$$(Y_n, Y_n)_{\mathbb{H}_{n,d}} = Y_{n-2}(\nabla)\overline{\Delta Y_n(\boldsymbol{x})} = 0.$$

Hence, $Y_n(\boldsymbol{x}) \equiv 0$.

Example 2.6. Obviously, $\mathbb{Y}_n(\mathbb{R}^d) = \mathbb{H}_n^d$ if $n = 0$ or 1.

For $d = 1$, $\mathbb{Y}_n(\mathbb{R}) = \emptyset$ for $n \geq 2$.

For $d = 2$, $\mathbb{Y}_2(\mathbb{R}^2)$ consists of all polynomials of the form $a\left(x_1^2 - x_2^2\right) + b\,x_1 x_2$, $a, b \in \mathbb{C}$. Polynomials of the form $(x_1 + i\,x_2)^n$ belong to $\mathbb{Y}_n(\mathbb{R}^2)$.

For $d = 3$, any polynomial of the form $(x_3 + i\,x_1 \cos\theta + i\,x_2 \sin\theta)^n$, $\theta \in \mathbb{R}$ being fixed, belongs to $\mathbb{Y}_n(\mathbb{R}^3)$.   □

Let us determine the dimension $N_{n,d} := \dim \mathbb{Y}_n(\mathbb{R}^d)$. The number $N_{n,d}$ will appear at various places in this text. Any polynomial $H_n \in \mathbb{H}_n^d$ can be written in the form

$$H_n(x_1, \cdots, x_d) = \sum_{j=0}^n (x_d)^j h_{n-j}(x_1, \cdots, x_{d-1}), \quad h_{n-j} \in \mathbb{H}_{n-j}^{d-1}. \tag{2.7}$$

Apply the Laplacian operator to this polynomial,

$$\Delta_{(d)} H_n(\boldsymbol{x}_{(d)}) = \sum_{j=0}^{n-2} (x_d)^j \left[\Delta_{(d-1)} h_{n-j}(\boldsymbol{x}_{(d-1)}) \right.$$

$$\left. + (j+2)(j+1) h_{n-j-2}(\boldsymbol{x}_{(d-1)}) \right].$$

Thus, if $H_n \in \mathbb{Y}_n(\mathbb{R}^d)$ so that $\Delta_{(d)} H_n(\boldsymbol{x}_{(d)}) \equiv 0$, then

$$h_{n-j-2} = -\frac{1}{(j+2)(j+1)} \Delta_{(d-1)} h_{n-j}, \quad 0 \leq j \leq n-2. \qquad (2.8)$$

Consequently, a homogeneous harmonic $H_n \in \mathbb{Y}_n(\mathbb{R}^d)$ is uniquely determined by $h_n \in \mathbb{H}_n^{d-1}$ and $h_{n-1} \in \mathbb{H}_{n-1}^{d-1}$ in the expansion (2.7). From this, we get the following relation on the polynomial space dimensions:

$$N_{n,d} = \dim \mathbb{H}_n^{d-1} + \dim \mathbb{H}_{n-1}^{d-1}. \qquad (2.9)$$

Using the formula (2.2) for $\dim \mathbb{H}_n^{d-1}$ and $\dim \mathbb{H}_{n-1}^{d-1}$, we have, for $d \geq 2$,

$$N_{n,d} = \frac{(2n+d-2)(n+d-3)!}{n!(d-2)!}, \quad n \in \mathbb{N}. \qquad (2.10)$$

In particular, with $n \in \mathbb{N}$, for $d = 2$, $N_{n,2} = 2$, and for $d = 3$, $N_{n,3} = 2n + 1$. It can be verified directly that $N_{0,d} = 1$ for any $d \geq 1$, and

$$N_{0,1} = N_{1,1} = 1, \quad N_{n,1} = 0 \; \forall n \geq 2. \qquad (2.11)$$

Note the asymptotic behavior

$$N_{n,d} = \mathcal{O}(n^{d-2}) \quad \text{for } n \text{ sufficiently large.} \qquad (2.12)$$

For the generating function of the sequence $\{N_{n,d}\}_n$, we apply the relation (2.9) for $n \geq 1$,

$$\sum_{n=0}^{\infty} N_{n,d} z^n = 1 + \sum_{n=1}^{\infty} N_{n,d} z^n$$

$$= 1 + \sum_{n=1}^{\infty} \left(\dim \mathbb{H}_n^{d-1} \right) z^n + \sum_{n=1}^{\infty} \left(\dim \mathbb{H}_{n-1}^{d-1} \right) z^n$$

$$= \sum_{n=0}^{\infty} \left(\dim \mathbb{H}_n^{d-1} \right) z^n + z \sum_{n=0}^{\infty} \left(\dim \mathbb{H}_n^{d-1} \right) z^n$$

$$= (1+z) \sum_{n=0}^{\infty} \left(\dim \mathbb{H}_n^{d-1} \right) z^n.$$

Thus, using the formula (2.5), we get a compact formula for the generating function of the sequence $\{N_{n,d}\}_n$:

$$\sum_{n=0}^{\infty} N_{n,d} z^n = \frac{1+z}{(1-z)^{d-1}}, \quad |z| < 1. \tag{2.13}$$

We can use (2.13) to derive a recursion formula for $N_{n,d}$ with respect to the dimension parameter d. Write

$$\frac{1+z}{(1-z)^{d-1}} = \frac{1+z}{(1-z)^{d-2}} \cdot \frac{1}{1-z} = \left(\sum_{m=0}^{\infty} N_{m,d-1} z^m\right) \left(\sum_{k=0}^{\infty} z^k\right).$$

We have

$$\frac{1+z}{(1-z)^{d-1}} = \sum_{n=0}^{\infty} \left(\sum_{m=0}^{n} N_{m,d-1}\right) z^n.$$

Comparing this formula with (2.13), we obtain

$$N_{n,d} = \sum_{m=0}^{n} N_{m,d-1}. \tag{2.14}$$

2.1.2 Legendre Harmonic and Legendre Polynomial

We now introduce a special homogeneous harmonic, the Legendre harmonic of degree n in d dimensions, $L_{n,d} : \mathbb{R}^d \to \mathbb{R}$, by the following three conditions:

$$L_{n,d} \in \mathbb{Y}_n(\mathbb{R}^d), \tag{2.15}$$

$$L_{n,d}(A\boldsymbol{x}) = L_{n,d}(\boldsymbol{x}) \quad \forall A \in \mathbb{O}^d(\boldsymbol{e}_d), \ \forall \boldsymbol{x} \in \mathbb{R}^d, \tag{2.16}$$

$$L_{n,d}(\boldsymbol{e}_d) = 1. \tag{2.17}$$

The condition (2.16) expresses the isotropical symmetry of $L_{n,d}$ with respect to the x_d-axis, whereas the condition (2.17) is a normalizing condition. Write $L_{n,d}$ in the form (2.7) and $A \in \mathbb{O}^d(\boldsymbol{e}_d)$ in the form (2.1). Then the condition (2.16) implies

$$h_{n-j}(A_1 \boldsymbol{x}_{(d-1)}) = h_{n-j}(\boldsymbol{x}_{(d-1)}) \quad \forall A_1 \in \mathbb{O}^{d-1}, \ \boldsymbol{x}_{(d-1)} \in \mathbb{R}^{d-1}, \ 0 \le j \le n.$$

From Proposition 2.1, h_{n-j} depends on $\boldsymbol{x}_{(d-1)}$ through $|\boldsymbol{x}_{(d-1)}|$. Since h_{n-j} is a homogeneous polynomial, this is possible only if $(n - j)$ is even and we have

$$h_{n-j}(\boldsymbol{x}_{(d-1)}) = \begin{cases} c_k |\boldsymbol{x}_{(d-1)}|^{2k} & \text{if } n-j = 2k, \\ 0 & \text{if } n-j = 2k+1, \end{cases} \quad c_k \in \mathbb{R}.$$

Hence,

$$L_{n,d}(\boldsymbol{x}) = \sum_{k=0}^{[n/2]} c_k |\boldsymbol{x}_{(d-1)}|^{2k} (x_d)^{n-2k},$$

where $[n/2]$ denotes the integer part of $n/2$. To determine the coefficients $\{c_k\}_{k=0}^{[n/2]}$, we apply the relation (2.8) to obtain

$$c_k = -\frac{(n - 2k + 2)(n - 2k + 1)}{2k(2k + d - 3)} c_{k-1}, \quad 1 \le k \le [n/2].$$

The normalization condition (2.17) implies $c_0 = 1$. Then

$$c_k = (-1)^k \frac{n! \, \Gamma(\frac{d-1}{2})}{4^k k! \, (n - 2k)! \, \Gamma(k + \frac{d-1}{2})}, \quad 0 \le k \le [n/2].$$

Therefore, we have derived the following formula for the Legendre harmonic

$$L_{n,d}(\boldsymbol{x}) = n! \, \Gamma\left(\frac{d-1}{2}\right) \sum_{k=0}^{[n/2]} (-1)^k \frac{|\boldsymbol{x}_{(d-1)}|^{2k} (x_d)^{n-2k}}{4^k k! \, (n - 2k)! \, \Gamma(k + \frac{d-1}{2})}. \tag{2.18}$$

Using the polar coordinates

$$\boldsymbol{x}_{(d)} = r\boldsymbol{\xi}_{(d)}, \quad \boldsymbol{\xi}_{(d)} = t\,\boldsymbol{e}_d + (1 - t^2)^{1/2}\boldsymbol{\xi}_{(d-1)},$$

we define the Legendre polynomial of degree n in d dimensions, $P_{n,d}(t) := L_{n,d}(\boldsymbol{\xi}_{(d)})$, as the restriction of the Legendre harmonic on the unit sphere. Then from the formula (2.18), we have

$$P_{n,d}(t) = n! \, \Gamma\left(\frac{d-1}{2}\right) \sum_{k=0}^{[n/2]} (-1)^k \frac{(1 - t^2)^k t^{n-2k}}{4^k k! \, (n - 2k)! \, \Gamma(k + \frac{d-1}{2})}. \tag{2.19}$$

Corresponding to (2.17), we have

$$P_{n,d}(1) = 1. \tag{2.20}$$

This property can be deduced straightforward from the formula (2.19). Note the relation

$$L_{n,d}(\boldsymbol{x}) = L_{n,d}(r\boldsymbol{\xi}_{(d)}) = r^n P_{n,d}(t). \tag{2.21}$$

The polynomial $P_{n,3}(t)$ is the standard Legendre polynomial of degree n. Following [85], we also call $P_{n,d}(t)$ of (2.19) Legendre polynomial.

Detailed discussion of the Legendre polynomials $P_{n,d}(t)$ is given in Sects. 2.6 and 2.7.

2.1.3 Spherical Harmonics

We are now ready to introduce spherical harmonics.

Definition 2.7. $\mathbb{Y}_n^d := \mathbb{Y}_n(\mathbb{R}^d)|_{\mathbb{S}^{d-1}}$ is called the spherical harmonic space of order n in d dimensions. Any function in $\mathbb{Y}_n^d$ is called a spherical harmonic of order n in d dimensions.

By the definition, we see that any spherical harmonic $Y_n \in \mathbb{Y}_n^d$ is related to a homogeneous harmonic $H_n \in \mathbb{Y}_n(\mathbb{R}^d)$ as follows:

$$H_n(r\boldsymbol{\xi}) = r^n Y_n(\boldsymbol{\xi}).$$

Thus the dimension of $\mathbb{Y}_n^d$ is the same as that of $\mathbb{Y}_n(\mathbb{R}^d)$:

$$\dim \mathbb{Y}_n^d = N_{n,d}$$

and $N_{n,d}$ is given by (2.10).

Take the case of $d = 2$ as an example. The complex-valued function $(x_1 + i\,x_2)^n$ is a homogeneous harmonic of degree n, and so are the real part and the imaginary part of the function. In polar coordinates (r, θ), $\boldsymbol{\xi} = (\cos\theta, \sin\theta)^T$ and the restriction of the function $(x_1 + i\,x_2)^n$ on the unit circle is

$$(\cos\theta + i\,\sin\theta)^n = e^{in\theta} = \cos(n\theta) + i\,\sin(n\theta).$$

Thus,

$$y_{n,1}(\boldsymbol{\xi}) = \cos(n\theta), \quad y_{n,2}(\boldsymbol{\xi}) = \sin(n\theta) \tag{2.22}$$

are elements of the space $\mathbb{Y}_n^2$.

Let $\boldsymbol{\xi} \in \mathbb{S}^{d-1}$ be fixed. A function $f : \mathbb{S}^{d-1} \to \mathbb{C}$ is said to be invariant with respect to $\mathbb{O}^d(\boldsymbol{\xi})$ if

$$f(A\boldsymbol{\eta}) = f(\boldsymbol{\eta}) \quad \forall\, A \in \mathbb{O}^d(\boldsymbol{\xi}), \forall\, \boldsymbol{\eta} \in \mathbb{S}^{d-1}.$$

We have the following result, which will be useful later on several occasions.

Theorem 2.8. Let $Y_n \in \mathbb{Y}_n^d$ and $\boldsymbol{\xi} \in \mathbb{S}^{d-1}$. Then Y_n is invariant with respect to $\mathbb{O}^d(\boldsymbol{\xi})$ if and only if

$$Y_n(\boldsymbol{\eta}) = Y_n(\boldsymbol{\xi}) \, P_{n,d}(\boldsymbol{\xi} \cdot \boldsymbol{\eta}) \quad \forall \boldsymbol{\eta} \in \mathbb{S}^{d-1}. \tag{2.23}$$

Proof. ($\Longrightarrow$) Since $\boldsymbol{\xi}$ is a unit vector, we can find an $A_1 \in \mathbb{O}^d$ such that $\boldsymbol{\xi} = A_1 \boldsymbol{e}_d$. Consider the function

$$\tilde{Y}_n(\boldsymbol{\eta}) := Y_n(A_1 \boldsymbol{\eta}), \quad \boldsymbol{\eta} \in \mathbb{S}^{d-1}.$$

Then $\tilde{Y}_n$ is invariant with respect to $\mathbb{O}^d(\boldsymbol{e}_d)$. From the definition of the Legendre harmonic $L_{n,d}(\boldsymbol{x})$, we know that the homogeneous harmonic $r^n \tilde{Y}_n(\boldsymbol{\eta})$ is a multiple of $L_{n,d}(r^n \boldsymbol{\eta})$,

$$r^n \tilde{Y}_n(\boldsymbol{\eta}) = c_1 L_{n,d}(r^n \boldsymbol{\eta}), \quad r \geq 0, \; \boldsymbol{\eta} \in \mathbb{S}^{d-1}$$

with some constant c_1. Thus,

$$\tilde{Y}_n(\boldsymbol{\eta}) = c_1 L_{n,d}(\boldsymbol{\eta}), \quad \boldsymbol{\eta} \in \mathbb{S}^{d-1}.$$

Choosing $\boldsymbol{\eta} = \boldsymbol{e}_d$, we find
$$c_1 = \tilde{Y}_n(\boldsymbol{e}_d).$$

Hence,

$$\tilde{Y}_n(\boldsymbol{\eta}) = \tilde{Y}_n(\boldsymbol{e}_d) \, L_{n,d}(\boldsymbol{\eta}) = \tilde{Y}_n(\boldsymbol{e}_d) \, P_{n,d}(\boldsymbol{\eta} \cdot \boldsymbol{e}_d), \quad \boldsymbol{\eta} \in \mathbb{S}^{d-1}.$$

Then,

$$\begin{aligned} Y_n(\boldsymbol{\eta}) &= \tilde{Y}_n(A_1^T \boldsymbol{\eta}) \\ &= Y_n(A_1 \boldsymbol{e}_d) \, P_{n,d}(A_1^T \boldsymbol{\eta} \cdot \boldsymbol{e}_d) \\ &= Y_n(A_1 \boldsymbol{e}_d) \, P_{n,d}(\boldsymbol{\eta} \cdot A_1 \boldsymbol{e}_d) \\ &= Y_n(\boldsymbol{\xi}) \, P_{n,d}(\boldsymbol{\eta} \cdot \boldsymbol{\xi}), \end{aligned}$$

i.e., the formula (2.23) holds.

($\Longleftarrow$) The function $Y_n(\boldsymbol{\eta})$ satisfying (2.23) is obviously invariant with respect to $\mathbb{O}^d(\boldsymbol{\xi})$. $\qquad \square$

Consequently, the subspaces of isotropically invariant functions from $\mathbb{Y}_n^d$ are one-dimensional.

2.2 Addition Theorem and Its Consequences

One important property regarding the spherical harmonics is the addition theorem.

Theorem 2.9 (Addition Theorem). *Let* $\{Y_{n,j} : 1 \leq j \leq N_{n,d}\}$ *be an orthonormal basis of* $\mathbb{Y}_n^d$, *i.e.*,

$$\int_{\mathbb{S}^{d-1}} Y_{n,j}(\boldsymbol{\eta})\overline{Y_{n,k}(\boldsymbol{\eta})}\, dS^{d-1}(\boldsymbol{\eta}) = \delta_{jk}, \quad 1 \leq j, k \leq N_{n,d}.$$

Then

$$\sum_{j=1}^{N_{n,d}} Y_{n,j}(\boldsymbol{\xi})\overline{Y_{n,j}(\boldsymbol{\eta})} = \frac{N_{n,d}}{|\mathbb{S}^{d-1}|} P_{n,d}(\boldsymbol{\xi}\cdot\boldsymbol{\eta}) \quad \forall\, \boldsymbol{\xi}, \boldsymbol{\eta} \in \mathbb{S}^{d-1}. \tag{2.24}$$

Proof. For any $A \in \mathbb{O}^d$ and $1 \leq k \leq N_{n,d}$, $Y_{n,k}(A\boldsymbol{\xi}) \in \mathbb{Y}_n^d$ and we can write

$$Y_{n,k}(A\boldsymbol{\xi}) = \sum_{j=1}^{N_{n,d}} c_{kj} Y_{n,j}(\boldsymbol{\xi}), \quad c_{kj} \in \mathbb{C}. \tag{2.25}$$

From

$$\int_{\mathbb{S}^{d-1}} Y_{n,j}(A\boldsymbol{\xi})\overline{Y_{n,k}(A\boldsymbol{\xi})}\, dS^{d-1}(\boldsymbol{\xi}) = \int_{\mathbb{S}^{d-1}} Y_{n,j}(\boldsymbol{\eta})\overline{Y_{n,k}(\boldsymbol{\eta})}\, dS^{d-1}(\boldsymbol{\eta}) = \delta_{jk},$$

we have

$$\delta_{jk} = \sum_{l,m=1}^{N_{n,d}} c_{jl}\overline{c_{km}}\, (Y_{n,l}, Y_{n,m}) = \sum_{l=1}^{N_{n,d}} c_{jl}\overline{c_{kl}}.$$

In matrix form, $CC^H = I$. Here C^H is the conjugate transpose of C. Thus, the matrix $C := (c_{jl})$ is unitary and so $C^H C = I$, i.e.,

$$\sum_{j=1}^{N_{n,d}} \overline{c_{jl}}c_{jk} = \delta_{lk}, \quad 1 \leq l, k \leq N_{n,d}. \tag{2.26}$$

Now consider the sum

$$Y(\boldsymbol{\xi}, \boldsymbol{\eta}) := \sum_{j=1}^{N_{n,d}} Y_{n,j}(\boldsymbol{\xi})\overline{Y_{n,j}(\boldsymbol{\eta})}, \quad \boldsymbol{\xi}, \boldsymbol{\eta} \in \mathbb{S}^{d-1}.$$

For any $A \in \mathbb{O}^d$, use the expansion (2.25),

$$Y(A\boldsymbol{\xi}, A\boldsymbol{\eta}) = \sum_{j=1}^{N_{n,d}} Y_{n,j}(A\boldsymbol{\xi})\overline{Y_{n,j}(A\boldsymbol{\eta})} = \sum_{j,k,l=1}^{N_{n,d}} c_{jk}\overline{c_{jl}}Y_{n,k}(\boldsymbol{\xi})\overline{Y_{n,l}(\boldsymbol{\eta})},$$

and then use the property (2.26),

$$Y(A\boldsymbol{\xi}, A\boldsymbol{\eta}) = \sum_{k=1}^{N_{n,d}} Y_{n,k}(\boldsymbol{\xi})\overline{Y_{n,k}(\boldsymbol{\eta})} = Y(\boldsymbol{\xi}, \boldsymbol{\eta}).$$

So for fixed $\boldsymbol{\xi}$, $Y(\boldsymbol{\xi}, \cdot) \in \mathbb{Y}_n^d$ and is invariant with respect to $\mathbb{O}^d(\boldsymbol{\xi})$. By Theorem 2.8,

$$Y(\boldsymbol{\xi}, \boldsymbol{\eta}) = Y(\boldsymbol{\xi}, \boldsymbol{\xi})\, P_{n,d}(\boldsymbol{\xi}\cdot\boldsymbol{\eta}).$$

Similarly, we have the equality

$$Y(\boldsymbol{\xi}, \boldsymbol{\eta}) = Y(\boldsymbol{\eta}, \boldsymbol{\eta})\, P_{n,d}(\boldsymbol{\xi}\cdot\boldsymbol{\eta}).$$

Thus, $Y(\boldsymbol{\xi}, \boldsymbol{\xi}) = Y(\boldsymbol{\eta}, \boldsymbol{\eta})$ and is a constant on $\mathbb{S}^{d-1}$. To determine this constant, we integrate the equality

$$Y(\boldsymbol{\xi}, \boldsymbol{\xi}) = \sum_{j=1}^{N_{n,d}} |Y_{n,j}(\boldsymbol{\xi})|^2$$

over $\mathbb{S}^{d-1}$ to obtain

$$Y(\boldsymbol{\xi}, \boldsymbol{\xi})\, |\mathbb{S}^{d-1}| = \sum_{j=1}^{N_{n,d}} \int_{\mathbb{S}^{d-1}} |Y_j(\boldsymbol{\xi})|^2 dS^{d-1} = N_{n,d}.$$

Therefore,

$$Y(\boldsymbol{\xi}, \boldsymbol{\xi}) = \frac{N_{n,d}}{|\mathbb{S}^{d-1}|}$$

and the equality (2.24) holds.   □

The equality (2.24) is, for $d = 3$,

$$\sum_{j=1}^{2n+1} Y_{n,j}(\boldsymbol{\xi})\overline{Y_{n,j}(\boldsymbol{\eta})} = \frac{2n+1}{4\pi}\, P_{n,3}(\boldsymbol{\xi}\cdot\boldsymbol{\eta}) \quad \forall \boldsymbol{\xi}, \boldsymbol{\eta} \in \mathbb{S}^2, \qquad (2.27)$$

and for $d = 2$,

$$\sum_{j=1}^{2} Y_{n,j}(\boldsymbol{\xi})\overline{Y_{n,j}(\boldsymbol{\eta})} = \frac{1}{\pi}\, P_{n,2}(\boldsymbol{\xi}\cdot\boldsymbol{\eta}) \quad \forall \boldsymbol{\xi}, \boldsymbol{\eta} \in \mathbb{S}^1. \qquad (2.28)$$

For the case $d = 2$, we write $\boldsymbol{\xi} = (\cos\theta, \sin\theta)^T$ and $\boldsymbol{\eta} = (\cos\psi, \sin\psi)^T$. Then, $\boldsymbol{\xi}\cdot\boldsymbol{\eta} = \cos(\theta - \psi)$. As an orthonormal basis for $\mathbb{Y}_n^2$, take (cf. (2.22))

$$Y_{n,1}(\boldsymbol{\xi}) = \frac{1}{\sqrt{\pi}}\, \cos(n\theta), \quad Y_{n,2}(\boldsymbol{\xi}) = \frac{1}{\sqrt{\pi}}\, \sin(n\theta).$$

By (2.28),

$$P_{n,2}(\cos(\theta - \psi)) = \cos(n\theta)\cos(n\psi) + \sin(n\theta)\sin(n\psi) = \cos(n(\theta - \psi)).$$

Thus,

$$P_{n,2}(t) = \cos(n\arccos t), \quad |t| \le 1, \tag{2.29}$$

i.e., $P_{n,2}$ is the ordinary Chebyshev polynomial of degree n.

We note that for $d = 2$,

$$\sum_{k=0}^{n} \frac{1}{\pi} P_{k,2}(\boldsymbol{\xi}\cdot\boldsymbol{\eta}) = \frac{1}{2\pi} \frac{\sin((n + 1/2)\phi)}{\sin(\phi/2)}, \quad \cos\phi := \boldsymbol{\xi}\cdot\boldsymbol{\eta},$$

is the Dirichlet kernel, whereas for $d = 3$,

$$\sum_{k=0}^{n} \sum_{j=1}^{2k+1} Y_{k,j}(\boldsymbol{\xi})\overline{Y_{k,j}(\boldsymbol{\eta})} = \frac{n+1}{4\pi} P_n^{(1,0)}(\boldsymbol{\xi}\cdot\boldsymbol{\eta}) \quad \forall \boldsymbol{\xi}, \boldsymbol{\eta} \in \mathbb{S}^2. \tag{2.30}$$

Here $P_n^{(1,0)}(t)$ is the Jacobi polynomial of degree n on $[-1, 1]$, based on the weight function $w(t) = 1 - t$; and as a normalization, $P_n^{(1,0)}(1) = n + 1$. This identity is noted in [50]. See Sect. 4.3.1 for an introduction of the Jacobi polynomials.

We now discuss several applications of the addition theorem.

The addition theorem can be used to find a compact expression of the reproducing kernel of $\mathbb{Y}_n^d$. Any $Y_n \in \mathbb{Y}_n^d$ can be written in the form

$$Y_n(\boldsymbol{\xi}) = \sum_{j=1}^{N_{n,d}} (Y_n, Y_{n,j})_{\mathbb{S}^{d-1}} Y_{n,j}(\boldsymbol{\xi}). \tag{2.31}$$

Applying (2.24),

$$Y_n(\boldsymbol{\xi}) = \int_{\mathbb{S}^{d-1}} Y_n(\boldsymbol{\eta}) \sum_{j=1}^{N_{n,d}} Y_{n,j}(\boldsymbol{\xi})\overline{Y_{n,j}(\boldsymbol{\eta})}\, dS^{d-1}(\boldsymbol{\eta})$$

$$= \frac{N_{n,d}}{|\mathbb{S}^{d-1}|} \int_{\mathbb{S}^{d-1}} P_{n,d}(\boldsymbol{\xi}\cdot\boldsymbol{\eta}) Y_n(\boldsymbol{\eta})\, dS^{d-1}(\boldsymbol{\eta}).$$

Hence,

$$K_{n,d}(\boldsymbol{\xi}, \boldsymbol{\eta}) := \frac{N_{n,d}}{|\mathbb{S}^{d-1}|} P_{n,d}(\boldsymbol{\xi}\cdot\boldsymbol{\eta}) \tag{2.32}$$

is the reproducing kernel of $\mathbb{Y}_n^d$, i.e.,

$$Y_n(\boldsymbol{\xi}) = (Y_n, K_{n,d}(\boldsymbol{\xi}, \cdot))_{\mathbb{S}^{d-1}} \quad \forall Y_n \in \mathbb{Y}_n^d, \, \boldsymbol{\xi} \in \mathbb{S}^{d-1}. \tag{2.33}$$

Define

$$\mathbb{Y}_{0:m}^d := \bigoplus_{n=0}^m \mathbb{Y}_n^d$$

to be the space of all the spherical harmonics of order less than or equal to m. Then by (2.33),

$$K_{0:m,d}(\boldsymbol{\xi}, \boldsymbol{\eta}) := \frac{1}{|\mathbb{S}^{d-1}|} \sum_{n=0}^m N_{n,d} P_{n,d}(\boldsymbol{\xi} \cdot \boldsymbol{\eta}) \qquad (2.34)$$

is the reproducing kernel of $\mathbb{Y}_{0:m}^d$ in the sense that

$$Y(\boldsymbol{\xi}) = (Y, K_{0:m,d}(\boldsymbol{\xi}, \cdot))_{\mathbb{S}^{d-1}} \quad \forall Y \in \mathbb{Y}_{0:m}^d, \boldsymbol{\xi} \in \mathbb{S}^{d-1}.$$

We now derive some bounds for any spherical harmonic and for the Legendre polynomial, see (2.38) and (2.39) below, respectively.

Since $P_{n,d}(1) = 1$, we get from (2.24) that

$$\sum_{j=1}^{N_{n,d}} |Y_{n,j}(\boldsymbol{\xi})|^2 = \frac{N_{n,d}}{|\mathbb{S}^{d-1}|} \quad \forall \boldsymbol{\xi} \in \mathbb{S}^{d-1}. \qquad (2.35)$$

This provides an upper bound for the maximum value of any member of an orthonormal basis in $\mathbb{Y}_n^d$:

$$\max\left\{ |Y_{n,j}(\boldsymbol{\xi})| : \boldsymbol{\xi} \in \mathbb{S}^{d-1}, 1 \le j \le N_{n,d} \right\} \le \left(\frac{N_{n,d}}{|\mathbb{S}^{d-1}|} \right)^{1/2}. \qquad (2.36)$$

Consider an arbitrary $Y_n \in \mathbb{Y}_n^d$. From (2.31), we find

$$\int_{\mathbb{S}^{d-1}} |Y_n(\boldsymbol{\xi})|^2 dS^{d-1}(\boldsymbol{\xi}) = \sum_{j=1}^{N_{n,d}} |(Y_n, Y_{n,j})_{\mathbb{S}^{d-1}}|^2. \qquad (2.37)$$

By (2.31) again,

$$|Y_n(\boldsymbol{\xi})|^2 \le \sum_{j=1}^{N_{n,d}} |Y_{n,j}(\boldsymbol{\xi})|^2 \sum_{j=1}^{N_{n,d}} |(Y_n, Y_{n,j})_{\mathbb{S}^{d-1}}|^2.$$

Then using (2.35) and (2.37),

$$|Y_n(\boldsymbol{\xi})|^2 \le \frac{N_{n,d}}{|\mathbb{S}^{d-1}|} \|Y_n\|_{L^2(\mathbb{S}^{d-1})}^2.$$

Thus we have the inequality

$$\|Y_n\|_\infty \le \left(\frac{N_{n,d}}{|\mathbb{S}^{d-1}|}\right)^{1/2} \|Y_n\|_{L^2(\mathbb{S}^{d-1})} \quad \forall Y_n \in \mathbb{Y}_n^d, \tag{2.38}$$

which extends the bound (2.36).

By (2.24) and (2.35), we have

$$\frac{N_{n,d}}{|\mathbb{S}^{d-1}|}\,|P_{n,d}(\boldsymbol{\xi}\cdot\boldsymbol{\eta})| \le \left[\sum_{j=1}^{N_{n,d}} |Y_{n,j}(\boldsymbol{\xi})|^2\right]^{1/2} \left[\sum_{j=1}^{N_{n,d}} |Y_{n,j}(\boldsymbol{\eta})|^2\right]^{1/2} = \frac{N_{n,d}}{|\mathbb{S}^{d-1}|}.$$

Therefore,

$$|P_{n,d}(t)| \le 1 = P_{n,d}(1) \quad \forall n \in \mathbb{N},\, d \ge 2,\, t \in [-1,1]. \tag{2.39}$$

We have an integral formula

$$\int_{\mathbb{S}^{d-1}} |P_{n,d}(\boldsymbol{\xi}\cdot\boldsymbol{\eta})|^2 dS^{d-1}(\boldsymbol{\eta}) = \frac{|\mathbb{S}^{d-1}|}{N_{n,d}}. \tag{2.40}$$

This formula is proved as follows. First we use (2.24) to get

$$\int_{\mathbb{S}^{d-1}} |P_{n,d}(\boldsymbol{\xi}\cdot\boldsymbol{\eta})|^2 dS^{d-1}(\boldsymbol{\eta})$$

$$= \left(\frac{|\mathbb{S}^{d-1}|}{N_{n,d}}\right)^2 \int_{\mathbb{S}^{d-1}} \left|\sum_{j=1}^{N_{n,d}} Y_{n,j}(\boldsymbol{\xi})\overline{Y_{n,j}(\boldsymbol{\eta})}\right|^2 dS^{d-1}(\boldsymbol{\eta})$$

$$= \left(\frac{|\mathbb{S}^{d-1}|}{N_{n,d}}\right)^2 \sum_{j=1}^{N_{n,d}} |Y_{n,j}(\boldsymbol{\xi})|^2.$$

Then we apply the identity (2.35).

As one more application of the addition theorem, we have the following result.

Theorem 2.10. *For any $n \in \mathbb{N}_0$ and any $d \in \mathbb{N}$, the spherical harmonic space $\mathbb{Y}_n^d$ is irreducible.*

Proof. We argue by contradiction. Suppose $\mathbb{Y}_n^d$ is reducible so that it is possible to write $\mathbb{Y}_n^d = \mathbb{V}_1 + \mathbb{V}_2$ with $\mathbb{V}_1 \ne \emptyset$, $\mathbb{V}_2 \ne \emptyset$, and $\mathbb{V}_1 \perp \mathbb{V}_2$. Choose an orthonormal basis of $\mathbb{Y}_n^d$ in such a way that the first N_1 functions span $\mathbb{V}_1$ and the remaining $N_2 = N_{n,d} - N_1$ functions span $\mathbb{V}_2$. For both $\mathbb{V}_1$ and $\mathbb{V}_2$, we can apply the addition theorem with the corresponding Legendre functions $P_{n,d,1}$ and $P_{n,d,2}$. Since $\mathbb{V}_1 \perp \mathbb{V}_2$,

$$\int_{\mathbb{S}^{d-1}} P_{n,d,1}(\boldsymbol{\xi}\cdot\boldsymbol{\eta})P_{n,d,2}(\boldsymbol{\xi}\cdot\boldsymbol{\eta})\,dS^{d-1}(\boldsymbol{\eta}) = 0 \quad \forall\, \boldsymbol{\xi} \in \mathbb{S}^{d-1}. \qquad (2.41)$$

For an arbitrary but fixed $\boldsymbol{\xi} \in \mathbb{S}^{d-1}$, consider the function $\boldsymbol{\eta} \mapsto P_{n,d,1}(\boldsymbol{\xi}\cdot\boldsymbol{\eta})$. For any $A \in \mathbb{O}^d(\boldsymbol{\xi})$, we have $A^T A = I$ and $A\boldsymbol{\xi} = \boldsymbol{\xi}$, implying $A^T\boldsymbol{\xi} = \boldsymbol{\xi}$. Then

$$P_{n,d,1}(\boldsymbol{\xi}\cdot A\boldsymbol{\eta}) = P_{n,d,1}(A^T\boldsymbol{\xi}\cdot\boldsymbol{\eta}) = P_{n,d,1}(\boldsymbol{\xi}\cdot\boldsymbol{\eta}),$$

i.e., the function $\boldsymbol{\eta} \mapsto P_{n,d,1}(\boldsymbol{\xi}\cdot\boldsymbol{\eta})$ is invariant with respect to $\mathbb{O}^d(\boldsymbol{\xi})$. By Theorem 2.8,

$$P_{n,d,1}(\boldsymbol{\xi}\cdot\boldsymbol{\eta}) = P_{n,d,1}(\boldsymbol{\xi}\cdot\boldsymbol{\xi})\,P_{n,d}(\boldsymbol{\xi}\cdot\boldsymbol{\eta}) = P_{n,d}(\boldsymbol{\xi}\cdot\boldsymbol{\eta}).$$

Similarly,

$$P_{n,d,2}(\boldsymbol{\xi}\cdot\boldsymbol{\eta}) = P_{n,d}(\boldsymbol{\xi}\cdot\boldsymbol{\eta}).$$

But then the integral in (2.41) equals $|\mathbb{S}^{d-1}|/N_{n,d}$ by (2.40) and we reach a contradiction. $\qquad\qquad\square$

2.3 A Projection Operator

Consider the problem of finding the best approximation in $\mathbb{Y}_n^d$ of a function $f \in L^2(\mathbb{S}^{d-1})$:

$$\inf\left\{\|f - Y_n\|_{L^2(\mathbb{S}^{d-1})} : Y_n \in \mathbb{Y}_n^d\right\}. \qquad (2.42)$$

In terms of an orthonormal basis $\{Y_{n,j} : 1 \le j \le N_{n,d}\}$ of $\mathbb{Y}_n^d$, the solution of the problem (2.42) is

$$(\mathcal{P}_{n,d}f)(\boldsymbol{\xi}) = \sum_{j=1}^{N_{n,d}} (f, Y_{n,j})_{\mathbb{S}^{d-1}} Y_{n,j}(\boldsymbol{\xi}). \qquad (2.43)$$

This is the projection of any f into $\mathbb{Y}_n^d$ and it is defined for $f \in L^1(\mathbb{S}^{d-1})$. The disadvantage of using this formula is the requirement of explicit knowledge of an orthonormal basis. We can circumvent this weakness by applying (2.24) to rewrite the right side of (2.43).

Definition 2.11. The projection of $f \in L^1(\mathbb{S}^{d-1})$ into $\mathbb{Y}_n^d$ is

$$(\mathcal{P}_{n,d}f)(\boldsymbol{\xi}) := \frac{N_{n,d}}{|\mathbb{S}^{d-1}|} \int_{\mathbb{S}^{d-1}} P_{n,d}(\boldsymbol{\xi}\cdot\boldsymbol{\eta})\,f(\boldsymbol{\eta})\,dS^{d-1}(\boldsymbol{\eta}), \quad \boldsymbol{\xi} \in \mathbb{S}^{d-1}. \qquad (2.44)$$

The operator $\mathcal{P}_{n,d}$ is obviously linear. Let us derive some bounds for the operator $\mathcal{P}_{n,d}$. First, we obtain from (2.39) that

$$|(\mathcal{P}_{n,d}f)(\boldsymbol{\xi})| \leq \frac{N_{n,d}}{|\mathbb{S}^{d-1}|}\|f\|_{L^1(\mathbb{S}^{d-1})}, \quad \boldsymbol{\xi} \in \mathbb{S}^{d-1}.$$

Then, for all $f \in L^1(\mathbb{S}^{d-1})$,

$$\|\mathcal{P}_{n,d}f\|_{C(\mathbb{S}^{d-1})} \leq \frac{N_{n,d}}{|\mathbb{S}^{d-1}|}\|f\|_{L^1(\mathbb{S}^{d-1})}, \tag{2.45}$$

$$\|\mathcal{P}_{n,d}f\|_{L^1(\mathbb{S}^{d-1})} \leq N_{n,d}\|f\|_{L^1(\mathbb{S}^{d-1})}. \tag{2.46}$$

Next, assume $f \in L^2(\mathbb{S}^{d-1})$. For any $\boldsymbol{\xi} \in \mathbb{S}^{d-1}$,

$$|(\mathcal{P}_{n,d}f)(\boldsymbol{\xi})|^2 \leq \left(\frac{N_{n,d}}{|\mathbb{S}^{d-1}|}\right)^2 \int_{\mathbb{S}^{d-1}} |P_{n,d}(\boldsymbol{\xi}\cdot\boldsymbol{\eta})|^2 dS^{d-1}(\boldsymbol{\eta})$$
$$\cdot \int_{\mathbb{S}^{d-1}} |f(\boldsymbol{\eta})|^2 dS^{d-1}(\boldsymbol{\eta}).$$

Use (2.40),

$$|(\mathcal{P}_{n,d}f)(\boldsymbol{\xi})|^2 \leq \frac{N_{n,d}}{|\mathbb{S}^{d-1}|}\|f\|_{L^2(\mathbb{S}^{d-1})}^2.$$

Hence, for all $f \in L^2(\mathbb{S}^{d-1})$,

$$\|\mathcal{P}_{n,d}f\|_{L^2(\mathbb{S}^{d-1})} \leq N_{n,d}^{1/2}\|f\|_{L^2(\mathbb{S}^{d-1})}, \tag{2.47}$$

$$\|\mathcal{P}_{n,d}f\|_{C(\mathbb{S}^{d-1})} \leq \left(\frac{N_{n,d}}{|\mathbb{S}^{d-1}|}\right)^{1/2}\|f\|_{L^2(\mathbb{S}^{d-1})}. \tag{2.48}$$

We remark that (2.47) can be improved to

$$\|\mathcal{P}_{n,d}f\|_{L^2(\mathbb{S}^{d-1})} \leq \|f\|_{L^2(\mathbb{S}^{d-1})};$$

see (2.134) later. Furthermore, if $f \in C(\mathbb{S}^{d-1})$, a similar argument leads to

$$\|\mathcal{P}_{n,d}f\|_{C(\mathbb{S}^{d-1})} \leq N_{n,d}^{1/2}\|f\|_{C(\mathbb{S}^{d-1})}. \tag{2.49}$$

Proposition 2.12. *The projection operator $\mathcal{P}_{n,d}$ and orthogonal transformations commute:*

$$\mathcal{P}_{n,d}f_A = (\mathcal{P}_{n,d}f)_A \quad \forall A \in \mathbb{O}^d.$$

Proof. We start with the left side of the equality,

$$(\mathcal{P}_{n,d}f_A)(\boldsymbol{\xi}) = \frac{N_{n,d}}{|\mathbb{S}^{d-1}|} \int_{\mathbb{S}^{d-1}} P_{n,d}(\boldsymbol{\xi}\cdot\boldsymbol{\eta}) \, f(A\boldsymbol{\eta}) \, dS^{d-1}(\boldsymbol{\eta})$$

$$= \frac{N_{n,d}}{|\mathbb{S}^{d-1}|} \int_{\mathbb{S}^{d-1}} P_{n,d}(A\boldsymbol{\xi}\cdot\boldsymbol{\zeta}) \, f(\boldsymbol{\zeta}) \, dS^{d-1}(\boldsymbol{\zeta}),$$

which is $(\mathcal{P}_{n,d}f)_A(\boldsymbol{\xi})$ by definition.   □

A useful consequence of Proposition 2.12 is the following result.

Corollary 2.13. *If* $\mathbb{V}$ *is an invariant space, then* $\mathcal{P}_{n,d}\mathbb{V} := \{\mathcal{P}_{n,d}f : f \in \mathbb{V}\}$ *is an invariant subspace of* $\mathbb{Y}_n^d$.

Since $\mathbb{Y}_n^d$ is irreducible, by Theorem 2.10, Corollary 2.13 implies that if $\mathbb{V}$ is an invariant space, then either $\mathbb{V}$ is orthogonal to $\mathbb{Y}_n^d$ or $\mathcal{P}_{n,d}\mathbb{V} = \mathbb{Y}_n^d$. Moreover, we have the next result.

Theorem 2.14. *If* $\mathbb{V}$ *is a primitive subspace of* $C(\mathbb{S}^{d-1})$, *then either* $\mathbb{V} \perp \mathbb{Y}_n^d$ *or* $\mathcal{P}_{n,d}$ *is a bijection from* $\mathbb{V}$ *to* $\mathbb{Y}_n^d$. *In the latter case,* $\mathbb{V} = \mathbb{Y}_n^d$.

Proof. We only need to prove that if $\mathcal{P}_{n,d} : \mathbb{V} \to \mathbb{Y}_n^d$ is a bijection, then $\mathbb{V} = \mathbb{Y}_n^d$. The two spaces are finite dimensional and have the same dimension $N_{n,d} = \dim(\mathbb{Y}_n^d)$. Let $\{V_j : 1 \le j \le N_{n,d}\}$ be an orthonormal basis of $\mathbb{V}$. Since $\mathbb{V}$ is primitive, for any $A \in \mathbb{O}^d$, we can write

$$V_j(A\boldsymbol{\xi}) = \sum_{k=1}^{N_{n,d}} c_{jk} V_k(\boldsymbol{\xi}), \quad c_{jk} \in \mathbb{C},$$

and the matrix (c_{jk}) is unitary as in the proof of Theorem 2.9. Consider the function

$$V(\boldsymbol{\xi}, \boldsymbol{\eta}) := \sum_{j=1}^{N_{n,d}} V_j(\boldsymbol{\xi})\overline{V_j(\boldsymbol{\eta})}.$$

Then again as in the proof of Theorem 2.9, we have

$$V(A\boldsymbol{\xi}, A\boldsymbol{\eta}) = V(\boldsymbol{\xi}, \boldsymbol{\eta}) \quad \forall A \in \mathbb{O}^d.$$

Given $\boldsymbol{\xi}, \boldsymbol{\eta} \in \mathbb{S}^{d-1}$, we can find an $A \in \mathbb{O}^d$ such that

$$A\boldsymbol{\xi} = e_d, \quad A\boldsymbol{\eta} = t\,e_d + (1-t^2)^{1/2}e_{d-1} \text{ with } t = \boldsymbol{\xi}\cdot\boldsymbol{\eta}.$$

Then

$$V(\boldsymbol{\xi}, \boldsymbol{\eta}) = V(e_d, t\,e_d + (1-t^2)^{1/2}e_{d-1})$$

is a function of $t = \boldsymbol{\xi}\cdot\boldsymbol{\eta}$. Denote this function by $P_d(t)$. For fixed $\boldsymbol{\xi}$, the mapping $\boldsymbol{\eta} \mapsto \overline{P_d(\boldsymbol{\xi}\cdot\boldsymbol{\eta})}$ is a function in $\mathbb{V}$, whereas for fixed $\boldsymbol{\zeta}$, the mapping $\boldsymbol{\eta} \mapsto P_{n,d}(\boldsymbol{\zeta}\cdot\boldsymbol{\eta})$ is a function in $\mathbb{Y}_n^d$. Consider the function

$$\phi(\boldsymbol{\xi},\boldsymbol{\zeta}) = \int_{\mathbb{S}^{d-1}} \overline{P_d(\boldsymbol{\xi}\cdot\boldsymbol{\eta})}\,P_{n,d}(\boldsymbol{\zeta}\cdot\boldsymbol{\eta})\,dS^{d-1}(\boldsymbol{\eta}).$$

We have the property

$$\phi(A\boldsymbol{\xi}, A\boldsymbol{\zeta}) = \phi(\boldsymbol{\xi},\boldsymbol{\zeta}) \quad \forall A \in \mathbb{O}^d.$$

So $\phi(\boldsymbol{\xi},\boldsymbol{\zeta})$ depends on $\boldsymbol{\xi}\cdot\boldsymbol{\zeta}$ only. This function belongs to both $\mathbb{V}$ and $\mathbb{Y}_n^d$. Thus, either $\mathbb{V} = \mathbb{Y}_n^d$ or $\phi \equiv 0$. In the latter case, we have

$$\sum_{j,k=1}^{N_{n,d}} \overline{V_j(\boldsymbol{\xi})}Y_{n,k}(\boldsymbol{\zeta})\,(V_j, Y_{n,k})_{L^2(\mathbb{S}^{d-1})} = 0 \quad \forall \boldsymbol{\xi},\boldsymbol{\zeta} \in \mathbb{S}^{d-1},$$

where $\{Y_{n,k} : 1 \le k \le N_{n,d}\}$ is an orthonormal basis of $\mathbb{Y}_n^d$. Since each of the sets $\{V_j : 1 \le j \le N_{n,d}\}$ and $\{Y_{n,j} : 1 \le j \le N_{n,d}\}$ consists of linearly independent elements, we obtain from the above identity that

$$(V_j, Y_{n,k})_{L^2(\mathbb{S}^{d-1})} = 0, \quad 1 \le j, k \le N_{n,d}.$$

This implies $\mathbb{V} \perp \mathbb{Y}_n^d$. $\qquad\square$

We let $\mathbb{V} = \mathbb{Y}_m^d$, $m \ne n$, in Theorem 2.14 to obtain the following result concerning orthogonality of spherical harmonics of different order.

Corollary 2.15. *For $m \ne n$, $\mathbb{Y}_m^d \perp \mathbb{Y}_n^d$.*

This result can be proved directly as follows. Let $Y_m \in \mathbb{Y}_m^d$ and $Y_n \in \mathbb{Y}_n^d$ be the restrictions on $\mathbb{S}^{d-1}$ of $H_m \in \mathbb{Y}_m(\mathbb{R}^d)$ and $H_n \in \mathbb{Y}_n(\mathbb{R}^d)$, respectively. Since $\Delta H_m(\boldsymbol{x}) = \Delta H_n(\boldsymbol{x}) = 0$, we have

$$\int_{\|\boldsymbol{x}\|<1} (H_m \Delta H_n - H_n \Delta H_m)\,d\boldsymbol{x} = 0.$$

Apply Green's formula,

$$\int_{\mathbb{S}^{d-1}} \left(H_m \frac{\partial H_n}{\partial r} - H_n \frac{\partial H_m}{\partial r} \right) dS^{d-1} = 0. \qquad (2.50)$$

Since H_m is a homogeneous polynomial of degree m,

$$\frac{\partial H_m(\boldsymbol{x})}{\partial r}\bigg|_{\boldsymbol{x}=\boldsymbol{\xi}} = m\,Y_m(\boldsymbol{\xi}), \quad \boldsymbol{\xi} \in \mathbb{S}^{d-1}.$$

Similarly,

$$\frac{\partial H_n(\boldsymbol{x})}{\partial r}\bigg|_{\boldsymbol{x}=\boldsymbol{\xi}} = n\,Y_n(\boldsymbol{\xi}), \quad \boldsymbol{\xi} \in \mathbb{S}^{d-1}.$$

Thus, from (2.50),

$$\int_{\mathbb{S}^{d-1}} (n-m)\,Y_m(\boldsymbol{\xi})\,Y_n(\boldsymbol{\xi})\,dS^{d-1}(\boldsymbol{\xi}) = 0.$$

Hence, since $m \neq n$,

$$\int_{\mathbb{S}^{d-1}} Y_m(\boldsymbol{\xi})\,Y_n(\boldsymbol{\xi})\,dS^{d-1}(\boldsymbol{\xi}) = 0.$$

2.4   Relations Among Polynomial Spaces

We have introduced several polynomial spaces in the previous sections. Here we discuss some relations among these polynomial spaces.

Proposition 2.16. *The Laplacian operator* Δ *is surjective from* $\mathbb{H}_n^d$ *to* $\mathbb{H}_{n-2}^d$ *for* $n \geq 2$.

Proof. Obviously, the operator Δ maps $\mathbb{H}_n^d$ to $\mathbb{H}_{n-2}^d$. By (2.2) and (2.10), we have

$$\begin{aligned}
\dim \mathbb{H}_n^d - \dim \mathbb{Y}_n(\mathbb{R}^d) &= \frac{(n+d-1)!}{n!\,(d-1)!} - \frac{(2n+d-2)\,(n+d-3)!}{n!\,(d-2)!} \\
&= \frac{(n-2+d-1)!}{(n-2)!\,(d-1)!} \\
&= \dim \mathbb{H}_{n-2}^d.
\end{aligned}$$

Therefore, $\Delta : \mathbb{H}_n^d \to \mathbb{H}_{n-2}^d$ is surjective. $\square$

It is possible to give another proof of Proposition 2.16 using the inner product (2.6). Suppose $\Delta : \mathbb{H}_n^d \to \mathbb{H}_{n-2}^d$ is not surjective. Then there exists a non-zero function $H_{n-2} \in \mathbb{H}_{n-2}^d$ such that

$$(\Delta H_n, H_{n-2})_{\mathbb{H}_{n-2}^d} = 0 \quad \forall\, H_n \in \mathbb{H}_n^d.$$

Take $H_n(\boldsymbol{x}) = |\boldsymbol{x}|^2 H_{n-2}(\boldsymbol{x})$ to get

$$\begin{aligned}
(H_n, H_n)_{\mathbb{H}_n^d} &= H_n(\boldsymbol{\nabla})\overline{H_n(\boldsymbol{x})} = H_{n-2}(\boldsymbol{\nabla})\overline{\Delta H_n(\boldsymbol{x})} \\
&= (H_{n-2}, \Delta H_n)_{\mathbb{H}_{n-2}^d} = 0.
\end{aligned}$$

Hence, $H_n(\boldsymbol{x}) = 0$ and then $H_{n-2}(\boldsymbol{x}) = 0$. This contradicts the assumption that $H_{n-2} \neq 0$.

Lemma 2.17. *For $n \geq 2$, $\mathbb{H}_n^d = \mathbb{Y}_n(\mathbb{R}^d) \oplus |\cdot|^2 \mathbb{H}_{n-2}^d$, with respect to the inner product (2.6).*

Proof. It is shown in the proof of Proposition 2.16 that

$$\dim \mathbb{H}_n^d = \dim \mathbb{Y}_n(\mathbb{R}^d) + \dim \mathbb{H}_{n-2}^d.$$

Thus, it remains to show $\mathbb{Y}_n(\mathbb{R}^d) \perp |\cdot|^2 \mathbb{H}_{n-2}^d$. For any $Y_n \in \mathbb{Y}_n(\mathbb{R}^d)$ and any $H_{n-2} \in \mathbb{H}_{n-2}^d$, there holds

$$\left(Y_n, |\cdot|^2 H_{n-2}\right)_{\mathbb{H}_n^d} = (\Delta Y_n, H_{n-2})_{\mathbb{H}_{n-2}^d} = 0.$$

Therefore, the statement is valid.   □

The orthogonal decomposition stated in Lemma 2.17 can be applied repeatedly, leading to the next result.

Theorem 2.18. *With respect to the inner product (2.6), we have*

$$\mathbb{H}_n^d = \mathbb{Y}_n(\mathbb{R}^d) \oplus |\cdot|^2 \mathbb{Y}_{n-2}(\mathbb{R}^d) \oplus \cdots \oplus |\cdot|^{2\,[n/2]} \mathbb{Y}_{n-2\,[n/2]}(\mathbb{R}^d). \qquad (2.51)$$

Proof. For any $H_n \in \mathbb{H}_n^d$, by Lemma 2.17, we have

$$H_n(\boldsymbol{x}) = Y_n(\boldsymbol{x}) + |\boldsymbol{x}|^2 H_{n-2}(\boldsymbol{x})$$

with uniquely determined $Y_n \in \mathbb{Y}_n(\mathbb{R}^d)$ and $H_{n-2} \in \mathbb{H}_{n-2}^d$. Applying Lemma 2.17 to $H_{n-2} \in \mathbb{H}_{n-2}^d$, we can uniquely determine a pair of functions $Y_{n-2} \in \mathbb{Y}_{n-2}(\mathbb{R}^d)$ and $H_{n-4} \in \mathbb{H}_{n-4}^d$ such that

$$H_{n-2}(\boldsymbol{x}) = Y_{n-2}(\boldsymbol{x}) + |\boldsymbol{x}|^2 H_{n-4}(\boldsymbol{x}).$$

Hence,

$$H_n(\boldsymbol{x}) = Y_n(\boldsymbol{x}) + |\boldsymbol{x}|^2 Y_{n-2}(\boldsymbol{x}) + |\boldsymbol{x}|^4 H_{n-4}(\boldsymbol{x}).$$

Continue this process to obtain the unique decomposition

$$H_n(\boldsymbol{x}) = Y_n(\boldsymbol{x}) + |\boldsymbol{x}|^2 Y_{n-2}(\boldsymbol{x}) + \cdots + |\boldsymbol{x}|^{2\,[n/2]} Y_{n-2\,[n/2]}(\boldsymbol{x}), \qquad (2.52)$$

where $Y_{n-2j} \in \mathbb{Y}_{n-2j}(\mathbb{R}^d)$. Note that the terms on the right side of (2.52) are mutually orthogonal with respect to the inner product (2.6).   □

As consequences of Theorem 2.18, we have the following two results.

Corollary 2.19.

$$\left(\sum_{j=0}^{n}\mathbb{H}_{j}^{d}\right)\Bigg|_{\mathbb{S}^{d-1}}=\sum_{j=0}^{n}\mathbb{Y}_{j}^{d}.$$

So the restriction of any polynomial on $\mathbb{S}^{d-1}$ is a sum of some spherical harmonics and the restriction of the space of the polynomials of d variables on $\mathbb{S}^{d-1}$ is $\sum_{j=0}^{\infty}\mathbb{Y}_{j}^{d}$.

Corollary 2.20. *A polynomial $H_n \in \mathbb{H}_n^d$ is harmonic if and only if*

$$\int_{\mathbb{S}^{d-1}} H_n(\boldsymbol{\xi})\,\overline{H_{n-2}(\boldsymbol{\xi})}\,dS^{d-1}(\boldsymbol{\xi})=0 \quad \forall\, H_{n-2}\in\mathbb{H}_{n-2}^d. \qquad (2.53)$$

Proof. ($\Longleftarrow$) Use (2.52) to obtain

$$H_n(\boldsymbol{\xi})=Y_n(\boldsymbol{\xi})+Y_{n-2}(\boldsymbol{\xi})+\cdots+Y_{n-2\,[n/2]}(\boldsymbol{\xi}).$$

Then by (2.53) and the orthogonality of spherical harmonics of different order (Corollary 2.15), we obtain

$$0=\int_{\mathbb{S}^{d-1}} H_n(\boldsymbol{\xi})\,\overline{Y_{n-2j}(\boldsymbol{\xi})}\,dS^{d-1}(\boldsymbol{\xi})$$

$$=\int_{\mathbb{S}^{d-1}} |Y_{n-2j}(\boldsymbol{\xi})|^2 dS^{d-1}(\boldsymbol{\xi}), \quad 1\le j\le[n/2].$$

So $Y_{n-2j}\equiv 0$ for $1\le j\le[n/2]$ and $H_n(\boldsymbol{x})=Y_n(\boldsymbol{x})$ is harmonic.

($\Longrightarrow$) Assume $H_n \in \mathbb{Y}_n(\mathbb{S}^d)$ is harmonic. Recalling (2.52), we write an arbitrary $H_{n-2}\in\mathbb{H}_{n-2}^d$ as

$$H_{n-2}(\boldsymbol{x})=Y_{n-2}(\boldsymbol{x})+|\boldsymbol{x}|^2 Y_{n-4}(\boldsymbol{x})+\cdots+|\boldsymbol{x}|^{2\,[(n-2)/2]}Y_{n-2-2\,[(n-2)/2]}(\boldsymbol{x}).$$

Then,

$$\int_{\mathbb{S}^{d-1}} H_n(\boldsymbol{\xi})\,\overline{H_{n-2}(\boldsymbol{\xi})}\,dS^{d-1}(\boldsymbol{\xi})=\sum_{j=0}^{[(n-2)/2]}\int_{\mathbb{S}^{d-1}} H_n(\boldsymbol{\xi})\,\overline{Y_{n-2-2j}(\boldsymbol{\xi})}\,dS^{d-1}(\boldsymbol{\xi})$$

$$=0,$$

again using the fact that spherical harmonics of different order are orthogonal.
□

Now we discuss the question of how to determine the harmonic polynomials Y_n, Y_{n-2}, ..., $Y_{n-2\,[n/2]}$ in the decomposition (2.52) for an arbitrary homogeneous polynomial H_n of degree n. Since $H_n(\boldsymbol{x})$ is homogeneous of degree n,

$$H_n(\lambda\,\boldsymbol{x}) = \lambda^n H_n(\boldsymbol{x}) \quad \forall\,\lambda \in \mathbb{R},\ \boldsymbol{x} \in \mathbb{R}^d.$$

We differentiate this equality with respect to λ and then set $\lambda = 1$ to obtain

$$\sum_{i=1}^{d} x_i \frac{\partial H_n(\boldsymbol{x})}{\partial x_i} = n\,H_n(\boldsymbol{x}), \quad H_n \in \mathbb{H}_n^d. \tag{2.54}$$

Consider the function $r^m H_n(\boldsymbol{x})$ with $r = |\boldsymbol{x}|$ and $m \in \mathbb{N}_0$. Note that

$$\frac{\partial r}{\partial x_i} = \frac{x_i}{r}, \quad 1 \le i \le d.$$

We take derivatives of the function $r^m H_n(\boldsymbol{x})$ to obtain

$$\frac{\partial}{\partial x_i}\left(r^m H_n(\boldsymbol{x})\right) = m\,r^{m-2} x_i H_n(\boldsymbol{x}) + r^m \frac{\partial H_n(\boldsymbol{x})}{\partial x_i},$$

$$\frac{\partial^2}{\partial x_i^2}\left(r^m H_n(\boldsymbol{x})\right) = \left[m\,(m-2)\,r^{m-4}x_i^2 + m\,r^{m-2}\right] H_n(\boldsymbol{x})$$

$$+ 2\,m\,r^{m-2}x_i \frac{\partial H_n(\boldsymbol{x})}{\partial x_i} + r^m \frac{\partial^2 H_n(\boldsymbol{x})}{\partial x_i^2},$$

and hence, using (2.54),

$$\Delta\left(r^m H_n(\boldsymbol{x})\right) = m\,(d + 2n + m - 2)\,r^{m-2} H_n(\boldsymbol{x}) + r^m \Delta H_n(\boldsymbol{x})$$

$$\forall\,H_n \in \mathbb{H}_n^d. \tag{2.55}$$

In particular, if $H_n(\boldsymbol{x}) = Y_n(\boldsymbol{x})$ is harmonic, then

$$\Delta\left(r^m Y_n(\boldsymbol{x})\right) = m\,(d + 2n + m - 2)\,r^{m-2} Y_n(\boldsymbol{x}) \quad \forall\,Y_n \in \mathbb{Y}_n(\mathbb{R}^d). \tag{2.56}$$

For $H_n \in \mathbb{H}_n^d$, we write (2.52) in a compact form

$$H_n(\boldsymbol{x}) = \sum_{j=0}^{[n/2]} |\boldsymbol{x}|^{2j} Y_{n-2j}(\boldsymbol{x}). \tag{2.57}$$

Apply the Laplacian operator Δ to both sides of (2.57) and use the formula (2.56),

$$\Delta H_n(\boldsymbol{x}) = \sum_{j=1}^{[n/2]} 2j\,(d + 2n - 2j - 2)\,|\boldsymbol{x}|^{2(j-1)} Y_{n-2j}(\boldsymbol{x}).$$

In general, for $k \geq 1$ an integer, we have

$$\Delta^k H_n(\boldsymbol{x}) = \sum_{j=k}^{[n/2]} 2j \cdot 2(j-1) \cdots 2(j-(k-1)) \, (d+2n-2j-2)$$

$$\cdot (d+2n-2j-4) \cdots (d+2n-2j-2k) \, |\boldsymbol{x}|^{2(j-k)} Y_{n-2j}(\boldsymbol{x}).$$

Using the notation of double factorial,

$$\Delta^k H_n(\boldsymbol{x}) = \sum_{j=k}^{[n/2]} \frac{(2j)!! \, (d+2n-2j-2)!!}{(2j-2k)!! \, (d+2n-2j-2k-2)!!} |\boldsymbol{x}|^{2(j-k)} Y_{n-2j}(\boldsymbol{x}).$$

$$(2.58)$$

By taking $k = [n/2]$, $[n/2] - 1$, ..., 1, 0 in (2.58), we can obtain in turn $Y_{n-2[n/2]}$, ..., $Y_n(\boldsymbol{x})$. In particular, for n even,

$$\Delta^{n/2} H_n(\boldsymbol{x}) = \frac{n!! \, (d+n-2)!!}{(d-2)!!} Y_0(\boldsymbol{x}).$$

Hence,

$$Y_0(\boldsymbol{x}) = \frac{(d-2)!!}{n!! \, (d+n-2)!!} \Delta^{n/2} H_n(\boldsymbol{x}). \qquad (2.59)$$

Example 2.21. Write

$$x_i^2 = Y_2(\boldsymbol{x}) + |\boldsymbol{x}|^2 Y_0(\boldsymbol{x}).$$

We first apply (2.59) to get

$$Y_0(\boldsymbol{x}) = \frac{1}{d}.$$

We then use (2.58) with $n = 2$ and $k = 0$ to obtain

$$Y_2(\boldsymbol{x}) = x_i^2 - \frac{1}{d} |\boldsymbol{x}|^2.$$

Hence, we have the decomposition

$$x_i^2 = \left(x_i^2 - \frac{1}{d} |\boldsymbol{x}|^2 \right) + |\boldsymbol{x}|^2 \frac{1}{d}, \quad 1 \leq i \leq d.$$

The same technique can be applied for higher degree homogeneous polynomials. $\qquad\square$

2.5   The Funk–Hecke Formula

The Funk–Hecke formula is useful in simplifying calculations of certain integrals over $\mathbb{S}^{d-1}$, cf. Sect. 3.7 for some examples. Introduce a weighted L^1 space

$$L^1_{(d-3)/2}(-1,1) := \left\{ f \text{ measurable on } (-1,1) : \|f\|_{L^1_{(d-3)/2}(-1,1)} < \infty \right\}$$
$$(2.60)$$

with the norm

$$\|f\|_{L^1_{(d-3)/2}(-1,1)} := \int_{-1}^{1} |f(t)| (1-t^2)^{(d-3)/2} dt.$$

Note that for $d \geq 2$, $C[-1,1] \subset L^1_{(d-3)/2}(-1,1)$. In the rest of the section, we assume $d \geq 2$.

Recall the projection operator $\mathcal{P}_{n,d}$ defined in (2.44). Given $f \in L^1_{(d-3)/2}(-1,1)$ and $\boldsymbol{\xi} \in \mathbb{S}^{d-1}$, define $f_{\boldsymbol{\xi}}(\boldsymbol{\eta}) = f(\boldsymbol{\xi} \cdot \boldsymbol{\eta})$ for $\boldsymbol{\eta} \in \mathbb{S}^{d-1}$. Then $(\mathcal{P}_{n,d} f_{\boldsymbol{\xi}})_A = \mathcal{P}_{n,d} f_{\boldsymbol{\xi}}$ for any $A \in \mathbb{O}^d(\boldsymbol{\xi})$. Since $\mathcal{P}_{n,d} f_{\boldsymbol{\xi}} \in \mathbb{Y}_n^d$, by Theorem 2.8, it is a multiple of $P_{n,d}(\boldsymbol{\xi} \cdot)$:

$$(\mathcal{P}_{n,d} f_{\boldsymbol{\xi}})(\boldsymbol{\eta}) = \lambda_n \frac{N_{n,d}}{|\mathbb{S}^{d-1}|} P_{n,d}(\boldsymbol{\xi} \cdot \boldsymbol{\eta}).$$

This is rewritten as, following the definition (2.44),

$$\lambda_n P_{n,d}(\boldsymbol{\xi} \cdot \boldsymbol{\eta}) = \int_{\mathbb{S}^{d-1}} P_{n,d}(\boldsymbol{\zeta} \cdot \boldsymbol{\eta}) f(\boldsymbol{\xi} \cdot \boldsymbol{\zeta}) \, dS^{d-1}(\boldsymbol{\zeta}). \qquad (2.61)$$

We determine the constant λ_n by setting $\boldsymbol{\eta} = \boldsymbol{\xi}$ in (2.61):

$$\lambda_n = \int_{\mathbb{S}^{d-1}} P_{n,d}(\boldsymbol{\xi} \cdot \boldsymbol{\zeta}) f(\boldsymbol{\xi} \cdot \boldsymbol{\zeta}) \, dS^{d-1}(\boldsymbol{\zeta}).$$

The integral does not depend on $\boldsymbol{\xi}$ and we may take $\boldsymbol{\xi} = e_d$. Then using (1.16),

$$\lambda_n = |\mathbb{S}^{d-2}| \int_{-1}^{1} P_{n,d}(t) f(t) (1-t^2)^{\frac{d-3}{2}} dt. \qquad (2.62)$$

Let $Y_n \in \mathbb{Y}_n^d$ be arbitrary yet fixed. Multiply (2.61) by Y_n and integrate over $\mathbb{S}^{d-1}$ with respect to $\boldsymbol{\eta}$:

$$\lambda_n \int_{\mathbb{S}^{d-1}} P_{n,d}(\boldsymbol{\xi} \cdot \boldsymbol{\eta}) Y_n(\boldsymbol{\eta}) \, dS^{d-1}(\boldsymbol{\eta})$$
$$= \int_{\mathbb{S}^{d-1}} f(\boldsymbol{\xi} \cdot \boldsymbol{\zeta}) \left(\int_{\mathbb{S}^{d-1}} P_{n,d}(\boldsymbol{\zeta} \cdot \boldsymbol{\eta}) Y_n(\boldsymbol{\eta}) \, dS^{d-1}(\boldsymbol{\eta}) \right) dS^{d-1}(\boldsymbol{\zeta}). \qquad (2.63)$$

Applying the addition theorem, Theorem 2.9, we see that

$$\int_{\mathbb{S}^{d-1}} P_{n,d}(\boldsymbol{\eta} \cdot \boldsymbol{\zeta}) Y_n(\boldsymbol{\eta}) \, dS^{d-1}(\boldsymbol{\eta}) = \frac{|\mathbb{S}^{d-1}|}{N_{n,d}} \sum_{j=1}^{N_{n,d}} (Y_n, Y_{n,j})_{\mathbb{S}^{d-1}} Y_{n,j}(\boldsymbol{\zeta}),$$

i.e.,

$$\int_{\mathbb{S}^{d-1}} P_{n,d}(\boldsymbol{\eta}\cdot\boldsymbol{\zeta})Y_n(\boldsymbol{\eta})\,dS^{d-1}(\boldsymbol{\eta}) = \frac{|\mathbb{S}^{d-1}|}{N_{n,d}}\,Y_n(\boldsymbol{\zeta}).\qquad(2.64)$$

Hence, from (2.63),

$$\int_{\mathbb{S}^{d-1}} f(\boldsymbol{\xi}\cdot\boldsymbol{\eta})\,Y_n(\boldsymbol{\eta})\,dS^{d-1}(\boldsymbol{\eta}) = \lambda_n Y_n(\boldsymbol{\xi}).\qquad(2.65)$$

We summarize the result in the form of a theorem.

Theorem 2.22 (Funk–Hecke Formula). *Let $f \in L^1_{(d-3)/2}(-1,1)$, $\boldsymbol{\xi} \in \mathbb{S}^{d-1}$ and $Y_n \in \mathbb{Y}_n^d$. Then the Funk–Hecke formula (2.65) holds with the constant λ_n given by (2.62).*

From (2.65), we can deduce the following statement using the formula (2.24). Assume $f \in L^1_{(d-3)/2}(-1,1)$. Then

$$\int_{\mathbb{S}^{d-1}} f(\boldsymbol{\xi}\cdot\boldsymbol{\zeta})\,P_{n,d}(\boldsymbol{\eta}\cdot\boldsymbol{\zeta})\,dS^{d-1}(\boldsymbol{\zeta}) = \lambda_n P_{n,d}(\boldsymbol{\xi}\cdot\boldsymbol{\eta})\quad \forall\boldsymbol{\xi},\boldsymbol{\eta}\in\mathbb{S}^{d-1},\,n\in\mathbb{N}_0,$$
$$(2.66)$$

where λ_n is given by the formula (2.62).

Letting $f = P_{n,d}$ in (2.65) and comparing it with (2.64), we deduce the formula

$$\int_{-1}^1 [P_{n,d}(t)]^2\,(1-t^2)^{\frac{d-3}{2}}\,dt = \frac{|\mathbb{S}^{d-1}|}{N_{n,d}|\mathbb{S}^{d-2}|},\qquad(2.67)$$

which is equivalent to (2.40).

2.6   Legendre Polynomials: Representation Formulas

Further studies of spherical harmonics require a deeper knowledge of the Legendre polynomials. In this section, we present compact formulas for the Legendre polynomial $P_{n,d}$ defined in (2.19): one differential formula (Rodrigues representation formula) and some integral representation formulas. These formulas are used in proving properties of the Legendre polynomials in Sect. 2.7.

2.6.1   Rodrigues Representation Formula

By Corollary 2.15,

$$\int_{\mathbb{S}^{d-1}} P_{m,d}(\boldsymbol{\xi}\cdot\boldsymbol{\zeta})\,P_{n,d}(\boldsymbol{\xi}\cdot\boldsymbol{\zeta})\,dS^{d-1}(\boldsymbol{\xi}) = 0\quad\text{for }m \neq n.$$

By the formula (1.17), the left side integral equals

$$\int_{\mathbb{S}^{d-2}} \left(\int_{-1}^{1} P_{m,d}(t)\, P_{n,d}(t) \left(1 - t^2\right)^{\frac{d-3}{2}} dt \right) dS^{d-2}$$

$$= |\mathbb{S}^{d-2}| \int_{-1}^{1} P_{m,d}(t)\, P_{n,d}(t) \left(1 - t^2\right)^{\frac{d-3}{2}} dt.$$

So

$$\int_{-1}^{1} P_{m,d}(t)\, P_{n,d}(t) \left(1 - t^2\right)^{\frac{d-3}{2}} dt = 0 \quad \text{for } m \neq n. \tag{2.68}$$

Consequently, denoting P_m a polynomial of degree less than or equal to m, we have the orthogonality

$$\int_{-1}^{1} P_m(t)\, P_{n,d}(t) \left(1 - t^2\right)^{\frac{d-3}{2}} dt = 0, \quad m < n. \tag{2.69}$$

The Legendre polynomials are determined by the orthogonality relation (2.68) and the normalization condition $P_{n,d}(1) = 1$.

Theorem 2.23 (Rodrigues representation formula).

$$P_{n,d}(t) = (-1)^n R_{n,d}(1 - t^2)^{\frac{3-d}{2}} \left(\frac{d}{dt} \right)^n (1 - t^2)^{n + \frac{d-3}{2}} \quad \text{for } d \geq 2, \tag{2.70}$$

where the Rodrigues constant

$$R_{n,d} = \frac{\Gamma(\frac{d-1}{2})}{2^n \Gamma(n + \frac{d-1}{2})}. \tag{2.71}$$

Proof. The function

$$p_n(t) = (1 - t^2)^{\frac{3-d}{2}} \left(\frac{d}{dt} \right)^n (1 - t^2)^{n + \frac{d-3}{2}}$$

is easily seen to be a polynomial of degree n. Let us show that these polynomials are orthogonal with respect to the weight $\left(1 - t^2\right)^{\frac{d-3}{2}}$. For $n > m$,

$$\int_{-1}^{1} p_n(t)\, p_m(t) \left(1 - t^2\right)^{\frac{d-3}{2}} dt = \int_{-1}^{1} p_m(t) \left(\frac{d}{dt} \right)^n (1 - t^2)^{n + \frac{d-3}{2}} dt.$$

Performing integration by parts n times shows that the integral is zero.

The value $p_n(1)$ is calculated as follows:

$$p_n(1) = (1-t^2)^{\frac{3-d}{2}} \left(\frac{d}{dt}\right)^n \left[(1+t)^{n+\frac{d-3}{2}}(1-t)^{n+\frac{d-3}{2}}\right]\Big|_{t=1}$$

$$= (1-t^2)^{\frac{3-d}{2}}(1+t)^{n+\frac{d-3}{2}} \left(\frac{d}{dt}\right)^n (1-t)^{n+\frac{d-3}{2}}\Big|_{t=1}$$

$$= (-1)^n \left(\frac{d-1}{2}\right)_n (1+t)^n\Big|_{t=1}$$

$$= (-1)^n \frac{2^n \Gamma(n+\frac{d-1}{2})}{\Gamma(\frac{d-1}{2})},$$

where the formula (1.12) for Pochhammer's symbol $((d-1)/2)_n$ is used. Hence,

$$P_{n,d}(t) = (-1)^n R_{n,d} p_n(t),$$

which is the stated formula.   □

In the case $d = 3$, we recover the Rodrigues representation formula for the standard Legendre polynomials:

$$P_{n,3}(t) = \frac{1}{2^n n!} \left(\frac{d}{dt}\right)^n (t^2-1)^n, \quad n \in \mathbb{N}_0.$$

In the case $d = 2$, we use the relation

$$\Gamma\left(n+\frac{1}{2}\right) = \frac{(2n)!}{2^{2n}n!}\Gamma\left(\frac{1}{2}\right),$$

derived from a repeated application of (1.6), and obtain

$$P_{n,2}(t) = (-1)^n \frac{2^n n!}{(2n)!}(1-t^2)^{\frac{1}{2}} \left(\frac{d}{dt}\right)^n (1-t^2)^{n-\frac{1}{2}}, \quad n \in \mathbb{N}_0.$$

This formula is not convenient to use. A more familiar form is given by the Chebyshev polynomial:

$$P_{n,2}(t) = \cos(n \arccos t), \quad t \in [-1,1].$$

This result is verified by showing $\cos(n \arccos t)$ is a polynomial of degree n, has a value 1 at $t = 1$, and these polynomials satisfy the orthogonality condition (2.68) with $d = 2$. See also the derivation leading to (2.29).

In the case $d = 4$, we can similarly verify the formula

$$P_{n,4}(t) = \frac{1}{n+1} U_n(t), \quad t \in [-1,1],$$

where

$$U_n(t) = \frac{1}{n+1} \, P'_{n+1,2}(t)$$

is the nth degree Chebyshev polynomial of the second kind. For $-1 < t < 1$, we have the formula

$$U_n(t) = \frac{\sin((n+1) \arccos t)}{\sin(\arccos t)}.$$

We note that the Legendre polynomial $P_{n,d}(t)$ is proportional to the Jacobi polynomial $P_n^{(\alpha,\alpha)}(t)$ with $\alpha = (d-3)/2$. The Jacobi polynomials $P_n^{(\alpha,\beta)}(t)$ are introduced in Sect. 4.3.1.

2.6.2 Integral Representation Formulas

In addition to the Rodrigues representation formula (2.70), there are integral representation formulas for the Legendre polynomials which are useful in showing certain properties of the Legendre polynomials.

Let $d \geq 3$. For a fixed $\boldsymbol{\eta} \in \mathbb{S}^{d-2}$, the function $\boldsymbol{x} \mapsto (x_d + i\,\boldsymbol{x}_{(d-1)} \cdot \boldsymbol{\eta})^n$ is a homogeneous harmonic polynomial of degree n. Consider its average with respect to $\boldsymbol{\eta} \in \mathbb{S}^{d-2}$,

$$L_n(\boldsymbol{x}) = \frac{1}{|\mathbb{S}^{d-2}|} \int_{\mathbb{S}^{d-2}} \left(x_d + i\,\boldsymbol{x}_{(d-1)} \cdot \boldsymbol{\eta} \right)^n dS^{d-2}(\boldsymbol{\eta}).$$

This function is a homogeneous harmonic of degree n. For $A \in \mathbb{O}^d(\boldsymbol{e}_d)$, we recall (2.1) and write

$$A\boldsymbol{x} = \begin{pmatrix} A_1 \boldsymbol{x}_{(d-1)} \\ x_d \end{pmatrix}, \quad A_1 \in \mathbb{O}^{d-1}.$$

Note that here we view $\boldsymbol{x}_{(d-1)}$ as a vector in $\mathbb{S}^{d-2}$. Then

$$L_n(A\boldsymbol{x}) = \frac{1}{|\mathbb{S}^{d-2}|} \int_{\mathbb{S}^{d-2}} \left(x_d + i\,\boldsymbol{x}_{(d-1)} \cdot A_1^T \boldsymbol{\eta} \right)^n dS^{d-2}(\boldsymbol{\eta}).$$

With a change of variable $\boldsymbol{\zeta} = A_1^T \boldsymbol{\eta}$, we have

$$L_n(A\boldsymbol{x}) = \frac{1}{|\mathbb{S}^{d-2}|} \int_{\mathbb{S}^{d-2}} \left(x_d + i\,\boldsymbol{x}_{(d-1)} \cdot \boldsymbol{\zeta} \right)^n dS^{d-2}(\boldsymbol{\zeta}),$$

which coincides with $L_n(\boldsymbol{x})$. Moreover, $L_n(\boldsymbol{e}_d) = 1$. Thus, $L_n(\boldsymbol{x})$ is the Legendre harmonic of degree n in dimension d. By the relation (2.21), we see that

$$P_{n,d}(t) = \frac{1}{|\mathbb{S}^{d-2}|} \int_{\mathbb{S}^{d-2}} \left[t + i\,(1-t^2)^{1/2} \boldsymbol{\xi}_{(d-1)} \cdot \boldsymbol{\eta} \right]^n dS^{d-2}(\boldsymbol{\eta}), \quad t \in [-1,1].$$

In this formula, $\boldsymbol{\xi}_{(d-1)} \in \mathbb{S}^{d-2}$ is arbitrary. In particular, choosing $\boldsymbol{\xi}_{(d-1)} = (0,\cdots,0,1)^T$ in $\mathbb{S}^{d-2}$ and applying (1.17), we obtain the first integral representation formula for the Legendre polynomials.

Theorem 2.24. *For $n \in \mathbb{N}_0$ and $d \geq 3$,*

$$P_{n,d}(t) = \frac{|\mathbb{S}^{d-3}|}{|\mathbb{S}^{d-2}|} \int_{-1}^{1} \left[t + i\,(1-t^2)^{1/2} s \right]^n (1-s^2)^{\frac{d-4}{2}} ds, \quad t \in [-1,1]. \quad (2.72)$$

An easy consequence of the representation formula (2.72) is that $P_{n,d}(t)$ has the same parity as the integer n, i.e.,

$$P_{n,d}(-t) = (-1)^n P_{n,d}(t), \quad -1 \leq t \leq 1. \quad (2.73)$$

There is another useful integral representation formula that can be derived from (2.72). Recall definitions of hyper-trigonometric functions:

$$\sinh x := \frac{e^x - e^{-x}}{2}, \quad \cosh x := \frac{e^x + e^{-x}}{2},$$

$$\tanh x := \frac{\sinh x}{\cosh x} = \frac{e^x - e^{-x}}{e^x + e^{-x}}$$

and differentiation formulas

$$(\sinh x)' = \cosh x, \quad (\cosh x)' = \sinh x, \quad (\tanh x)' = \frac{1}{\cosh^2 x}.$$

Use the change of variable

$$s = \tanh u, \quad u \in \mathbb{R}. \quad (2.74)$$

We have $s \to 1-$ as $u \to \infty$, $s \to -1+$ as $u \to -\infty$, and

$$ds = \frac{1}{\cosh^2 u} du, \quad 1 - s^2 = \frac{1}{\cosh^2 u}. \quad (2.75)$$

Since $P_{n,d}(-t) = (-1)^n P_{n,d}(t)$ by (2.73), it is sufficient to consider the case $t \in (0,1]$ for the second integral representation formula. Write

$$t + i\,(1-t^2)^{1/2} = e^{i\theta}$$

for a uniquely determined $\theta \in [0, \pi/2)$. Then $t = \cos\theta$ and

$$t + i\,(1-t^2)^{1/2} s = \cos\theta + i\tanh u \sin\theta.$$

The hyper-trigonometric functions are defined for complex variables and it can be verified that

$$\cos\theta + i\tanh u \sin\theta = \frac{\cosh(u + i\theta)}{\cosh u}.$$

Thus,

$$\int_{-1}^{1}\left[t + i\,(1 - t^2)^{1/2}s\right]^n (1 - s^2)^{\frac{d-4}{2}}ds = \int_{-\infty}^{\infty}\frac{\cosh^n(u + i\theta)}{\cosh^{n+d-2}u}\,du.$$

The integrand is a meromorphic function of u with poles at $u = i\pi\,(k + 1/2)$, $k \in \mathbb{Z}$. We then apply the Cauchy integral theorem in complex analysis [2] to obtain

$$\int_{-\infty}^{\infty}\frac{\cosh^n(u + i\theta)}{\cosh^{n+d-2}u}\,du = \int_{-\infty}^{\infty}\frac{\cosh^n u}{\cosh^{n+d-2}(u - i\theta)}\,du.$$

Return back to the variable s, using the relation

$$\cosh(u - i\theta) = \cosh u\left[t - i(1 - t^2)^{1/2}s\right]$$

together with (2.74) and (2.75),

$$P_{n,d}(t) = \frac{|\mathbb{S}^{d-3}|}{|\mathbb{S}^{d-2}|}\int_{-1}^{1}\frac{(1 - s^2)^{\frac{d-4}{2}}}{\left[t - i\,(1 - t^2)^{1/2}s\right]^{n+d-2}}\,ds.$$

Note that changing s to $-s$ for the integrand leads to another integral representation formula for $P_{n,d}(t)$. In summary, the following result holds.

Theorem 2.25. *For $n \in \mathbb{N}_0$ and $d \geq 3$,*

$$P_{n,d}(t) = \frac{|\mathbb{S}^{d-3}|}{|\mathbb{S}^{d-2}|}\int_{-1}^{1}\frac{(1 - s^2)^{\frac{d-4}{2}}}{\left[t \pm i\,(1 - t^2)^{1/2}s\right]^{n+d-2}}\,ds, \quad t \in (0, 1]. \tag{2.76}$$

2.7 Legendre Polynomials: Properties

In this section, we explore properties of the Legendre polynomials by using the compact presentation formulas given in Sect. 2.6.

2.7.1   Integrals, Orthogonality

The following result is useful in computing integrals involving the Legendre polynomials.

Proposition 2.26. *If $f \in C^n([-1,1])$, then*

$$\int_{-1}^{1} f(t)\, P_{n,d}(t)\, (1-t^2)^{\frac{d-3}{2}}\, dt = R_{n,d} \int_{-1}^{1} f^{(n)}(t)\, (1-t^2)^{n+\frac{d-3}{2}}\, dt, \quad (2.77)$$

where the constant $R_{n,d}$ is given in (2.71).

Proof. By the Rodrigues representation formula (2.70), the left side of (2.77) is

$$(-1)^n R_{n,d} \int_{-1}^{1} f(t) \left(\frac{d}{dt}\right)^n (1-t^2)^{n+\frac{d-3}{2}}\, dt.$$

Performing integration by parts n times on this integral leads to (2.77). $\square$

Recall the formula (2.40) or (2.67),

$$\int_{-1}^{1} [P_{n,d}(t)]^2\, (1-t^2)^{\frac{d-3}{2}}\, dt = \frac{|\mathbb{S}^{d-1}|}{N_{n,d}\, |\mathbb{S}^{d-2}|}. \quad (2.78)$$

Combining (2.68) and (2.78), we have the orthogonality relation

$$\int_{-1}^{1} P_{m,d}(t)\, P_{n,d}(t)\, (1-t^2)^{\frac{d-3}{2}}\, dt = \frac{|\mathbb{S}^{d-1}|}{N_{n,d}\, |\mathbb{S}^{d-2}|}\, \delta_{mn}. \quad (2.79)$$

Using (1.18), we can rewrite (2.78) as

$$\int_{-1}^{1} [P_{n,d}(t)]^2\, (1-t^2)^{\frac{d-3}{2}}\, dt = \frac{\sqrt{\pi}\,\Gamma(\frac{d-1}{2})}{N_{n,d}\,\Gamma(\frac{d}{2})}.$$

In particular, for $d=3$, $N_{n,3} = 2n+1$ and

$$\int_{-1}^{1} [P_{n,3}(t)]^2\, dt = \frac{2}{2n+1}.$$

For $d=2$, $N_{n,2} = 2$ and

$$\int_{-1}^{1} [P_{n,2}(t)]^2\, (1-t^2)^{-\frac{1}{2}}\, dt = \frac{\pi}{2}.$$

We can verify this result easily by a direct calculation using the formula

$$P_{n,2}(t) = \cos(n \arccos t).$$

2.7.2 Differential Equation and Distribution of Roots

First we derive a differential equation satisfied by the Legendre polynomial $P_{n,d}(t)$. Introduce a second-order differential operator L_d defined by

$$L_d g(t) := \left(1 - t^2\right)^{\frac{3-d}{2}} \frac{d}{dt}\left[\left(1 - t^2\right)^{\frac{d-1}{2}} \frac{d}{dt} g(t)\right], \quad g \in C^2[-1,1].$$

Also introduce a weighted inner product

$$(f,g)_d := \int_{-1}^{1} f(t)\, g(t) \left(1 - t^2\right)^{\frac{d-3}{2}} dt.$$

Then through integration by parts, we have

$$(L_d f, g)_d = (f, L_d g)_d \quad \forall\, f, g \in C^2[-1,1]. \tag{2.80}$$

Thus, the operator L_d is self-adjoint with respect to the weighted inner product $(\cdot, \cdot)_d$.

Consider the function $L_d P_{n,d}(t)$. Since

$$L_d g(t) = (1 - t^2)\, g''(t) - (d - 1)\, t\, g'(t),$$

we see that if $p_n(t)$ is a polynomial of degree n, then so is $L_d p_n(t)$. Let $0 \le m \le n - 1$. By the weighted orthogonality relation (2.69), we have

$$(P_{n,d}, L_d P_{m,d})_d = 0.$$

Then by (2.80),

$$(P_{m,d}, L_d P_{n,d})_d = 0, \quad 0 \le m \le n - 1.$$

Thus, the polynomial $L_d P_{n,d}(t)$ must be a multiple of $P_{n,d}(t)$. Writing

$$P_{n,d}(t) = a_{n,d}^0 t^n + \text{l.d.t.} \tag{2.81}$$

Here l.d.t. stands for the lower degree terms. We have

$$L_d P_{n,d}(t) = -n(n+d-2) a_{n,d}^0 t^n + \text{l.d.t.}$$

Hence,

$$L_d P_{n,d}(t) + n(n+d-2) P_{n,d}(t) = 0.$$

So $P_{n,d}$ is an eigenfunction for the differential operator $-L_d$ corresponding to the eigenvalue $n(n+d-2)$. In other words, the Legendre polynomial $P_{n,d}(t)$ satisfies the differential equation

$$(1-t^2)^{\frac{3-d}{2}} \frac{d}{dt}\left[(1-t^2)^{\frac{d-1}{2}} \frac{d}{dt} P_{n,d}(t)\right] + n(n+d-2) P_{n,d}(t) = 0, \quad (2.82)$$

which can also be written as

$$(1-t^2) P_{n,d}''(t) - (d-1) t\, P_{n,d}'(t) + n(n+d-2) P_{n,d}(t) = 0. \quad (2.83)$$

Next, we present a result regarding distributions of the roots of the Legendre polynomials. This result plays an important role in the theory of Gaussian quadratures. From the differential equation (2.83), we deduce that $P_{n,d}(t)$ and $P_{n,d}'(t)$ cannot both vanish at any point in $(-1,1)$; in other words, $P_{n,d}(t)$ has no multiple roots in $(-1,1)$. Assume $P_{n,d}(t)$ has k distinct roots $t_1, \cdots, t_k$ in the interval $(-1,1)$, and $k < n$. Then

$$p_k(t) = (t-t_1)\cdots(t-t_k)$$

is a polynomial of degree k, $p_k(1) > 0$, and $P_{n,d}(t) = q_{n-k}(t) p_k(t)$ with a polynomial q_{n-k} of degree $n-k$. Since the polynomial $q_{n-k}(t)$ does not change sign in $(-1,1)$ and is positive at 1, it is positive in $(-1,1)$. So

$$\int_{-1}^1 P_{n,d}(t) p_k(t) \left(1-t^2\right)^{\frac{d-3}{2}} dt = \int_{-1}^1 q_{n-k}(t) p_k(t)^2 \left(1-t^2\right)^{\frac{d-3}{2}} dt > 0.$$

However, since $k < n$, the integral on the left side is zero and this leads to contradiction. We summarize the result in the form of a proposition.

Proposition 2.27. *The Legendre polynomial $P_{n,d}(t)$ has exactly n distinct roots in $(-1,1)$.*

For n even, $P_{n,d}(t)$ is an even function so that its roots can be written as $\pm t_1, \ldots, \pm t_{n/2}$ with $0 < t_1 < \cdots < t_{n/2} < 1$. For n odd, $P_{n,d}(t)$ is an odd function so that its roots can be written as $0, \pm t_1, \ldots, \pm t_{(n-1)/2}$ with $0 < t_1 < \cdots < t_{(n-1)/2} < 1$.

In the particular case $d = 2$, it is easy to find the n roots of the equation

$$P_{n,2}(t) = \cos(n \arccos t) = 0$$

to be

$$t_j = \cos \frac{(2j+1)\,\pi}{2\,n}, \quad 0 \le j \le n-1.$$

For $n = 2k$ even, noting that $t_{2k-1-j} = -t_j$, we can list the roots as

$$\pm t_0, \pm t_1, \cdots, \pm t_{k-1} \text{ with } t_j = \cos \frac{(2j+1)\,\pi}{4k}, \; 0 \le j \le k-1.$$

For $n = 2k+1$ odd, noting that $t_k = 0$ and $t_{2k-j} = -t_j$, we can list the roots as

$$0, \pm t_0, \pm t_1, \cdots, \pm t_{k-1} \text{ where } t_j = \cos \frac{(2j+1)\,\pi}{2(2k+1)}, \; 0 \le j \le k-1.$$

2.7.3 Recursion Formulas

Recursion formulas are useful in computing values of the Legendre polynomials, especially those of a higher degree.

Let us first determine the leading coefficient $a_{n,d}^0$ of $P_{n,d}(t)$ (see (2.81)). We start with the equality

$$\int_{-1}^{1} [P_{n,d}(t)]^2 \left(1-t^2\right)^{\frac{d-3}{2}} dt = a_{n,d}^0 \int_{-1}^{1} t^n P_{n,d}(t) \left(1-t^2\right)^{\frac{d-3}{2}} dt, \quad (2.84)$$

obtained by an application of the orthogonality property (2.69). By (2.78), the left side of (2.84) equals

$$\frac{|\mathbb{S}^{d-1}|}{N_{n,d}\,|\mathbb{S}^{d-2}|}.$$

Applying Proposition 2.26, we see that the right side of (2.84) equals

$$a_{n,d}^0 R_{n,d} n! \int_{-1}^{1} \left(1-t^2\right)^{n+\frac{d-3}{2}} dt.$$

To compute the integral, we let $s = t^2$:

$$\int_{-1}^{1} \left(1-t^2\right)^{n+\frac{d-3}{2}} dt = \int_{0}^{1} s^{\frac{1}{2}-1}(1-s)^{n+\frac{d-1}{2}-1} ds$$

$$= \frac{\Gamma(\frac{1}{2})\,\Gamma(n+\frac{d-1}{2})}{\Gamma(n+\frac{d}{2})}.$$

Hence,

$$\frac{|\mathbb{S}^{d-1}|}{N_{n,d}\,|\mathbb{S}^{d-2}|} = a_{n,d}^0 R_{n,d} n! \frac{\Gamma(\frac{1}{2})\,\Gamma(n+\frac{d-1}{2})}{\Gamma(n+\frac{d}{2})}.$$

Therefore, the leading coefficient of the Legendre polynomial $P_{n,d}(t)$ is

$$a_{n,d}^0 = \frac{2^{n-1}\Gamma(d-1)\,\Gamma(n+\frac{d-2}{2})}{\Gamma(\frac{d}{2})\Gamma(n+d-2)}. \tag{2.85}$$

As an application of the formula (2.85), we note that

$$\frac{a_{n,d}^0}{a_{n-1,d}^0} = \frac{2n+d-4}{n+d-3}.$$

So

$$(n+d-3)\,P_{n,d}(t) - (2n+d-4)\,t\,P_{n-1,d}(t)$$

is a polynomial of degree $\leq n-1$ and is orthogonal to $P_{k,d}(t)$ with respect to the weighted inner product $(\cdot,\cdot)_d$ for $0 \leq k \leq n-3$. Thus, when this polynomial is expressed as a linear combination of $P_{j,d}(t)$, $0 \leq j \leq n-1$, only the two terms involving $P_{n-2,d}(t)$ and $P_{n-1,d}(t)$ remain. In other words, for two suitable constants c_1 and c_2,

$$(n+d-3)\,P_{n,d}(t) - (2n+d-4)\,t\,P_{n-1,d}(t) = c_1 P_{n-1,d}(t) + c_2 P_{n-2,d}(t).$$

The constants c_1 and c_2 can be found from the above equality at $t = \pm 1$, since $P_{k,d}(1) = 1$ and $P_{k,d}(-1) = (-1)^k$ (cf. (2.73)):

$$c_1 + c_2 = 1 - n,$$

$$c_1 - c_2 = n - 1.$$

The solution of this system is $c_1 = 0$, $c_2 = 1 - n$. Thus, the Legendre polynomials satisfy the recursion relation

$$P_{n,d}(t) = \frac{2n+d-4}{n+d-3}\,t\,P_{n-1,d}(t) - \frac{n-1}{n+d-3}P_{n-2,d}(t), \quad n \geq 2,\ d \geq 2. \tag{2.86}$$

The initial conditions for the recursion formula (2.86) are

$$P_{0,d}(t) = 1, \quad P_{1,d}(t) = t. \tag{2.87}$$

It is convenient to use the recursion formula (2.86) to derive expressions of the Legendre polynomials. The following are some examples. Note that in any dimension d, the first two Legendre polynomials are the same, given by (2.87).

For $d = 2$,

$$P_{2,2}(t) = 2\,t^2 - 1,$$

$$P_{3,2}(t) = 4\,t^3 - 3\,t,$$

$$P_{4,2}(t) = 8\,t^4 - 8\,t^2 + 1,$$

$$P_{5,2}(t) = 16\,t^5 - 20\,t^3 + 5\,t.$$

For $d = 3$,

$$P_{2,3}(t) = \frac{1}{2}\left(3\,t^2 - 1\right),$$

$$P_{3,3}(t) = \frac{1}{2}\left(5\,t^3 - 3\,t\right),$$

$$P_{4,3}(t) = \frac{1}{8}\left(35\,t^4 - 30\,t^2 + 3\right),$$

$$P_{5,3}(t) = \frac{1}{8}\left(63\,t^5 - 70\,t^3 + 15\,t\right).$$

For $d = 4$,

$$P_{2,4}(t) = \frac{1}{3}\left(4\,t^2 - 1\right),$$

$$P_{3,4}(t) = 2\,t^3 - t,$$

$$P_{4,4}(t) = \frac{1}{5}\left(16\,t^4 - 12\,t^2 + 1\right),$$

$$P_{5,4}(t) = \frac{1}{3}\left(16\,t^5 - 16\,t^3 + 3\,t\right).$$

For $d = 5$,

$$P_{2,5}(t) = \frac{1}{4}\left(5\,t^2 - 1\right),$$

$$P_{3,5}(t) = \frac{1}{4}\left(7\,t^3 - 3\,t\right),$$

$$P_{4,5}(t) = \frac{1}{8}\left(21\,t^4 - 14\,t^2 + 1\right),$$

$$P_{5,5}(t) = \frac{1}{8}\left(33\,t^5 - 30\,t^3 + 5\,t\right).$$

Graphs of these Legendre polynomials are found in Figs. 2.1–2.4.

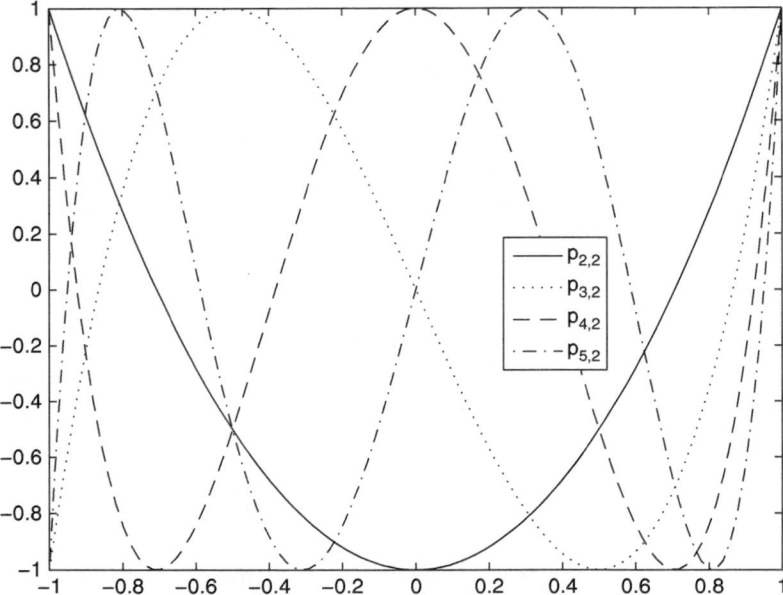

Fig. 2.1 Legendre polynomials for dimension 2

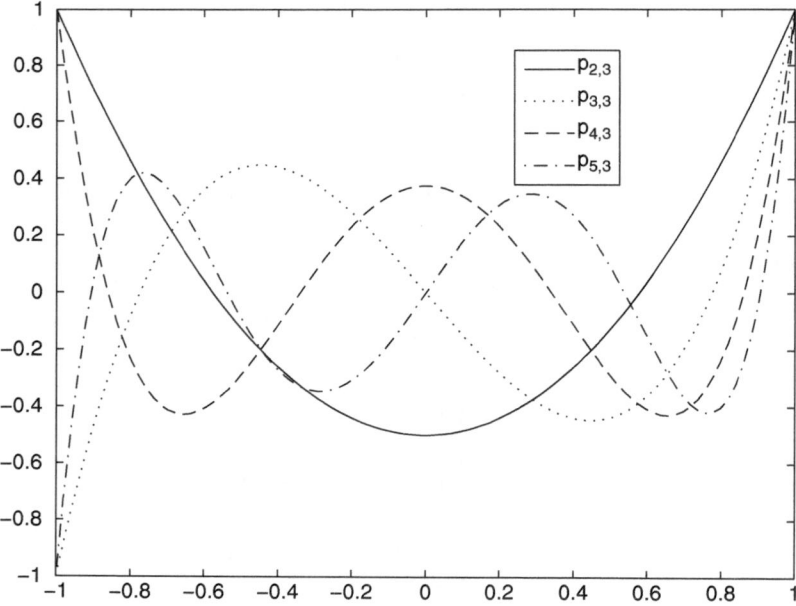

Fig. 2.2 Legendre polynomials for dimension 3

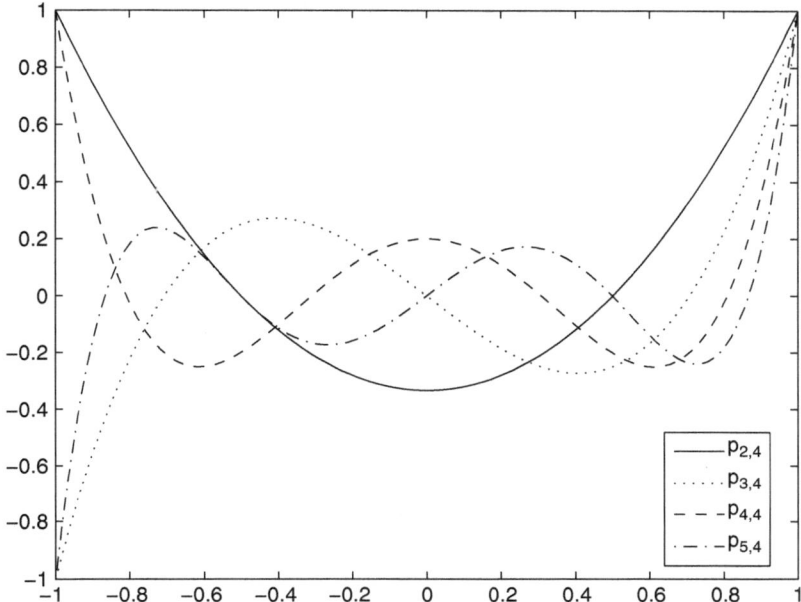

Fig. 2.3 Legendre polynomials for dimension 4

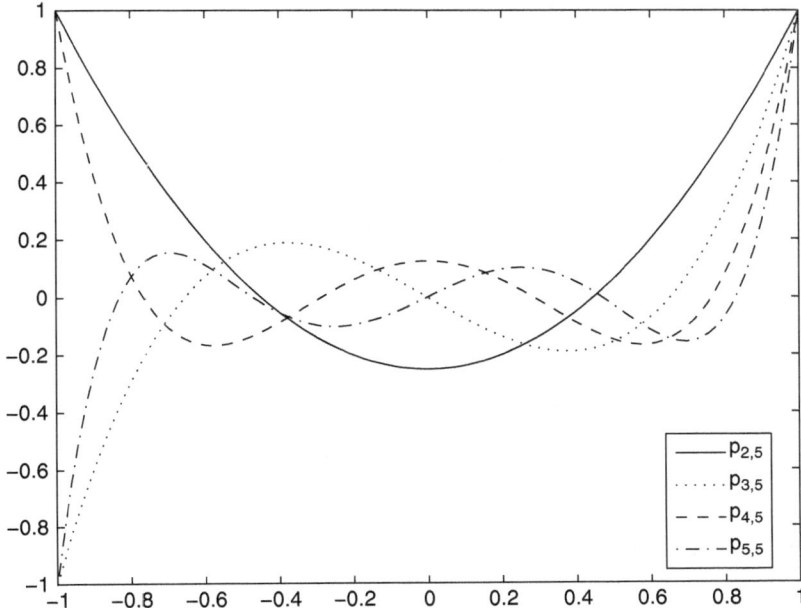

Fig. 2.4 Legendre polynomials for dimension 5

As another application of the formula (2.85), we derive a formula for derivatives of the Legendre polynomials in terms of the polynomials themselves. Note that

$$\frac{a_{n,d}^0}{a_{n-1,d+2}^0} = \frac{n+d-2}{d-1}.$$

So

$$(d-1)\, P_{n,d}'(t) - n\,(n+d-2)\, P_{n-1,d+2}(t) \qquad (2.88)$$

is a polynomial of degree $\leq n-2$. For $k \leq n-2$,

$$\int_{-1}^{1} P_{n,d}'(t)\, P_{k,d+2}(t) \left(1-t^2\right)^{\frac{d-1}{2}} dt$$

$$= -\int_{-1}^{1} P_{n,d}(t)\, \frac{d}{dt}\left[P_{k,d+2}(t)\left(1-t^2\right)^{\frac{d-1}{2}}\right] dt$$

$$= -\int_{-1}^{1} P_{n,d}(t)\left[\left(1-t^2\right) P_{k,d+2}'(t) - (d-1)\,t\,P_{k,d+2}(t)\right]\left(1-t^2\right)^{\frac{d-3}{2}} dt.$$

Since

$$\left(1-t^2\right) P_{k,d+2}'(t) - (d-1)\,t\,P_{k,d+2}(t)$$

is a polynomial of degree $\leq n-1$,

$$\int_{-1}^{1} P_{n,d}'(t)\, P_{k,d+2}(t) \left(1-t^2\right)^{\frac{d-1}{2}} dt = 0, \quad 0 \leq k \leq n-2.$$

Thus, the polynomial (2.88) is of degree $\leq n-2$ and is orthogonal to all the polynomials of degree $\leq n-2$ with respect to the weighted inner product $(\cdot,\cdot)_{d+2}$. Then the polynomial (2.88) must be zero. Summarizing, we have shown the following relation

$$P_{n,d}'(t) = \frac{n\,(n+d-2)}{d-1}\, P_{n-1,d+2}(t), \quad n \geq 1,\ d \geq 2. \qquad (2.89)$$

Applying (2.89) recursively, we see that

$$P_{n,d}^{(j)}(t) = c_{n,d,j}\, P_{n-j,d+2j}(t)$$

where the constant $c_{n,d,j}$ is

$$\frac{n\,(n-1)\cdots(n-(j-1))\cdot(n+d-2)\,(n+d-1)\cdots(n+j+d-3)}{(d-1)\,(d+1)\cdots(d+2j-3)}.$$

The denominator of the above fraction can be rewritten as

$$2^j \left(\frac{d-1}{2} \right)_j = \frac{2^j \Gamma(j + \frac{d-1}{2})}{\Gamma(\frac{d-1}{2})},$$

where (1.12) is applied. Thus,

$$P_{n,d}^{(j)}(t) = \frac{n!\,(n+j+d-3)!\,\Gamma(\frac{d-1}{2})}{2^j\,(n-j)!\,(n+d-3)!\,\Gamma(j + \frac{d-1}{2})}\, P_{n-j,d+2j}(t), \quad n \geq j,\ d \geq 2.$$
(2.90)

Note that for $n < j$, $P_{n,d}^{(j)}(t) = 0$.

The formula (2.90) provides one way to compute the Legendre polynomials in higher dimensions $d \geq 4$ through differentiating the Legendre polynomials for $d = 3$ and $d = 2$. This is done as follows. First, rewrite (2.90) as

$$P_{n,d}(t) = \frac{2^j n!\,(n+d-j-3)!\,\Gamma(\frac{d-1}{2})}{(n+j)!\,(n+d-3)!\,\Gamma(\frac{d-1}{2} - j)}\, P_{n+j,d-2j}^{(j)}(t). \tag{2.91}$$

For $d = 2k$ even, take $j = k - 1$. Then from (2.91),

$$P_{n,2k}(t) = \frac{2^{k-1} n!\,(n+k-2)!\,\Gamma(k - \frac{1}{2})}{(n+k-1)!\,(n+2k-3)!\,\Gamma(\frac{1}{2})}\, P_{n+k-1,2}^{(k-1)}(t).$$

Applying (1.10), we have

$$P_{n,2k}(t) = \frac{(2k-2)!\,n!}{2^{k-1}(n+k-1)\,(k-1)!\,(n+2k-3)!}\, P_{n+k-1,2}^{(k-1)}(t).$$

For $d = 2k + 1$ odd, take $j = k - 1$. Then from (2.91),

$$P_{n,2k+1}(t) = \frac{2^{k-1} n!\,(k-1)!\,(n+k-1)!}{(n+k-1)!\,(n+2k-2)!}\, P_{n+k-1,3}^{(k-1)}(t)$$

$$= \frac{2^{k-1} n!\,(k-1)!}{(n+2k-2)!}\, P_{n+k-1,3}^{(k-1)}(t).$$

Let us derive some recursion formulas for the computation of the derivative $P_{n,d}'(t)$. First, we differentiate (2.76) to obtain

$$(1 - t^2)\, P_{n,d}'(t) = -(n+d-2)\,[P_{n+1,d}(t) - t\,P_{n,d}(t)]. \tag{2.92}$$

Since (2.76) is valid for $d \geq 3$ and $t \in (0,1]$, the relation (2.92) is proved for $d \geq 3$ and $t \in (0,1)$. For $d = 2$, $P_{n,2}(t) = \cos(n\theta)$ with $\theta = \arccos\theta$, and it is easy to verify that both sides of (2.92) are equal to $n \sin\theta \sin(n\theta)$. Then the relation (2.92) is valid for $d \geq 2$ and $t \in (0,1)$. Since $P_{n,d}(-t) = (-1)^n P_{n,d}(t)$, we know that (2.92) holds for $t \in (-1, 0)$ as well. Finally, since both sides of (2.92) are polynomials, we conclude that the relation remains true for $t = \pm 1$ and 0, i.e.,

$$(1 - t^2) \, P'_{n,d}(t) = -(n + d - 2) \, [P_{n+1,d}(t) - t \, P_{n,d}(t)] \,,$$

$$n \in \mathbb{N}_0, \ d \geq 2, \ t \in [-1, 1]. \qquad (2.93)$$

Then, from (2.86), we have

$$t \, P_{n,d}(t) = \frac{1}{2n + d - 2} \, [(n + d - 2) \, P_{n+1,d}(t) + n \, P_{n-1,d}(t)] \,.$$

Using this equality in (2.93) we obtain another relation

$$(1 - t^2) \, P'_{n,d}(t) = \frac{n \, (n + d - 2)}{2n + d - 2} \, [P_{n-1,d}(t) - P_{n+1,d}(t)] \,,$$

$$n \in \mathbb{N}, \ d \geq 2, \ t \in [-1, 1]. \qquad (2.94)$$

Finally, we differentiate the integral representation formula (2.72),

$$P'_{n,d}(t) = \frac{|\mathbb{S}^{d-3}|}{|\mathbb{S}^{d-2}|} \int_{-1}^{1} n \left[t + i \, (1 - t^2)^{1/2} s \right]^{n-1}$$
$$\cdot \left[1 - i \, t \, (1 - t^2)^{-1/2} s \right] (1 - s^2)^{\frac{d-4}{2}} \, ds.$$

Then we find out

$$(1 - t^2) \, P'_{n,d}(t) = n \, [P_{n-1,d}(t) - t \, P_{n,d}(t)] \,.$$

This equality is proved for $d \geq 3$. For $d = 2$, $P_{n,2}(t) = \cos(n \arccos t)$ and one can verify directly the equality. So we have the relation

$$(1 - t^2) \, P'_{n,d}(t) = n \, [P_{n-1,d}(t) - t \, P_{n,d}(t)] \,, \quad n \geq 1, \ d \geq 2, \ t \in [-1, 1]. \qquad (2.95)$$

2.7.4 Generating Function

Consider the following generating function of the Legendre polynomials

$$\phi(r) = \sum_{n=0}^{\infty} \binom{n + d - 3}{d - 3} P_{n,d}(t) \, r^n, \quad |t| \leq 1, \ |r| < 1. \qquad (2.96)$$

Let us first derive a compact formula for $\phi(r)$.

Since $|P_{n,d}(t)| \leq 1$ for any n, d and t, it is easy to verify that the series converges absolutely for any r with $|r| < 1$. We differentiate (2.96) with respect to r to find

$$\phi'(r) = \sum_{n=1}^{\infty} n \binom{n+d-3}{d-3} P_{n,d}(t) \, r^{n-1} \tag{2.97}$$

$$= \sum_{n=0}^{\infty} (n+1) \binom{n+d-2}{d-3} P_{n+1,d}(t) \, r^n. \tag{2.98}$$

Using (2.97) and (2.98), we can write

$$\left(1 + r^2 - 2\,r\,t\right) \phi'(r) = \sum_{n=0}^{\infty} (n+1) \binom{n+d-2}{d-3} P_{n+1,d}(t) \, r^n$$

$$+ \sum_{n=1}^{\infty} n \binom{n+d-3}{d-3} P_{n,d}(t) \, r^{n+1}$$

$$- 2t \sum_{n=1}^{\infty} n \binom{n+d-3}{d-3} P_{n,d}(t) \, r^n. \tag{2.99}$$

In the first sum of (2.99), for $n \geq 1$, use the following relation from (2.86):

$$P_{n+1,d}(t) = \frac{2n+d-2}{n+d-2} \, t \, P_{n,d}(t) - \frac{n}{n+d-2} P_{n-1,d}(t).$$

Then after some straightforward algebraic manipulations, we obtain from (2.99) that

$$\left(1 + r^2 - 2\,r\,t\right) \phi'(r) = (d-2)\,(t-r)\,\phi(r). \tag{2.100}$$

The unique solution of the differential equation (2.100) with the initial condition

$$\phi(0) = P_{0,d}(0) = 1$$

is

$$\phi(r) = \left(1 + r^2 - 2\,r\,t\right)^{-\frac{d-2}{2}}.$$

Therefore, we have the following compact formula for the generating function of the Legendre polynomials:

$$\sum_{n=0}^{\infty} \binom{n+d-3}{d-3} P_{n,d}(t) \, r^n = \left(1 + r^2 - 2\,r\,t\right)^{-\frac{d-2}{2}}, \quad |t| \leq 1, \; |r| < 1. \tag{2.101}$$

In particular, we have, for $P_n(t) := P_{n,3}(t)$,

$$\sum_{n=0}^{\infty} r^n P_n(t) = \frac{1}{(1 + r^2 - 2rt)^{1/2}}, \quad |t| \leq 1, \; |r| < 1. \tag{2.102}$$

The Legendre polynomials $P_{n,3}(t)$ were originally introduced as coefficients of the expansion (2.102).

For $d \geq 3$, we differentiate (2.101) with respect to r for $|r| < 1$:

$$\sum_{n=1}^{\infty} n \binom{n+d-3}{d-3} P_{n,d}(t) \, r^{n-1} = \frac{(d-2)\,(t-r)}{(1+r^2-2\,r\,t)^{\frac{d}{2}}}. \tag{2.103}$$

Note that

$$\frac{1}{(1+r^2-2\,r\,t)^{\frac{d-2}{2}}} + \frac{2\,r\,(t-r)}{(1+r^2-2\,r\,t)^{\frac{d}{2}}} = \frac{1-r^2}{(1+r^2-2\,r\,t)^{\frac{d}{2}}}.$$

Multiply both sides by $(d-2)$ and apply (2.101) and (2.103). Then we obtain

$$\sum_{n=0}^{\infty} (2\,n+d-2) \binom{n+d-3}{d-3} P_{n,d}(t)\, r^n = \frac{(d-2)\,(1-r^2)}{(1+r^2-2\,r\,t)^{\frac{d}{2}}}.$$

This identity can be rewritten as

$$\sum_{n=0}^{\infty} N_{n,d} r^n P_{n,d}(t) = \frac{1-r^2}{(1+r^2-2rt)^{\frac{d}{2}}}$$

and has been proved for $d \geq 3$. It can be verified that the identity holds also for $d = 2$. Therefore, we have the next result.

Proposition 2.28. (Poisson identity) *For $d \geq 2$,*

$$\sum_{n=0}^{\infty} N_{n,d} r^n P_{n,d}(t) = \frac{1-r^2}{(1+r^2-2rt)^{\frac{d}{2}}}, \quad |r| < 1, \ t \in [-1,1]. \tag{2.104}$$

Consider the special case $d = 2$. Then $P_{n,2}(t) = \cos(n \arccos t)$. With $t = \cos\theta$, the Poisson identity (2.104) is

$$1 + 2\sum_{n=1}^{\infty} r^n \cos(n\theta) = \frac{1-r^2}{1+r^2-2r\cos\theta} \quad |r| < 1, \ 0 \leq \theta \leq \pi. \tag{2.105}$$

With $d = 3$, the Poisson identity (2.104) is

$$\sum_{n=0}^{\infty} (2n+1)\, r^n P_{n,3}(t) = \frac{1-r^2}{(1+r^2-2rt)^{\frac{3}{2}}}, \quad |r| < 1, \ t \in [-1,1].$$

This Poisson identity provides the expansion of the Henyey–Greenstein phase function (1.2) with respect to the Legendre polynomials.

We now use (2.101) to derive a few more recursive relations involving the first order derivative of the Legendre polynomials. Differentiate (2.101) with respect to t,

$$\sum_{n=0}^{\infty} \binom{n+d-3}{d-3} P'_{n,d}(t)\, r^n = (d-2)\, r \left(1+r^2-2\,r\,t\right)^{-\frac{d}{2}}.$$

Differentiate (2.101) with respect to r,

$$\sum_{n=1}^{\infty} n \binom{n+d-3}{d-3} P_{n,d}(t)\, r^{n-1} = (d-2)\, (t-r) \left(1+r^2-2\,r\,t\right)^{-\frac{d}{2}}.$$

Combining these two equalities we have

$$(t-r) \sum_{n=1}^{\infty} \binom{n+d-3}{d-3} P'_{n,d}(t)\, r^n = \sum_{n=1}^{\infty} n \binom{n+d-3}{d-3} P_{n,d}(t)\, r^n,$$

i.e.,

$$t \sum_{n=1}^{\infty} \binom{n+d-3}{d-3} P'_{n,d}(t)\, r^n - \sum_{n=2}^{\infty} \binom{n+d-4}{d-3} P'_{n-1,d}(t)\, r^n$$

$$= \sum_{n=1}^{\infty} n \binom{n+d-3}{d-3} P_{n,d}(t)\, r^n.$$

Thus, for $n \geq 2$,

$$t \binom{n+d-3}{d-3} P'_{n,d}(t) - \binom{n+d-4}{d-3} P'_{n-1,d}(t) = n \binom{n+d-3}{d-3} P_{n,d}(t),$$

which can be simplified to

$$(n+d-3)\, t\, P'_{n,d}(t) - n\, P'_{n-1,d}(t) = n\, (n+d-3)\, P_{n,d}(t). \qquad (2.106)$$

Differentiate (2.86) with respect to t,

$$(n+d-2)\, P'_{n+1,d}(t) = (2\,n+d-2) \left[t\, P'_{n,d}(t) + P_{n,d}(t)\right] - n\, P'_{n-1,d}(t). \qquad (2.107)$$

Add (2.106) and (2.107) to obtain

$$(n+d-2)\, P'_{n+1,d}(t) - (n+1)\, t\, P'_{n,d}(t) = \left[n^2 + (d-1)\, n + d - 2\right] P_{n,d}(t). \qquad (2.108)$$

We can use either (2.106) or (2.108) to express a Legendre polynomial in terms of derivatives of Legendre polynomials:

$$P_{n,d}(t) = \frac{1}{n} t\, P'_{n,d}(t) - \frac{1}{n+d-3}\, P'_{n-1,d}(t), \qquad (2.109)$$

and

$$P_{n,d}(t) = \frac{n+d-2}{n^2 + (d-1)\, n + d - 2}\, P'_{n+1,d}(t)$$
$$- \frac{n+1}{n^2 + (d-1)\, n + d - 2}\, t\, P'_{n,d}(t). \qquad (2.110)$$

Replace n by $(n-1)$ in (2.108),

$$(n+d-3)\, P'_{n,d}(t) - n\, t\, P'_{n-1,d}(t) = n\, (n+d-3)\, P_{n-1,d}(t).$$

Then subtract from this relation the identity obtained from (2.106) multiplied by t,

$$(1 - t^2)\, P'_{n,d}(t) = n\, [P_{n-1,d}(t) - t\, P_{n,d}(t)]\,.$$

This is the formula (2.95).

From (2.86),

$$P_{n-1,d}(t) = \frac{2n+d-2}{n}\, t\, P_{n,d}(t) - \frac{n+d-2}{n} P_{n+1,d}(t).$$

We can use this relation in (2.95) to recover (2.93).

2.7.5 Values and Bounds

First, we recall the parity property (2.73),

$$P_{n,d}(-t) = (-1)^n P_{n,d}(t), \qquad -1 \le t \le 1. \qquad (2.111)$$

We know from (2.20) that

$$P_{n,d}(1) = 1.$$

Using the property (2.111), we further have

$$P_{n,d}(-1) = (-1)^n. \qquad (2.112)$$

This result also follows from the value

$$p_n(-1) = \frac{2^n \Gamma(n + \frac{d-1}{2})}{\Gamma(\frac{d-1}{2})},$$

computed with a similar technique used in evaluating $p_n(1)$ in the proof of Theorem 2.23.

We use (2.72) to compute $P_{n,d}(0)$ for $d \geq 3$.

$$P_{n,d}(0) = \frac{|\mathbb{S}^{d-3}|}{|\mathbb{S}^{d-2}|} \int_{-1}^1 i^n s^n (1-s^2)^{\frac{d-4}{2}} ds.$$

For n odd, $n = 2k+1$, $k \in \mathbb{N}_0$, obviously,

$$P_{2k+1,d}(0) = 0. \tag{2.113}$$

For n even, $n = 2k$, $k \in \mathbb{N}_0$,

$$P_{2k,d}(0) = (-1)^k \frac{|\mathbb{S}^{d-3}|}{|\mathbb{S}^{d-2}|} 2 \int_0^1 s^{2k}(1-s^2)^{\frac{d-4}{2}} ds.$$

Use the change of variable $t = s^2$,

$$P_{2k,d}(0) = (-1)^k \frac{|\mathbb{S}^{d-3}|}{|\mathbb{S}^{d-2}|} \int_0^1 t^{k-1/2}(1-t)^{\frac{d-4}{2}} dt.$$

Therefore,

$$P_{2k,d}(0) = (-1)^k \frac{|\mathbb{S}^{d-3}|}{|\mathbb{S}^{d-2}|} \frac{\Gamma(\frac{d-2}{2})\Gamma(k+\frac{1}{2})}{\Gamma(k+\frac{d-1}{2})}. \tag{2.114}$$

As an example,

$$P_{2k,3}(0) = (-1)^k \frac{(2k-1)!!}{2^k k!}.$$

Alternatively, we may use the generating function formula (2.101) to compute the values. For example, take $x = -1$ in (2.101):

$$\sum_{n=0}^{\infty} \binom{n+d-3}{d-3} P_{n,d}(-1) r^n = (1+r)^{-(d-2)}.$$

Apply (2.4) to expand the right side to obtain

$$\sum_{n=0}^{\infty} \binom{n+d-3}{d-3} P_{n,d}(-1) r^n = \sum_{n=0}^{\infty} \binom{n+d-3}{d-3} (-r)^n, \quad |r| < 1.$$

Hence,

$$P_{n,d}(-1) = (-1)^n.$$

We may also apply (2.90) to find derivative values at particular points. For instance, since

$$P_{n-j,d+2j}(1) = 1,$$

$$P_{n-j,d+2j}(-1) = (-1)^{n-j},$$

we have for $n \geq j$ and $d \geq 2$,

$$P_{n,d}^{(j)}(1) = \frac{n!(n+j+d-3)!\Gamma(\frac{d-1}{2})}{2^j(n-j)!(n+d-3)!\Gamma(j+\frac{d-1}{2})},$$

$$P_{n,d}^{(j)}(-1) = \frac{(-1)^n n!(n+j+d-3)!\Gamma(\frac{d-1}{2})}{2^j(n-j)!(n+d-3)!\Gamma(j+\frac{d-1}{2})}.$$

In particular, for $d = 3$,

$$P_{n,3}^{(j)}(1) = \frac{(n+j)!}{2^j j!(n-j)!},$$

from which,

$$P_{n,3}'(1) = \frac{1}{2}n(n+1), \qquad P_{n,3}''(1) = \frac{1}{8}(n-1)n(n+1)(n+2).$$

Next we provide some bounds for the Legendre polynomials and their derivatives. We use (2.72) to bound $P_{n,d}(t)$. For $s, t \in [-1, 1]$,

$$\left| t + i(1-t^2)^{1/2}s \right| = \left[t^2 + (1-t^2)s^2 \right]^{1/2} \leq (t^2 + 1 - t^2)^{1/2} = 1. \quad (2.115)$$

So for $d \geq 3$,

$$|P_{n,d}(t)| \leq \frac{|\mathbb{S}^{d-3}|}{|\mathbb{S}^{d-2}|} \int_{-1}^{1} (1-s^2)^{\frac{d-4}{2}} ds = 1, \quad t \in [-1, 1].$$

This bound is valid also for $d = 2$. Thus,

$$|P_{n,d}(t)| \leq 1, \quad n \in \mathbb{N}_0, \ d \geq 2, \ t \in [-1, 1]. \quad (2.116)$$

Instead of (2.115), we can use the bound

$$\left| t + i(1-t^2)^{1/2}s \right| = \left[1 - (1-t^2)(1-s^2) \right]^{1/2} \leq e^{-(1-t^2)(1-s^2)/2}$$

for $s, t \in [-1, 1]$. Then,

$$|P_{n,d}(t)| \leq \frac{|\mathbb{S}^{d-3}|}{|\mathbb{S}^{d-2}|} \int_{-1}^{1} e^{-n\,(1-t^2)\,(1-s^2)/2}(1 - s^2)^{\frac{d-4}{2}}\, ds$$

$$= 2\frac{|\mathbb{S}^{d-3}|}{|\mathbb{S}^{d-2}|} \int_{0}^{1} e^{-n\,(1-t^2)\,(1-s^2)/2}(1 - s^2)^{\frac{d-4}{2}}\, ds.$$

Let $t \in (-1, 1)$. Use the change of variable $s = 1 - u$ and the relation $u \leq 1 - s^2 \leq 2u$ for $s \in [0, 1]$,

$$|P_{n,d}(t)| < 2^{\frac{d-2}{2}}\frac{|\mathbb{S}^{d-3}|}{|\mathbb{S}^{d-2}|} \int_{0}^{\infty} e^{-n\,(1-t^2)\,u/2}u^{\frac{d-4}{2}}\, du.$$

For the integral, we apply the formula (1.4),

$$\int_{0}^{\infty} e^{-n\,(1-t^2)\,u/2}u^{\frac{d-4}{2}}\, du = \left[\frac{2}{n\,(1-t^2)}\right]^{\frac{d-2}{2}}\Gamma\!\left(\frac{d-2}{2}\right).$$

Then,

$$|P_{n,d}(t)| < 2^{d-2}\frac{|\mathbb{S}^{d-3}|}{|\mathbb{S}^{d-2}|}\frac{\Gamma(\frac{d-2}{2})}{[n\,(1-t^2)]^{\frac{d-2}{2}}}$$

$$= \frac{\Gamma(\frac{d-1}{2})}{\sqrt{\pi}}\left[\frac{4}{n\,(1-t^2)}\right]^{\frac{d-2}{2}}.$$

This inequality is valid also for $d = 2$. Therefore,

$$|P_{n,d}(t)| < \frac{\Gamma(\frac{d-1}{2})}{\sqrt{\pi}}\left[\frac{4}{n\,(1-t^2)}\right]^{\frac{d-2}{2}}, \quad n \in \mathbb{N}_0,\ d \geq 2,\ t \in (-1, 1). \quad (2.117)$$

From (2.90), we have bounds for derivatives of $P_{n,d}(t)$ of any order:

$$\left|P_{n,d}^{(j)}(t)\right| \leq P_{n,d}^{(j)}(1) = \frac{n!(n+j+d-3)!\Gamma(\frac{d-1}{2})}{2^j(n-j)!(n+d-3)!\Gamma(j+\frac{d-1}{2})}.$$

In particular,

$$\max_{t\in[-1,1]}\left|P_{n,d}^{(j)}(t)\right| = \mathcal{O}(n^{2j}). \quad (2.118)$$

As an application of (2.118), we observe that for any $t, s \in [-1, 1]$,

$$P_{n,d}(t) - P_{n,d}(s) = P_{n,d}'(\tau)\,(t - s)$$

for some τ between t and s. Applying (2.118) with $j = 1$, we have

$$|P_{n,d}(t) - P_{n,d}(s)| \leq c n^2 |t - s| \quad \forall t, s \in [-1, 1]. \tag{2.119}$$

Hence,

$$|P_{n,d}(\boldsymbol{\xi} \cdot \boldsymbol{\zeta}) - P_{n,d}(\boldsymbol{\eta} \cdot \boldsymbol{\zeta})| \leq c n^2 |\boldsymbol{\xi} - \boldsymbol{\eta}| \quad \forall \boldsymbol{\xi}, \boldsymbol{\eta}, \boldsymbol{\zeta} \in \mathbb{S}^{d-1}. \tag{2.120}$$

2.8 Completeness

In this section, we show in a constructive way that the spherical harmonics are complete in $C(\mathbb{S}^{d-1})$ and in $L^2(\mathbb{S}^{d-1})$, i.e., linear combinations of the spherical harmonics are dense in $C(\mathbb{S}^{d-1})$ and in $L^2(\mathbb{S}^{d-1})$.

2.8.1 Completeness in $C(\mathbb{S}^{d-1})$

Let $f \in C(\mathbb{S}^{d-1})$. Formally,

$$f(\boldsymbol{\xi}) = \int_{\mathbb{S}^{d-1}} \delta(1 - \boldsymbol{\xi} \cdot \boldsymbol{\eta}) \, f(\boldsymbol{\eta}) \, dS^{d-1}(\boldsymbol{\eta}), \quad \boldsymbol{\xi} \in \mathbb{S}^{d-1}$$

using a Dirac delta function $\delta(t)$ whose value is 0 at $t \neq 0$, $+\infty$ at $t = 0$, and which satisfies formally

$$\int_{\mathbb{S}^{d-1}} \delta(1 - \boldsymbol{\xi} \cdot \boldsymbol{\eta}) \, dS^{d-1}(\boldsymbol{\eta}) = 1 \quad \forall \boldsymbol{\xi} \in \mathbb{S}^{d-1}.$$

The idea to demonstrate the completeness of the spherical harmonics in $C(\mathbb{S}^{d-1})$ is to construct a sequence of kernel functions $\{k_n(t)\}$ such that $k_n(\boldsymbol{\xi} \cdot \boldsymbol{\eta})$ approaches $\delta(1 - \boldsymbol{\xi} \cdot \boldsymbol{\eta})$ and is such that for each $n \in \mathbb{N}$, the function $\int_{\mathbb{S}^{d-1}} k_n(\boldsymbol{\xi} \cdot \boldsymbol{\eta}) \, f(\boldsymbol{\eta}) \, dS^{d-1}(\boldsymbol{\eta})$ is a linear combination of spherical harmonics of order less than or equal to n. One possibility is to choose $k_n(t)$ proportional to $(1 + t)^n / 2^n$. Thus, we let

$$k_n(t) = E_{n,d} \left(\frac{1 + t}{2} \right)^n,$$

where $E_{n,d}$ is a scaling constant so that

$$\int_{\mathbb{S}^{d-1}} k_n(\boldsymbol{\xi} \cdot \boldsymbol{\eta}) \, dS^{d-1}(\boldsymbol{\eta}) = 1 \quad \forall \boldsymbol{\xi} \in \mathbb{S}^{d-1}. \tag{2.121}$$

To satisfy the condition (2.121), we have

$$E_{n,d} = \frac{(n+d-2)!}{(4\pi)^{\frac{d-1}{2}}\Gamma(n+\frac{d-1}{2})}.$$
(2.122)

This formula is derived as follows. First,

$$\int_{\mathbb{S}^{d-1}}\left(\frac{1+\boldsymbol{\xi}\cdot\boldsymbol{\eta}}{2}\right)^n dS^{d-1}(\boldsymbol{\eta}) = |\mathbb{S}^{d-2}|\int_{-1}^{1}\left(\frac{1+t}{2}\right)^n(1-t^2)^{\frac{d-3}{2}}\,dt.$$

Use the change of variable $s = (1+t)/2$,

$$\int_{\mathbb{S}^{d-1}}\left(\frac{1+\boldsymbol{\xi}\cdot\boldsymbol{\eta}}{2}\right)^n dS^{d-1}(\boldsymbol{\eta}) = 2^{d-2}|\mathbb{S}^{d-2}|\int_{0}^{1} s^{n+\frac{d-3}{2}}(1-s)^{\frac{d-3}{2}}\,ds.$$

By (1.19),

$$|\mathbb{S}^{d-2}| = \frac{2\pi^{\frac{d-1}{2}}}{\Gamma(\frac{d-1}{2})}.$$

Moreover,

$$\int_{0}^{1} s^{n+\frac{d-3}{2}}(1-s)^{\frac{d-3}{2}}\,ds = B\left(n+\frac{d-1}{2},\frac{d-1}{2}\right) = \frac{\Gamma(n+\frac{d-1}{2})\Gamma(\frac{d-1}{2})}{\Gamma(n+d-1)}.$$

Thus

$$\int_{\mathbb{S}^{d-1}}\left(\frac{1+\boldsymbol{\xi}\cdot\boldsymbol{\eta}}{2}\right)^n dS^{d-1}(\boldsymbol{\eta}) = (4\pi)^{\frac{d-1}{2}}\frac{\Gamma(n+\frac{d-1}{2})}{\Gamma(n+d-1)}.$$

Hence, (2.122) holds.

Now we introduce an operator $\Pi_{n,d}$ by the following formula

$$(\Pi_{n,d}f)(\boldsymbol{\xi}) := E_{n,d}\int_{\mathbb{S}^{d-1}}\left(\frac{1+\boldsymbol{\xi}\cdot\boldsymbol{\eta}}{2}\right)^n f(\boldsymbol{\eta})\,dS^{d-1}(\boldsymbol{\eta}),\quad f\in C(\mathbb{S}^{d-1}).$$
(2.123)

Let us express $(\Pi_{n,d}f)(\boldsymbol{\xi})$ as a linear combination of spherical harmonics of order less than or equal to n. For this purpose, we write

$$E_{n,d}\left(\frac{1+t}{2}\right)^n = \sum_{k=0}^{n}\mu_{n,k,d}\frac{N_{k,d}}{|\mathbb{S}^{d-1}|}P_{k,d}(t).$$
(2.124)

To determine the coefficients $\{\mu_{n,k,d}\}_{k=0}^{n}$, multiply both sides by the function $P_{l,d}(t)(1-t^2)^{\frac{d-3}{2}}$, $0 \le l \le n$, integrate from $t = -1$ to $t = 1$ and use the orthogonality relation (2.79) to obtain

$$\mu_{n,l,d} = |\mathbb{S}^{d-2}| E_{n,d} \int_{-1}^{1} \left(\frac{1+t}{2}\right)^n P_{l,d}(t)\,(1-t^2)^{\frac{d-3}{2}}\,dt.$$

Applying Proposition 2.26, we have

$$\mu_{n,l,d} = |\mathbb{S}^{d-2}| E_{n,d} R_{l,d} \int_{-1}^{1} \left(\frac{d}{dt}\right)^l \left(\frac{1+t}{2}\right)^n (1-t^2)^{l+\frac{d-3}{2}}\,dt$$

$$= |\mathbb{S}^{d-2}| E_{n,d} R_{l,d} \frac{n!}{2^n(n-l)!} \int_{-1}^{1} (1+t)^{n-l}(1-t^2)^{l+\frac{d-3}{2}}\,dt.$$

To compute the integral, we let $t = 2\,s - 1$. Then

$$\int_{-1}^{1} (1+t)^{n-l}(1-t^2)^{l+\frac{d-3}{2}}\,dt = 2^{n+l+d-2} \int_{0}^{1} s^{n+\frac{d-3}{2}}(1-s)^{l+\frac{d-3}{2}}\,ds$$

$$= 2^{n+l+d-2} \frac{\Gamma(n+\frac{d-1}{2})\Gamma(l+\frac{d-1}{2})}{\Gamma(n+l+d-1)}.$$

Hence, using the formulas (1.19), (2.71), and (2.122), we have

$$\mu_{n,l,d} = \frac{n!(n+d-2)!}{(n-l)!(n+l+d-2)!}.$$

It is easy to see that $\mu_{n,l,d} < \mu_{n+1,l,d}$ and $\mu_{n,l,d} \to 1$ as $n \to \infty$. From the expansion (2.124), we get, by making use of the projection operator $\mathcal{P}_{n,d}$ defined in Definition 2.11,

$$(\Pi_{n,d}f)(\boldsymbol{\xi}) = \sum_{k=0}^{n} \mu_{n,k,d}(\mathcal{P}_{k,d}f)(\boldsymbol{\xi}). \tag{2.125}$$

In other words, $\Pi_{n,d}f$ is a linear combination of spherical harmonics of order less than or equal to n.

To prove the completeness, we note the following property.

Lemma 2.29. *If $t \in [-1,1)$, then*

$$\lim_{n\to\infty} E_{n,d}\left(\frac{1+t}{2}\right)^n = 0.$$

Proof. By Stirling's formula (1.11),

$$\Gamma(x) \sim \sqrt{2\pi}\,x^{x-1/2}e^{-x} \quad \text{for } x \to \infty.$$

Then,

$$E_{n,d} \sim \frac{n^{\frac{d}{2}}}{(4\pi)^{\frac{d-1}{2}}}$$

and the statement holds.   □

Now we state and prove a completeness result.

Theorem 2.30.

$$\lim_{n\to\infty} \|\Pi_{n,d}f - f\|_{C(\mathbb{S}^{d-1})} = 0 \quad \forall f \in C(\mathbb{S}^{d-1}). \qquad (2.126)$$

Proof. Use the modulus of continuity

$$\omega(f;\delta) = \sup\{|f(\boldsymbol{\xi}) - f(\boldsymbol{\eta})| : \boldsymbol{\xi}, \boldsymbol{\eta} \in \mathbb{S}^{d-1}, |\boldsymbol{\xi} - \boldsymbol{\eta}| \leq \delta\}, \quad \delta > 0,$$

and recall that since $f \in C(\mathbb{S}^{d-1})$,

$$\omega(f;\delta) \to 0 \quad \text{as } \delta \to 0.$$

Denote

$$M := \sup\{|f(\boldsymbol{\xi}) - f(\boldsymbol{\eta})| : \boldsymbol{\xi}, \boldsymbol{\eta} \in \mathbb{S}^{d-1}\} < \infty.$$

Let $\boldsymbol{\xi} \in \mathbb{S}^{d-1}$ be arbitrary but fixed. Using (2.121), we have

$$(\Pi_{n,d}f)(\boldsymbol{\xi}) - f(\boldsymbol{\xi}) = E_{n,d} \int_{\mathbb{S}^{d-1}} \left(\frac{1+\boldsymbol{\xi}\cdot\boldsymbol{\eta}}{2}\right)^n [f(\boldsymbol{\eta}) - f(\boldsymbol{\xi})] \, dS^{d-1}(\boldsymbol{\eta})$$

$$\equiv I_1(\boldsymbol{\xi}) + I_2(\boldsymbol{\xi}),$$

where

$$I_1(\boldsymbol{\xi}) = E_{n,d} \int_{\{\boldsymbol{\eta}\in\mathbb{S}^{d-1}:|\boldsymbol{\xi}-\boldsymbol{\eta}|\leq\delta\}} \left(\frac{1+\boldsymbol{\xi}\cdot\boldsymbol{\eta}}{2}\right)^n [f(\boldsymbol{\eta}) - f(\boldsymbol{\xi})] \, dS^{d-1}(\boldsymbol{\eta}),$$

$$I_2(\boldsymbol{\xi}) = E_{n,d} \int_{\{\boldsymbol{\eta}\in\mathbb{S}^{d-1}:|\boldsymbol{\xi}-\boldsymbol{\eta}|>\delta\}} \left(\frac{1+\boldsymbol{\xi}\cdot\boldsymbol{\eta}}{2}\right)^n [f(\boldsymbol{\eta}) - f(\boldsymbol{\xi})] \, dS^{d-1}(\boldsymbol{\eta}).$$

We bound each term as follows:

$$|I_1(\boldsymbol{\xi})| \leq \omega(f;\delta) \, E_{n,d} \int_{\mathbb{S}^{d-1}} \left(\frac{1+\boldsymbol{\xi}\cdot\boldsymbol{\eta}}{2}\right)^n dS^{d-1}(\boldsymbol{\eta}) = \omega(f;\delta),$$

$$|I_2(\boldsymbol{\xi})| \leq M \, E_{n,d} |\mathbb{S}^{d-1}| \left(1 - \frac{\delta^2}{2}\right)^n.$$

In bounding $I_2(\boldsymbol{\xi})$, we used the relation

$$|\boldsymbol{\xi} - \boldsymbol{\eta}| > \delta \quad \Longrightarrow \quad \boldsymbol{\xi} \cdot \boldsymbol{\eta} < 1 - \frac{\delta^2}{2}$$

for $\boldsymbol{\xi}, \boldsymbol{\eta} \in \mathbb{S}^{d-1}$. Thus, for any $\delta \in (0,1)$, applying Lemma 2.29, we have

$$\limsup_{n \to \infty} \|\Pi_{n,d} f - f\|_{C(\mathbb{S}^{d-1})} \leq \omega(f; \delta).$$

Note that $\omega(f; \delta) \to 0$ as $\delta \to 0$. So the stated result holds. $\square$

Using the formula (2.125), we can restate Theorem 2.30 as follows.

Theorem 2.31. *For any $f \in C(\mathbb{S}^{d-1})$,*

$$f(\boldsymbol{\xi}) = \lim_{n \to \infty} \sum_{k=0}^{n} \mu_{n,k,d} (\mathcal{P}_{k,d} f)(\boldsymbol{\xi}) \text{ uniformly in } \boldsymbol{\xi} \in \mathbb{S}^{d-1}.$$

If $\mathcal{P}_{k,d} f = 0$ for all $n \in \mathbb{N}_0$, then $f = 0$.

Theorem 2.31 combined with Theorem 2.14 implies that $\{\mathbb{Y}_n^d : n \in \mathbb{N}_0\}$ is the only system of primitive spaces in $C(\mathbb{S}^{d-1})$ since any primitive space not identical with one of $\mathbb{Y}_n^d$, $n \in \mathbb{N}_0$, is orthogonal to all and is therefore trivial.

2.8.2 Completeness in $C(\mathbb{S}^{d-1})$ via the Poisson Identity

We now use the Poisson identity (2.104) to give another constructive proof of the completeness of the spherical harmonics. First we introduce a lemma.

Lemma 2.32. *The function*

$$G_d(r,t) := \frac{|\mathbb{S}^{d-2}|}{|\mathbb{S}^{d-1}|} \frac{1-r^2}{(1+r^2-2rt)^{\frac{d}{2}}}, \quad |r| < 1, \ t \in [-1,1] \qquad (2.127)$$

is positive and has the properties:

$$\int_{-1}^{1} G_d(r,t) (1-t^2)^{\frac{d-3}{2}} dt = 1, \qquad (2.128)$$

$$\lim_{r \to 1-} G_d(r,t) = 0 \text{ uniformly for } t \in [-1, t_0]$$

$$\text{with any fixed } t_0 \in (-1,1). \qquad (2.129)$$

Proof. For (2.128),

$$\int_{-1}^{1} G_d(r,t)\,(1-t^2)^{\frac{d-3}{2}}\,dt = \frac{|\mathbb{S}^{d-2}|}{|\mathbb{S}^{d-1}|} \int_{-1}^{1} \sum_{n=0}^{\infty} N_{n,d} r^n P_{n,d}(t)(1-t^2)^{\frac{d-3}{2}}\,dt$$

$$= \frac{|\mathbb{S}^{d-2}|}{|\mathbb{S}^{d-1}|} \int_{-1}^{1} (1-t^2)^{\frac{d-3}{2}}\,dt$$

$$= 1.$$

For (2.129), note the bound

$$\frac{1-r^2}{(1+r^2-2rt)^{\frac{d}{2}}} = \frac{1-r^2}{[(1-r)^2+2r(1-t)]^{\frac{d}{2}}} \le \frac{1-r^2}{[2r(1-t_0)]^{\frac{d}{2}}}$$

which is valid for $t \in [-1,t_0]$. $\qquad\square$

Define an operator $\mathcal{G}_d(r)$ by

$$(\mathcal{G}_d(r)f)(\boldsymbol{\xi}) = \frac{1}{|\mathbb{S}^{d-2}|} \int_{\mathbb{S}^{d-1}} G_d(r,\boldsymbol{\xi}\cdot\boldsymbol{\eta})\,f(\boldsymbol{\eta})\,dS^{d-1}(\boldsymbol{\eta}).$$

Note that for $|r| < 1$,

$$(\mathcal{G}_d(r)f)(\boldsymbol{\xi}) = \frac{1}{|\mathbb{S}^{d-2}|} \sum_{n=0}^{\infty} N_{n,d} r^n \int_{\mathbb{S}^{d-1}} P_{n,d}(\boldsymbol{\xi}\cdot\boldsymbol{\eta})\,f(\boldsymbol{\eta})\,dS^{d-1}(\boldsymbol{\eta}),$$

i.e.,

$$(\mathcal{G}_d(r)f)(\boldsymbol{\xi}) = \sum_{n=0}^{\infty} r^n (\mathcal{P}_{n,d}f)(\boldsymbol{\xi}). \tag{2.130}$$

Thus, $\mathcal{G}_d(r)f$ is the limit of a sequence of finite linear combinations of the spherical harmonics.

Theorem 2.33 (Completeness).

$$\lim_{r\to 1-} \|\mathcal{G}_d(r)f - f\|_{C(\mathbb{S}^{d-1})} = 0 \quad \forall f \in C(\mathbb{S}^{d-1}). \tag{2.131}$$

Proof. The proof is similar to that of Theorem 2.30. Using (2.128),

$$(\mathcal{G}_d(r)f)(\boldsymbol{\xi}) - f(\boldsymbol{\xi}) = \int_{\mathbb{S}^{d-1}} G_d(r,\boldsymbol{\xi}\cdot\boldsymbol{\eta})\,[f(\boldsymbol{\eta}) - f(\boldsymbol{\xi})]\,dS^{d-1}(\boldsymbol{\eta})$$

$$\equiv I_1(\boldsymbol{\xi}) + I_2(\boldsymbol{\xi}),$$

where

$$\mathrm{I}_1(\boldsymbol{\xi}) = \int_{|\boldsymbol{\xi}-\boldsymbol{\eta}|\geq\delta} G_d(r,\boldsymbol{\xi}\cdot\boldsymbol{\eta})\,[f(\boldsymbol{\eta})-f(\boldsymbol{\xi})]\,dS^{d-1}(\boldsymbol{\eta}),$$

$$\mathrm{I}_2(\boldsymbol{\xi}) = \int_{|\boldsymbol{\xi}-\boldsymbol{\eta}|<\delta} G_d(r,\boldsymbol{\xi}\cdot\boldsymbol{\eta})\,[f(\boldsymbol{\eta})-f(\boldsymbol{\xi})]\,dS^{d-1}(\boldsymbol{\eta}).$$

For any $\delta > 0$, by (2.129),

$$|\mathrm{I}_1(\boldsymbol{\xi})| \to 0 \text{ uniformly as } r \to 1-.$$

Also,

$$|\mathrm{I}_2(\boldsymbol{\xi})| \leq \omega(f;\delta).$$

So

$$\limsup_{r\to1-} \|\mathcal{G}_d(r)f - f\|_{C(\mathbb{S}^{d-1})} \leq \omega(f;\delta)$$

and (2.131) follows.   □

2.8.3 Convergence of Fourier–Laplace Series

We now consider convergence in average and uniform convergence of the Fourier–Laplace series. For a given function f, the series

$$\sum_{k=0}^{\infty} \mathcal{P}_{k,d}f$$

is called the Fourier–Laplace series of the function f. Recall Definition 2.11 for the projection $\mathcal{P}_{k,d}f$.

First, we present a result for convergence in average.

Theorem 2.34. *We have the convergence in average of the Fourier–Laplace series:*

$$\lim_{n\to\infty} \left\| f - \sum_{k=0}^{n} \mathcal{P}_{k,d}f \right\|_{L^2(\mathbb{S}^{d-1})} = 0 \quad \forall\, f \in L^2(\mathbb{S}^{d-1}). \tag{2.132}$$

Proof. Note that the operator $\mathcal{P}_{k,d}$ is self-adjoint:

$$(f, \mathcal{P}_{k,d}g) = (\mathcal{P}_{k,d}f, g) \quad \forall\, f, g \in L^2(\mathbb{S}^{d-1}).$$

Also, $(\mathcal{P}_{k,d})^2 = \mathcal{P}_{k,d}$. Therefore,

$$(f, \mathcal{P}_{k,d}f) = (f, (\mathcal{P}_{k,d})^2 f) = (\mathcal{P}_{k,d}f, \mathcal{P}_{k,d}f) = \|\mathcal{P}_{k,d}f\|_{L^2(\mathbb{S}^{d-1})}^2$$

and
$$(\mathcal{P}_{k,d}f, \mathcal{P}_{n,d}f) = \delta_{kn}\|\mathcal{P}_{k,d}f\|^2_{L^2(\mathbb{S}^{d-1})}.$$

Apply the above two equalities to obtain

$$\left\|f - \sum_{k=0}^{n}\mathcal{P}_{k,d}f\right\|^2_{L^2(\mathbb{S}^{d-1})} = \|f\|^2_{L^2(\mathbb{S}^{d-1})} - \sum_{k=0}^{n}\|\mathcal{P}_{k,d}f\|^2_{L^2(\mathbb{S}^{d-1})}. \qquad (2.133)$$

Hence,

$$\sum_{k=0}^{n}\|\mathcal{P}_{k,d}f\|^2_{L^2(\mathbb{S}^{d-1})} \le \|f\|^2_{L^2(\mathbb{S}^{d-1})} \quad \forall n \in \mathbb{N}_0.$$

Then,

$$\sum_{k=0}^{\infty}\|\mathcal{P}_{k,d}f\|^2_{L^2(\mathbb{S}^{d-1})} \le \|f\|^2_{L^2(\mathbb{S}^{d-1})} \quad \forall f \in L^2(\mathbb{S}^{d-1}). \qquad (2.134)$$

First we assume $f \in C(\mathbb{S}^{d-1})$. From the formula (2.130),

$$\|\mathcal{G}_d(r)f\|^2_{L^2(\mathbb{S}^{d-1})} = \sum_{k=0}^{\infty}r^{2k}\|\mathcal{P}_{k,d}f\|^2_{L^2(\mathbb{S}^{d-1})}. \qquad (2.135)$$

By Theorem 2.33, $\mathcal{G}_d(r)f$ converges uniformly to f on $\mathbb{S}^{d-1}$ as $r \to 1-$. Take the limit $r \to 1-$ in (2.135) to obtain

$$\|f\|^2_{L^2(\mathbb{S}^{d-1})} = \sum_{k=0}^{\infty}\|\mathcal{P}_{k,d}f\|^2_{L^2(\mathbb{S}^{d-1})}. \qquad (2.136)$$

Then by (2.133) we obtain (2.132) for $f \in C(\mathbb{S}^{d-1})$.

Extension of the result from a $C(\mathbb{S}^{d-1})$ function to an $L^2(\mathbb{S}^{d-1})$ function is achieved by using the density of $C(\mathbb{S}^{d-1})$ in $L^2(\mathbb{S}^{d-1})$, by noticing that since spherical harmonics of different order are orthogonal,

$$\left\|\sum_{k=0}^{n}\mathcal{P}_{k,d}f\right\|^2_{L^2(\mathbb{S}^{d-1})} = \sum_{k=0}^{n}\|\mathcal{P}_{k,d}f\|^2_{L^2(\mathbb{S}^{d-1})}$$

and by applying the bound (2.134).   □

Then we turn to a study of uniform convergence of the Fourier–Laplace series.

Define $\mathcal{S}_n : C(\mathbb{S}^{d-1}) \to C(\mathbb{S}^{d-1})$ to be the linear operator given by the partial sum of the spherical harmonic expansion

$$\mathcal{S}_n f(\boldsymbol{\xi}) := \sum_{k=0}^{n} (\mathcal{P}_{k,d} f)(\boldsymbol{\xi}), \quad f \in C(\mathbb{S}^{d-1}). \tag{2.137}$$

Denote by $\|\mathcal{S}_n\|$ the norm of the operator. To answer the question when do the partial sums $\{\mathcal{S}_n f\}$ converge uniformly to f, an important tool is the following result, due to Lebesgue.

Theorem 2.35. *For* $f \in C(\mathbb{S}^{d-1})$,

$$\|f - \mathcal{S}_n f\|_{C(\mathbb{S}^{d-1})} \le (1 + \|\mathcal{S}_n\|) E_{n,\infty}(f), \tag{2.138}$$

where

$$E_{n,\infty}(f) := \inf \left\{ \|f - p_n\|_{C(\mathbb{S}^{d-1})} : p_n \in \mathbb{Y}_{0:n}^d \right\} \tag{2.139}$$

and

$$\mathbb{Y}_{0:n}^d := \bigoplus_{j=0}^{n} \mathbb{Y}_j^d.$$

Proof. Note that

$$\mathcal{S}_n p_n = p_n \quad \forall p_n \in \mathbb{Y}_{0:n}^d.$$

Thus,

$$f - \mathcal{S}_n f = (f - p_n) - \mathcal{S}_n(f - p_n) \quad \forall p_n \in \mathbb{Y}_{0:n}^d.$$

Apply the $C(\mathbb{S}^{d-1})$-norm,

$$\|f - \mathcal{S}_n f\|_{C(\mathbb{S}^{d-1})} \le (1 + \|\mathcal{S}_n\|) \|f - p_n\|_{C(\mathbb{S}^{d-1})}.$$

Then take the infimum with respect to p_n over the subspace $\mathbb{Y}_{0:n}^d$ to get (2.138). $\qquad\square$

The operator norm $\|\mathcal{S}_n\|$ is called the "Lebesgue constant". In [94], it is shown that

$$\|\mathcal{S}_n\| = \mathcal{O}\left(n^{(d-2)/2}\right), \quad d \ge 3.$$

Based on this bound, the next result regarding the uniform convergence of the Fourier–Laplace series can be proved.

Theorem 2.36. *Let* $d \ge 3$ *and* $f \in C^{k,\alpha}(\mathbb{S}^{d-1})$ *for some* $k \ge 0$ *and* $\alpha \in (0, 1]$. *Assume* $k + \alpha > d/2 - 1$. *Then* $\mathcal{S}_n f$ *converges uniformly to* f *over* $\mathbb{S}^{d-1}$.

The spaces $C^k(\mathbb{S}^{d-1})$ and $C^{k,\alpha}(\mathbb{S}^{d-1})$ can be defined in a variety of ways, some of which are discussed in Sects. 4.2.1 and 4.2.2. We say $f \in C^{k,\alpha}(\mathbb{S}^{d-1})$ if all of its kth-order derivatives are Hölder continuous with exponent $\alpha \in (0, 1]$.

This theorem is proven in [94], based on results from [93] and [54]. Results from these papers are discussed in greater detail in Sect. 4.2 for the special case of $\mathbb{S}^2$.

In the case $d = 2$, the Fourier–Laplace series reduces to the ordinary Fourier series. The Lebesgue constant is [123, Chap. 2, p. 67]

$$\|\mathcal{S}_n\| = \frac{4}{\pi^2} \ln n + \mathcal{O}(1).$$

The following uniform convergence result on the Fourier series holds (see, e.g., [13, Sect. 3.7]).

Theorem 2.37. *Let $f : \mathbb{R} \to \mathbb{R}$ be a periodic function, with 2π being an integer multiple of its period. If $f \in C^{k,\alpha}(\mathbb{R})$ with $k \in \mathbb{N}_0$ and $\alpha \in (0,1]$, then for the nth order partial sum $\mathcal{S}_n f$ of the Fourier series of the function f,*

$$\|f - \mathcal{S}_n f\|_{C[0,2\pi]} \le c \, \frac{\ln(n+2)}{n^{k+\alpha}}.$$

In particular, this implies the uniform convergence of the Fourier series of the function f.

2.8.4 Completeness in $L^2(\mathbb{S}^{d-1})$

Theorem 2.34 implies the completeness of spherical harmonics in $L^2(\mathbb{S}^{d-1})$, i.e., the subspace of linear combinations of spherical harmonics is dense in $L^2(\mathbb{S}^{d-1})$.

An alternative way to show the completeness of spherical harmonics in $L^2(\mathbb{S}^{d-1})$ is through using the operator $\Pi_{n,d}$ defined in (2.123). First, we show the operator $\Pi_{n,d}$ is bounded as a mapping from $L^2(\mathbb{S}^{d-1})$ to $L^2(\mathbb{S}^{d-1})$:

$$\|\Pi_{n,d}f\|_{L^2(\mathbb{S}^{d-1})} \le \|f\|_{L^2(\mathbb{S}^{d-1})} \quad \forall f \in L^2(\mathbb{S}^{d-1}). \tag{2.140}$$

This is proved as follows:

$$\|\Pi_{n,d}f\|^2_{L^2(\mathbb{S}^{d-1})} = \int_{\mathbb{S}^{d-1}} E_{n,d}^2 \left[\int_{\mathbb{S}^{d-1}} \left(\frac{1+\boldsymbol{\xi}\cdot\boldsymbol{\eta}}{2} \right)^n f(\boldsymbol{\eta}) \, dS^{d-1}(\boldsymbol{\eta}) \right]^2 dS^{d-1}(\boldsymbol{\xi})$$

$$\le \int_{\mathbb{S}^{d-1}} E_{n,d}^2 \left[\int_{\mathbb{S}^{d-1}} \left(\frac{1+\boldsymbol{\xi}\cdot\boldsymbol{\eta}}{2} \right)^n dS^{d-1}(\boldsymbol{\eta}) \right.$$

$$\left. \int_{\mathbb{S}^{d-1}} \left(\frac{1+\boldsymbol{\xi}\cdot\boldsymbol{\eta}}{2} \right)^n |f(\boldsymbol{\eta})|^2 dS^{d-1}(\boldsymbol{\eta}) \right] dS^{d-1}(\boldsymbol{\xi}).$$

Apply (2.121),

$$\|\Pi_{n,d}f\|^2_{L^2(\mathbb{S}^{d-1})} \leq \int_{\mathbb{S}^{d-1}} E_{n,d} \left[\int_{\mathbb{S}^{d-1}} \left(\frac{1+\boldsymbol{\xi}\cdot\boldsymbol{\eta}}{2} \right)^n |f(\boldsymbol{\eta})|^2 dS^{d-1}(\boldsymbol{\eta}) \right] dS^{d-1}(\boldsymbol{\xi})$$

$$= \int_{\mathbb{S}^{d-1}} |f(\boldsymbol{\eta})|^2 \left[E_{n,d} \int_{\mathbb{S}^{d-1}} \left(\frac{1+\boldsymbol{\xi}\cdot\boldsymbol{\eta}}{2} \right)^n dS^{d-1}(\boldsymbol{\xi}) \right] dS^{d-1}(\boldsymbol{\eta}).$$

Apply (2.121) again to obtain

$$\|\Pi_{n,d}f\|^2_{L^2(\mathbb{S}^{d-1})} \leq \|f\|^2_{L^2(\mathbb{S}^{d-1})},$$

i.e., (2.140) holds.

Let $f \in L^2(\mathbb{S}^{d-1})$. For any $\varepsilon > 0$, by the density of $C(\mathbb{S}^{d-1})$ in $L^2(\mathbb{S}^{d-1})$, we can find a function $f_\varepsilon \in C(\mathbb{S}^{d-1})$ such that

$$\|f - f_\varepsilon\|_{L^2(\mathbb{S}^{d-1})} < \frac{\varepsilon}{3}.$$

Choose n sufficiently large so that, following Theorem 2.30,

$$\|\Pi_{n,d}f_\varepsilon - f_\varepsilon\|_{L^2(\mathbb{S}^{d-1})} < \frac{\varepsilon}{3}.$$

Then,

$$\|\Pi_{n,d}f - f\|_{L^2(\mathbb{S}^{d-1})} \leq \|\Pi_{n,d}(f - f_\varepsilon)\|_{L^2(\mathbb{S}^{d-1})} + \|\Pi_{n,d}f_\varepsilon - f_\varepsilon\|_{L^2(\mathbb{S}^{d-1})}$$

$$+ \|f - f_\varepsilon\|_{L^2(\mathbb{S}^{d-1})}$$

$$\leq 2\|f - f_\varepsilon\|_{L^2(\mathbb{S}^{d-1})} + \|\Pi_{n,d}f_\varepsilon - f_\varepsilon\|_{L^2(\mathbb{S}^{d-1})}$$

$$< \varepsilon.$$

Thus, the spherical harmonics are dense in $L^2(\mathbb{S}^{d-1})$.

Since spherical harmonics of different orders are orthogonal, we can also deduce the next result.

Theorem 2.38. *We have the orthogonal decomposition*

$$L^2(\mathbb{S}^{d-1}) = \bigoplus_{n=0}^{\infty} \mathbb{Y}_n^d.$$

Thus, any function $f \in L^2(\mathbb{S}^{d-1})$ can be uniquely represented as

$$f(\boldsymbol{\xi}) = \sum_{n=0}^{\infty} f_n(\boldsymbol{\xi}) \text{ in } L^2(\mathbb{S}^{d-1}), \ f_n \in \mathbb{Y}_n^d, \ n \geq 0. \tag{2.141}$$

We call $f_n \in \mathbb{Y}_n^d$ the n-spherical harmonic component of f and have the following formula

$$f_n(\boldsymbol{\xi}) = \frac{N_{n,d}}{|\mathbb{S}^{d-1}|} \int_{\mathbb{S}^{d-1}} f(\boldsymbol{\eta}) \, P_{n,d}(\boldsymbol{\xi} \cdot \boldsymbol{\eta}) \, dS^{d-1}(\boldsymbol{\eta}), \quad n \geq 0. \qquad (2.142)$$

This formula is derived from (2.141) as follows. Replace $\boldsymbol{\xi}$ by $\boldsymbol{\eta}$ in (2.141), multiply both sides by $P_{n,d}(\boldsymbol{\xi} \cdot \boldsymbol{\eta})$ and integrate with respect to $\boldsymbol{\eta} \in \mathbb{S}^{d-1}$ to obtain

$$\int_{\mathbb{S}^{d-1}} f(\boldsymbol{\eta}) \, P_{n,d}(\boldsymbol{\xi} \cdot \boldsymbol{\eta}) \, dS^{d-1}(\boldsymbol{\eta}) = \int_{\mathbb{S}^{d-1}} \sum_{j=0}^{\infty} f_j(\boldsymbol{\eta}) \, P_{n,d}(\boldsymbol{\xi} \cdot \boldsymbol{\eta}) \, dS^{d-1}(\boldsymbol{\eta})$$

$$= \sum_{j=0}^{\infty} \int_{\mathbb{S}^{d-1}} f_j(\boldsymbol{\eta}) \, P_{n,d}(\boldsymbol{\xi} \cdot \boldsymbol{\eta}) \, dS^{d-1}(\boldsymbol{\eta}).$$

By the orthogonality of spherical harmonics of different orders,

$$\int_{\mathbb{S}^{d-1}} f_j(\boldsymbol{\eta}) \, P_{n,d}(\boldsymbol{\xi} \cdot \boldsymbol{\eta}) \, dS^{d-1}(\boldsymbol{\eta}) = 0 \quad \forall j \neq n.$$

Moreover, by (2.33),

$$\int_{\mathbb{S}^{d-1}} f_n(\boldsymbol{\eta}) \, P_{n,d}(\boldsymbol{\xi} \cdot \boldsymbol{\eta}) \, dS^{d-1}(\boldsymbol{\eta}) = \frac{|\mathbb{S}^{d-1}|}{N_{n,d}} \, f_n(\boldsymbol{\xi}).$$

Hence,

$$\int_{\mathbb{S}^{d-1}} f(\boldsymbol{\eta}) \, P_{n,d}(\boldsymbol{\xi} \cdot \boldsymbol{\eta}) \, dS^{d-1}(\boldsymbol{\eta}) = \frac{|\mathbb{S}^{d-1}|}{N_{n,d}} \, f_n(\boldsymbol{\xi})$$

and the formula (2.142) is proved. Notice that $f_n(\boldsymbol{\xi}) = (\mathcal{P}_{n,d} f)(\boldsymbol{\xi})$ with the projection operator $\mathcal{P}_{n,d}$ defined in (2.44).

As a consequence of (2.141), we have the Parseval equality on $L^2(\mathbb{S}^{d-1})$:

$$\|f\|_{L^2(\mathbb{S}^{d-1})}^2 = \sum_{n=0}^{\infty} \|f_n\|_{L^2(\mathbb{S}^{d-1})}^2 \quad \forall f \in L^2(\mathbb{S}^{d-1}), \qquad (2.143)$$

where f_n is given by (2.142). This equality extends (2.136) from $C(\mathbb{S}^{d-1})$ functions to $L^2(\mathbb{S}^{d-1})$ functions.

2.9 The Gegenbauer Polynomials

The Gegenbauer polynomials are useful in generalizing the expansion (2.102). Recall the integral representation formula (2.72) for the Legendre polynomials.

Definition 2.39. For $\nu > 0$, $n \in \mathbb{N}_0$,

$$C_{n,\nu}(t) := \binom{n + 2\nu - 1}{n} \frac{\Gamma(\nu + \frac{1}{2})}{\sqrt{\pi}\,\Gamma(\nu)} \int_{-1}^{1} \left[t + i\,(1 - t^2)^{1/2} s \right]^n (1 - s^2)^{\nu - 1} ds \tag{2.144}$$

is called the Gegenbauer polynomial of degree n with index ν.

Note that for an arbitrary number a, the binomial coefficient

$$\binom{a}{n} := \frac{a\,(a - 1) \cdots (a - (n - 1))}{n!}, \quad n \in \mathbb{N}.$$

Why $C_{n,\nu}(t)$ is a polynomial of degree n? First,

$$\left[t + i\,(1 - t^2)^{1/2} s \right]^n = \sum_{j=0}^{n} \binom{n}{j} t^{n-j} (1 - t^2)^{j/2} (is)^j.$$

For $j = 2k + 1$ odd, the integral of the corresponding term is

$$\int_{-1}^{1} s^{2k+1} (1 - s^2)^{\nu - 1} ds = 0.$$

So $C_{n,\nu}(t)$ is real valued and

$$C_{n,\nu}(t) = \binom{n + 2\nu - 1}{n} \frac{\Gamma(\nu + \frac{1}{2})}{\sqrt{\pi}\,\Gamma(\nu)} \sum_{k=0}^{[n/2]} \binom{n}{2k} t^{n-2k} (-1)^k (1 - t^2)^k$$

$$\cdot \int_{-1}^{1} s^{2k} (1 - s^2)^{\nu - 1} ds$$

is a polynomial of degree $\leq n$. The coefficient of t^n in $C_{n,\nu}(t)$ is

$$\binom{n + 2\nu - 1}{n} \frac{\Gamma(\nu + \frac{1}{2})}{\sqrt{\pi}\,\Gamma(\nu)} \sum_{k=0}^{[n/2]} \binom{n}{2k} \int_{-1}^{1} s^{2k} (1 - s^2)^{\nu - 1} ds > 0.$$

Hence, $C_{n,\nu}(t)$ is a polynomial of degree n.
   Observe that, recalling the formula (2.72),

$$C_{n, \frac{d-2}{2}}(t) = \binom{n + d - 3}{n} P_{n,d}(t), \quad d \geq 3. \tag{2.145}$$

Proposition 2.40. (Gegenbauer identity)

$$\sum_{n=0}^{\infty} r^n C_{n,\nu}(t) = \frac{1}{(1 + r^2 - 2rt)^\nu}, \quad |r| < 1,\ t \in [-1, 1]. \tag{2.146}$$

Proof. First we calculate $C_{n,\nu}(1)$:

$$C_{n,\nu}(1) = \binom{n+2\nu-1}{n} \frac{\Gamma(\nu+\frac{1}{2})}{\sqrt{\pi}\,\Gamma(\nu)} \int_{-1}^{1} (1-s^2)^{\nu-1} ds.$$

Let $t = s^2$. Then,

$$C_{n,\nu}(1) = \binom{n+2\nu-1}{n} \frac{\Gamma(\nu+\frac{1}{2})}{\sqrt{\pi}\,\Gamma(\nu)} \int_{0}^{1} t^{-\frac{1}{2}}(1-t)^{\nu-1} dt.$$

Since

$$\int_{0}^{1} t^{-\frac{1}{2}}(1-t)^{\nu-1} dt = \frac{\Gamma(\frac{1}{2})\Gamma(\nu)}{\Gamma(\nu+\frac{1}{2})},$$

we have

$$C_{n,\nu}(1) = \binom{n+2\nu-1}{n}. \tag{2.147}$$

From the power series (2.4),

$$\sum_{n=0}^{\infty} \binom{n+2\nu-1}{n} z^n = \frac{1}{(1-z)^{2\nu}}, \quad |z| < 1.$$

For $|r| < 1$ and $|t| \leq 1$,

$$\sum_{n=0}^{\infty} r^n C_{n,\nu}(t) = \frac{\Gamma(\nu+\frac{1}{2})}{\sqrt{\pi}\,\Gamma(\nu)} \int_{-1}^{1} \frac{(1-s^2)^{\nu-1}}{[1-rt-ir(1-t^2)^{1/2}s]^{2\nu}} ds. \tag{2.148}$$

Write

$$1 - rt - ir(1-t^2)^{1/2} = (1+r^2-2rt)^{1/2} e^{-i\alpha}$$

for some $\alpha \in [0, \frac{\pi}{2})$. Use the substitution (2.74), recall the relations (2.75), and note that

$$1 - rt - ir(1-t^2)^{1/2}s = (1+r^2-2rt)^{1/2}(\cos\alpha - i\tanh u \sin\alpha)$$

$$= (1+r^2-2rt)^{1/2}\frac{\cosh(u-i\alpha)}{\cosh u}.$$

So from (2.148), we have

$$\sum_{n=0}^{\infty} r^n C_{n,\nu}(t) = \frac{\Gamma(\nu+\frac{1}{2})}{\sqrt{\pi}\,\Gamma(\nu)} \frac{1}{(1+r^2-2rt)^{\nu}} \int_{-\infty}^{\infty} \frac{1}{\cosh^{2\nu}(u-i\alpha)} du.$$

Since the poles of the function $(\cosh u)^{-2\nu}$ are $u = i\pi(k+1/2)$, $k \in \mathbb{Z}$, and since $0 \leq \alpha < \pi/2$, we can apply the Cauchy integral theorem in complex analysis to get

$$\int_{-\infty}^{\infty} \frac{1}{\cosh^{2\nu}(u-i\alpha)}\, du = \int_{-\infty}^{\infty} \frac{1}{\cosh^{2\nu} u}\, du,$$

which is a constant. Thus, for some constant c,

$$\sum_{n=0}^{\infty} r^n C_{n,\nu}(t) = \frac{c}{(1+r^2-2rt)^\nu}.$$

Let $t=1$ and use the value (2.147):

$$\frac{c}{(1-r)^{2\nu}} = \sum_{n=0}^{\infty} \binom{n+2\nu-1}{n} r^n = \frac{1}{(1-r)^{2\nu}}.$$

So the constant $c=1$.   □

Obviously, (2.102) is a special case of (2.146) by taking $\nu = 1/2$.

2.10   The Associated Legendre Functions

We have seen that the Legendre polynomials play an important role in the study of spherical harmonics. In an increasing order of complexity, we next introduce associated Legendre functions which are useful in constructing spherical harmonics from those in a lower dimension.

2.10.1   Definition and Representation Formulas

Recall the first integral representation formula (2.72) for the Legendre polynomials. We then introduce the following definition.

Definition 2.41. For $d \geq 3$ and $n,j \in \mathbb{N}_0$,

$$P_{n,d,j}(t) = \frac{|\mathbb{S}^{d-3}|}{|\mathbb{S}^{d-2}|} i^{-j} \int_{-1}^{1} \left[t + i(1-t^2)^{1/2}s\right]^n P_{j,d-1}(s)(1-s^2)^{\frac{d-4}{2}}\, ds,$$

$$t \in [-1,1]. \tag{2.149}$$

is called the associated Legendre function of degree n with order j in dimension d.

When $j=0$, $P_{n,d,0}(t) = P_{n,d}(t)$ is the Legendre polynomial of degree n in d-dimensions. The factor i^{-j} is included in (2.149) to make $P_{n,d,j}(t)$ real-valued. To see this, note that

$$\left[t + i\,(1-t^2)^{1/2}s\right]^n = \sum_{k=0}^{n} \binom{n}{k} t^{n-k}(1-t^2)^{k/2} i^k s^k.$$

Thus,

$$P_{n,d,j}(t) = \frac{|\mathbb{S}^{d-3}|}{|\mathbb{S}^{d-2}|} \sum_{k=0}^{n} \binom{n}{k} t^{n-k}(1-t^2)^{k/2} i^{k-j} \int_{-1}^{1} s^k P_{j,d-1}(s)\,(1-s^2)^{\frac{d-4}{2}} ds.$$

By the parity property (2.111) for the Legendre polynomials, when $|k - j|$ is odd, $s^k P_{j,d-1}(s)$ is an odd function and then

$$\int_{-1}^{1} s^k P_{j,d-1}(s)\,(1-s^2)^{\frac{d-4}{2}} ds = 0.$$

Consequently, $P_{n,d,j}(t)$ is real-valued.

The associated Legendre functions can be used to generate orthonormal systems of spherical harmonics on $\mathbb{S}^{d-1}$; see Sect. 2.11.

Applying Proposition 2.26, we have

$$P_{n,d,j}(t) = R_{j,d-1} \frac{|\mathbb{S}^{d-3}|}{|\mathbb{S}^{d-2}|} \frac{n!}{(n-j)!} (1-t^2)^{\frac{j}{2}}$$

$$\cdot \int_{-1}^{1} \left[t + i\,(1-t^2)^{1/2}s\right]^{n-j} (1-s^2)^{j+\frac{d-4}{2}} ds,$$

where by (2.71),

$$R_{j,d-1} = \frac{\Gamma(\frac{d-2}{2})}{2^j \Gamma(j + \frac{d-2}{2})}.$$

Since

$$\int_{-1}^{1} \left[t + i\,(1-t^2)^{1/2}s\right]^{n-j} (1-s^2)^{j+\frac{d-4}{2}} ds = \frac{\pi^{\frac{1}{2}}\Gamma(j + \frac{d-2}{2})}{\Gamma(j + \frac{d-1}{2})} P_{n-j,d+2j}(t)$$

by an application of the integral representation formula (2.72), we have thus shown the following result.

Proposition 2.42. *For $d \geq 3$ and $0 \leq j \leq n$,*

$$P_{n,d,j}(t) = \frac{n!\,\Gamma(\frac{d-1}{2})}{2^j (n-j)!\,\Gamma(j + \frac{d-1}{2})} (1-t^2)^{\frac{j}{2}} P_{n-j,d+2j}(t), \quad t \in [-1,1].$$

In some references, the associated Legendre functions are also called the associated Legendre polynomials. From Proposition 2.42, it is evident that the associated Legendre function $P_{n,d,j}(t)$ is a polynomial in t if and only if j is even.

Combining Theorem 2.23 and Proposition 2.42, we obtain the formula

$$P_{n,d,j}(t) = \frac{(-1)^{n-j} n! \Gamma(\frac{d-1}{2})}{2^n (n-j)! \Gamma(n + \frac{d-1}{2})} (1-t^2)^{\frac{3-d-j}{2}} \left(\frac{d}{dt}\right)^{n-j} (1-t^2)^{n+\frac{d-3}{2}}$$

for $d \geq 3$, $0 \leq j \leq n$ and $t \in [-1,1]$. For the particular case $d = 3$, with $0 \leq j \leq n$ and $t \in [-1,1]$,

$$P_{n,3,j}(t) = \frac{(-1)^{n-j}}{2^n (n-j)!} (1-t^2)^{-\frac{j}{2}} \left(\frac{d}{dt}\right)^{n-j} (1-t^2)^n. \tag{2.150}$$

Furthermore, by the formula (2.90), we obtain the next result.

Proposition 2.43. *For $d \geq 3$ and $0 \leq j \leq n$,*

$$P_{n,d,j}(t) = \frac{(n+d-3)!}{(n+j+d-3)!} (1-t^2)^{\frac{j}{2}} P_{n,d}^{(j)}(t), \quad t \in [-1,1].$$

Thus, the associated Legendre functions can be computed through differentiating the Legendre polynomials.

Combining Theorem 2.23 and Proposition 2.43, we obtain the formula

$$P_{n,d,j}(t) = \frac{(-1)^n (n+d-3)! \Gamma(\frac{d-1}{2})}{2^n (n+j+d-3)! \Gamma(n + \frac{d-1}{2})} (1-t^2)^{\frac{j}{2}}$$

$$\cdot \left(\frac{d}{dt}\right)^j \left[(1-t^2)^{\frac{3-d}{2}} \left(\frac{d}{dt}\right)^n (1-t^2)^{n+\frac{d-3}{2}}\right]$$

for $d \geq 3$, $0 \leq j \leq n$ and $t \in [-1,1]$. For $d = 3$, with $0 \leq j \leq n$ and $t \in [-1,1]$,

$$P_{n,3,j}(t) = \frac{(-1)^n}{2^n (n+j)!} (1-t^2)^{\frac{j}{2}} \left(\frac{d}{dt}\right)^{n+j} (1-t^2)^n. \tag{2.151}$$

From (2.150) and (2.151), we obtain an identity

$$(1-t^2)^j \left(\frac{d}{dt}\right)^{n+j} (1-t^2)^n = (-1)^j \frac{(n+j)!}{(n-j)!} \left(\frac{d}{dt}\right)^{n-j} (1-t^2)^n, \quad 0 \leq j \leq n.$$

For $d = 2$, we use the formulas given in Proposition 2.42 or Proposition 2.43 to define $P_{n,2,j}(t)$ for $0 \leq j \leq n$.

2.10.2 Properties

First we present an addition theorem for the associated Legendre functions. The function $\left[t - i\,(1 - t^2)^{1/2}s\right]^n$ is a polynomial of degree n in the variable s. Consider the expansion

$$\left[t + i\,(1 - t^2)^{1/2}s\right]^n = \sum_{j=0}^{n} c_j(t)\,P_{j,d-1}(s)$$

and let us determine $c_j(t)$. By Definition 2.41, for $0 \le k \le n$,

$$P_{n,d,k}(t) = \frac{\left|\mathbb{S}^{d-3}\right|}{\left|\mathbb{S}^{d-2}\right|}\,i^{-k}\sum_{j=0}^{n} c_j(t)\int_{-1}^{1} P_{k,d-1}(s)\,P_{j,d-1}(s)\,(1 - s^2)^{\frac{d-4}{2}}ds.$$

Using (2.79), we have

$$P_{n,d,k}(t) = \frac{1}{i^k N_{k,d-1}}\,c_k(t).$$

So

$$c_k(t) = i^k N_{k,d-1} P_{n,d,k}(t)$$

and then we can write the expansion as

$$\left[t + i\,(1 - t^2)^{1/2}s\right]^n = \sum_{j=0}^{n} i^j N_{j,d-1} P_{n,d,j}(t)\,P_{j,d-1}(s). \qquad (2.152)$$

Temporarily assume $m \ge n \ge 0$. We use the identity (2.152) to obtain

$$P_{m+n,d}(t) = \frac{\left|\mathbb{S}^{d-3}\right|}{\left|\mathbb{S}^{d-2}\right|}\int_{-1}^{1}\left[t + i\,(1 - t^2)^{1/2}s\right]^{m+n}(1 - s^2)^{\frac{d-4}{2}}ds$$

$$= \frac{\left|\mathbb{S}^{d-3}\right|}{\left|\mathbb{S}^{d-2}\right|}\int_{-1}^{1}\left[t + i\,(1 - t^2)^{1/2}s\right]^{m}\sum_{j=0}^{n} i^j N_{j,d-1} P_{n,d,j}(t)$$

$$\cdot\,P_{j,d-1}(s)\,(1 - s^2)^{\frac{d-4}{2}}ds$$

$$= \frac{\left|\mathbb{S}^{d-3}\right|}{\left|\mathbb{S}^{d-2}\right|}\sum_{j=0}^{n} i^j N_{j,d-1} P_{n,d,j}(t)\int_{-1}^{1}\left[t + i\,(1 - t^2)^{1/2}s\right]^{m}$$

$$\cdot\,P_{j,d-1}(s)\,(1 - s^2)^{\frac{d-4}{2}}ds$$

$$= \sum_{j=0}^{n}(-1)^j N_{j,d-1} P_{m,d,j}(t) P_{n,d,j}(t),$$

recalling the defining relation (2.149). Thus,

$$P_{m+n,d}(t) = \sum_{j=0}^{\min\{m,n\}} (-1)^j N_{j,d-1} P_{m,d,j}(t) P_{n,d,j}(t), \quad m,n \in \mathbb{N}_0. \quad (2.153)$$

This is an addition theorem for the associated Legendre functions.

For the case $d = 2$, $P_{n,2}(t) = \cos(n \arccos t)$. With the new variable $\theta = \arccos t$, we have $P_{n,2}(\cos \theta) = \cos(n\theta)$. Also, in this case, $N_{n,1}$ is given by (2.11). By Proposition 2.43,

$$P_{n,2,j}(t) = \frac{(n-1)!}{(n+j-1)!} (1-t^2)^{\frac{j}{2}} \left(\frac{d}{dt}\right)^j \cos(n \arccos t). \quad (2.154)$$

In particular, with $j = 1$, we obtain from (2.154) that

$$P_{n,2,1}(t) = \sin(n \arccos t).$$

The addition theorem formula (2.153) with $d = 2$

$$P_{m+n,2}(t) = \sum_{j=0}^{\min\{m,n\}} (-1)^j N_{j,1} P_{m,2,j}(t) P_{n,2,j}(t)$$

takes the following familiar form, with $\theta = \arccos t$,

$$\cos((m+n)\theta) = \cos(m\theta)\cos(n\theta) - \sin(m\theta)\sin(n\theta), \quad m,n \in \mathbb{N}_0.$$

Next, we derive a differential equation for $P_{n,d,j}(t)$. Differentiate (2.83) j times,

$$\left(\frac{d}{dt}\right)^j \left[(1-t^2) P''_{n,d}(t) - (d-1)t\, P'_{n,d}(t) + n(n+d-2) P_{n,d}(t)\right] = 0.$$

Since

$$\left(\frac{d}{dt}\right)^j \left[(1-t^2) P''_{n,d}(t)\right] = (1-t^2) P^{(j+2)}_{n,d}(t) - 2jt\, P^{(j+1)}_{n,d}(t)$$

$$- j(j-1) P^{(j)}_{n,d}(t),$$

$$\left(\frac{d}{dt}\right)^j \left[t\, P'_{n,d}(t)\right] = t\, P^{(j+1)}_{n,d}(t) + j\, P^{(j)}_{n,d}(t),$$

we get

$$(1 - t^2)\, P_{n,d}^{(j+2)}(t) - (2j + d - 1)\, t\, P_{n,d}^{(j+1)}(t)$$

$$+ \left[n\,(n + d - 2) - j\,(j + d - 2)\right] P_{n,d}^{(j)}(t) = 0. \tag{2.155}$$

By Proposition 2.43,

$$P_{n,d}^{(j)}(t) = c_0 (1 - t^2)^{-\frac{j}{2}} P_{n,d,j}(t), \qquad c_0 = \frac{(n + j + d - 3)!}{(n + d - 3)!}.$$

Then,

$$P_{n,d}^{(j+1)}(t) = c_0 (1 - t^2)^{-\frac{j}{2}} \left[P'_{n,d,j}(t) + j\, t\, (1 - t^2)^{-1} P_{n,d,j}(t)\right],$$

$$P_{n,d}^{(j+2)}(t) = c_0 (1 - t^2)^{-\frac{j}{2}-1} \big[(1 - t^2)\, P''_{n,d,j}(t) + 2\, j\, t\, P'_{n,d,j}(t)$$

$$+ j\left((j + 2)(1 - t^2)^{-1} - (j + 1)\right) P_{n,d,j}(t)\big].$$

Substitute these expressions in (2.155) and rearrange the terms to get the differential equation

$$(1 - t^2)\, P''_{n,d,j}(t) - (d - 1)\, t\, P'_{n,d,j}(t)$$

$$+ \left[n\,(n + d - 2) - \frac{j\,(j + d - 3)}{1 - t^2}\right] P_{n,d,j}(t) = 0. \tag{2.156}$$

Taking $j = 0$ in (2.156), we recover the differential equation (2.83) for the Legendre polynomials $P_{n,d}(t) = P_{n,d,0}(t)$.

We now use the differential equation (2.156) to prove the following orthogonality property.

Proposition 2.44.

$$\int_{-1}^{1} P_{m,d,j}(t)\, P_{n,d,j}(t)\, (1 - t^2)^{\frac{d-3}{2}}\, dt = 0, \quad m \neq n. \tag{2.157}$$

Proof. We rewrite (2.156) in the form

$$(1 - t^2)^{-\frac{d-3}{2}} \frac{d}{dt} \left[(1 - t^2)^{\frac{d-1}{2}} \frac{d}{dt} P_{n,d,j}(t)\right]$$

$$+ \left[n\,(n + d - 2) - \frac{j\,(j + d - 3)}{1 - t^2}\right] P_{n,d,j}(t) = 0.$$

From this equation, we deduce that

$$P_{m,d,j}(t) \frac{d}{dt}\left[(1-t^2)^{\frac{d-1}{2}} \frac{d}{dt} P_{n,d,j}(t)\right] - P_{n,d,j}(t) \frac{d}{dt}$$

$$\left[(1-t^2)^{\frac{d-1}{2}} \frac{d}{dt} P_{m,d,j}(t)\right] + (m-n)(m+n+d-2)$$

$$P_{m,d,j}(t) P_{n,d,j}(t)(1-t^2)^{\frac{d-3}{2}} = 0.$$

Integrate this equation for $t \in [-1,1]$ to get

$$(m-n)(m+n+d-2) \int_{-1}^{1} P_{m,d,j}(t) P_{n,d,j}(t)(1-t^2)^{\frac{d-3}{2}} dt = 0.$$

Thus, (2.157) holds.   □

Various recursion formulas for the associated Legendre functions exist; see [49, Sect. 3.12] in the case $d = 3$. The recursion formulas are useful for pointwise evaluation of the functions.

2.10.3 Normalized Associated Legendre Functions

In explicit calculations involving the associated Legendre functions, usually it is more convenient to use the normalized ones. From the formula given in Proposition 2.42,

$$\int_{-1}^{1} [P_{n,d,j}(t)]^2 (1-t^2)^{\frac{d-3}{2}} dt = \left[\frac{n!\Gamma(\frac{d-1}{2})}{2^j(n-j)!\Gamma(j+\frac{d-1}{2})}\right]^2$$

$$\int_{-1}^{1} [P_{n-j,d+2j}(t)]^2 (1-t^2)^{j+\frac{d-3}{2}} dt.$$

Use (2.79) for the integral,

$$\int_{-1}^{1} [P_{n-j,d+2j}(t)]^2 (1-t^2)^{j+\frac{d-3}{2}} dt = \frac{|\mathbb{S}^{d+2j-1}|}{N_{n-j,d+2j}|\mathbb{S}^{d+2j-2}|}.$$

Then

$$\int_{-1}^{1} [P_{n,d,j}(t)]^2 (1-t^2)^{\frac{d-3}{2}} dt = \frac{2^{d-2}(n!)^2 \Gamma(\frac{d-1}{2})^2}{(2n+d-2)(n-j)!(n+d+j-3)!}.$$

Thus, we define normalized associated Legendre functions

$$\tilde{P}_{n,d,j}(t) = \frac{\left[(2n + d - 2)\,(n - j)!\,(n + d + j - 3)!\right]^{\frac{1}{2}}}{2^{\frac{d-2}{2}}\,n!\,\Gamma(\frac{d-1}{2})}\,P_{n,d,j}(t),$$

$$t \in [-1, 1]. \tag{2.158}$$

We can also write, with the help of Proposition 2.43,

$$\tilde{P}_{n,d,j}(t) = \frac{(n + d - 3)!}{n!\,\Gamma(\frac{d-1}{2})}\left[\frac{(2n + d - 2)\,(n - j)!}{2^{d-2}(n + d + j - 3)!}\right]^{\frac{1}{2}}(1 - t^2)^{\frac{j}{2}}\,P_{n,d}^{(j)}(t),$$

$$t \in [-1, 1]. \tag{2.159}$$

These functions are normalized:

$$\int_{-1}^{1}\left[\tilde{P}_{n,d,j}(t)\right]^2 (1 - t^2)^{\frac{d-3}{2}}\,dt = 1.$$

Moreover, note that $\tilde{P}_{n,d,j}(t)$ is proportional to $P_{n,d,j}(t)$. Hence, these functions are orthonormal:

$$\int_{-1}^{1}\tilde{P}_{n,d,j}(t)\,\tilde{P}_{m,d,j}(t)\,(1 - t^2)^{\frac{d-3}{2}}\,dt = \delta_{nm}. \tag{2.160}$$

In the case $d = 3$,

$$\tilde{P}_{n,3,j}(t) = \left[\frac{(n + \frac{1}{2})\,(n - j)!}{(n + j)!}\right]^{\frac{1}{2}}(1 - t^2)^{\frac{j}{2}}\,P_{n,3}^{(j)}(t). \tag{2.161}$$

In the case $j = 0$, $\tilde{P}_{n,d,0}(t)$ is proportional to the Legendre polynomial $P_{n,d}(t)$,

$$\tilde{P}_{n,d,0}(t) = \frac{1}{\Gamma(\frac{d-1}{2})}\left[\frac{(2n + d - 2)\,(n + d - 3)!}{2^{d-2}n!}\right]^{\frac{1}{2}}P_{n,d}(t).$$

2.11 Generating Orthonormalized Bases for Spherical Harmonic Spaces

We now discuss a procedure to generate an orthonormal basis in $\mathbb{Y}_n^d$ from orthonormal bases in $(d - 1)$ dimensions, by making use of the associated Legendre functions introduced in Sect. 2.10.

Let $d \geq 3$. Consider a vector $\boldsymbol{\zeta} = \boldsymbol{\zeta}_{(d)} \in \mathbb{C}^d$ of the form $\boldsymbol{\zeta}_{(d)} = \boldsymbol{e}_d + i\left(\boldsymbol{\eta}^T, 0\right)^T$ with $\boldsymbol{\eta} \in \mathbb{S}^{d-2}$. A simple calculation shows $\boldsymbol{\zeta} \cdot \boldsymbol{\zeta} = 0$ and hence $\Delta_{\boldsymbol{x}}(\boldsymbol{\zeta} \cdot \boldsymbol{x})^n = 0$. So the function $\boldsymbol{x} \mapsto (\boldsymbol{\zeta} \cdot \boldsymbol{x})^n = (x_d + i\boldsymbol{x}_{(d-1)} \cdot \boldsymbol{\eta})^n$ is homogeneous and harmonic. Then

$$f(\boldsymbol{x}) := \frac{i^{-j}}{|\mathbb{S}^{d-2}|} \int_{\mathbb{S}^{d-2}} (x_d + i\boldsymbol{x}_{(d-1)} \cdot \boldsymbol{\eta})^n Y_{j,d-1}(\boldsymbol{\eta}) \, dS^{d-2}(\boldsymbol{\eta})$$

is a homogeneous harmonic polynomial of degree n, i.e., it is an element of $\mathbb{Y}_n(\mathbb{R}^d)$. Use the polar coordinates (1.15),

$$\boldsymbol{x} = |\boldsymbol{x}|\,\boldsymbol{\xi}, \quad \boldsymbol{\xi} = t\,\boldsymbol{e}_d + \sqrt{1-t^2}\,\boldsymbol{\xi}_{(d-1)}, \quad |t| \leq 1,\ \boldsymbol{\xi}_{(d-1)} \in \mathbb{S}^{d-1},$$

noting that $\boldsymbol{\xi}_{(d-1)}$ denotes a d-dimensional vector $(\xi_1, \cdots, \xi_{d-1}, 0)^T$. The restriction of the function $f(\boldsymbol{x})$ to $\mathbb{S}^{d-1}$ is

$$f(\boldsymbol{\xi}) = \frac{i^{-j}}{|\mathbb{S}^{d-2}|} \int_{\mathbb{S}^{d-2}} (t + i\,(1-t^2)^{\frac{1}{2}}\boldsymbol{\xi}_{(d-1)} \cdot \boldsymbol{\eta})^n Y_{j,d-1}(\boldsymbol{\eta}) \, dS^{d-2}(\boldsymbol{\eta}).$$

Applying the Funk–Hecke formula (Theorem 2.22), we have

$$\int_{\mathbb{S}^{d-2}} (t + i\,(1-t^2)^{\frac{1}{2}}\boldsymbol{\xi}_{(d-1)} \cdot \boldsymbol{\eta})^n Y_{j,d-1}(\boldsymbol{\eta}) \, dS^{d-2}(\boldsymbol{\eta}) = \lambda\, Y_{j,d-1}(\boldsymbol{\xi}_{(d-1)}),$$

where

$$\lambda = |\mathbb{S}^{d-3}| \int_{-1}^{1} P_{j,d-1}(s) \left(t + i\,(1-t^2)^{\frac{1}{2}}s\right)^j (1-t^2)^{\frac{d-4}{2}} \, dt.$$

Thus,

$$f(\boldsymbol{\xi}) = P_{n,d,j}(t)\, Y_{j,d-1}(\boldsymbol{\xi}_{(d-1)})$$

is a spherical harmonic of order n in dimension d. So we have shown the following result.

Proposition 2.45. *If $Y_{j,d-1} \in \mathbb{Y}_j^{d-1}$, then $P_{n,d,j}(t)Y_{j,d-1}(\boldsymbol{\xi}_{(d-1)}) \in \mathbb{Y}_n^d$ in polar coordinates* (1.15).

This result allows us to construct a basis for $\mathbb{Y}_n^d$ in d dimensions in terms of bases in $\mathbb{Y}_0^{d-1}, \ldots, \mathbb{Y}_n^{d-1}$ in $(d-1)$ dimensions. In the following we use the normalized associated functions $\tilde{P}_{n,d,j}$ since most formulas will then have a simpler form.

Definition 2.46. For $d \geq 3$ and $m \leq n$, define an operator

$$\tilde{\mathcal{P}}_{n,m} : \mathbb{Y}_m^{d-1} \to \mathbb{Y}_n^d$$

by the formula

$$(\tilde{\mathcal{P}}_{n,m} Y_{m,d-1})(\boldsymbol{\xi}) = \tilde{P}_{n,d,m}(t) Y_{m,d-1}(\boldsymbol{\xi}_{(d-1)}), \quad Y_{m,d-1} \in \mathbb{Y}_m^{d-1}.$$

Then define $\mathbb{Y}_{n,m}^d := \tilde{\mathcal{P}}_{n,m}(\mathbb{Y}_m^{d-1})$, called the associated space of order m in $\mathbb{Y}_n^d$.

The spherical harmonic space $\mathbb{Y}_n^d$ can be decomposed as an orthogonal sum of the associated spaces $\mathbb{Y}_{n,m}^d$, $0 \leq m \leq n$.

Theorem 2.47. *For $d \geq 3$ and $n \geq 0$,*

$$\mathbb{Y}_n^d = \mathbb{Y}_{n,0}^d \oplus \cdots \oplus \mathbb{Y}_{n,n}^d. \tag{2.162}$$

Proof. First we show that the subspaces on the right side of (2.162) are pairwise orthogonal. Let $0 \leq k, m \leq n$ with $k \neq m$. For any $Y_{k,d-1} \in \mathbb{Y}_k^{d-1}$ and any $Y_{m,d-1} \in \mathbb{Y}_m^{d-1}$,

$$(\tilde{\mathcal{P}}_{n,k} Y_{k,d-1}, \tilde{\mathcal{P}}_{n,m} Y_{m,d-1})_{L^2(\mathbb{S}^{d-1})}$$

$$= (Y_{k,d-1}, Y_{m,d-1})_{L^2(\mathbb{S}^{d-2})} \int_{-1}^{1} \tilde{P}_{n,d,k}(t)\, \tilde{P}_{n,d,m}(t)\, (1-t^2)^{\frac{d-3}{2}}\, dt$$

$$= 0$$

using the orthogonality (2.160). Thus, $\mathbb{Y}_{n,k}^d \perp \mathbb{Y}_{n,m}^d$ for $k \neq m$.

For each m, $0 \leq m \leq n$, $\mathbb{Y}_{n,m}^d$ is a subspace of $\mathbb{Y}_n^d$ and so

$$\mathbb{Y}_n^d \supset \mathbb{Y}_{n,0}^d \oplus \cdots \oplus \mathbb{Y}_{n,n}^d. \tag{2.163}$$

Since the mapping $\tilde{\mathcal{P}}_{n,m} : \mathbb{Y}_m^{d-1} \to \mathbb{Y}_{n,m}^d$ is a bijection,

$$\dim \mathbb{Y}_{n,m}^d = \dim \mathbb{Y}_m^{d-1} = N_{m,d-1}.$$

Hence, recalling the identity (2.14),

$$\sum_{m=0}^{n} \dim \mathbb{Y}_{n,m}^d = \sum_{m=0}^{n} N_{m,d-1} = N_{n,d} = \dim \mathbb{Y}_n^d.$$

In other words, the two sides of equality (2.162) are finite-dimensional spaces of equal dimension. Then the equality (2.162) holds in view of the relation (2.163). $\qquad \square$

From Theorem 2.47 and its proof, we see that if

$$\{Y_{m,d-1,j} : 1 \leq j \leq N_{m,d-1}\}$$

is an orthonormal basis for $\mathbb{Y}_m^{d-1}$, $0 \le m \le n$, then

$$\left\{ \tilde{P}_{n,d,m}(t) Y_{m,d-1,j}(\boldsymbol{\xi}_{(d-1)}) : 1 \le j \le N_{m,d-1}, 0 \le m \le n \right\} \qquad (2.164)$$

is an orthonormal basis for $\mathbb{Y}_n^d$.

Example 2.48. An orthonormal basis for $\mathbb{Y}_n^2$ is presented in Sect. 2.2. Let us apply the above result and use the orthonormal basis for $\mathbb{Y}_n^2$ to construct an orthonormal basis for $\mathbb{Y}_n^3$. We use the relation

$$\boldsymbol{\xi}_{(3)} = t\, e_3 + \sqrt{1 - t^2} \begin{pmatrix} \boldsymbol{\xi}_{(2)} \\ 0 \end{pmatrix},$$

where $t = \cos\theta$ for $0 \le \theta \le \pi$, $\boldsymbol{\xi}_{(2)} = (\cos\phi, \sin\phi)^T$ for $0 \le \phi \le 2\pi$. In the notation of the above discussion,

$$\left\{ Y_{m,2,1}(\boldsymbol{\xi}_{(2)}) = \frac{1}{\sqrt{\pi}} \cos(m\phi),\ Y_{m,2,2}(\boldsymbol{\xi}_{(2)}) = \frac{1}{\sqrt{\pi}} \sin(m\phi) \right\}$$

is an orthonormal basis for $\mathbb{Y}_m^2$. Recall the formula (2.161),

$$\tilde{P}_{n,3,m}(t) = \left[\frac{(n + \frac{1}{2})(n - m)!}{(n + m)!} \right]^{\frac{1}{2}} (1 - t^2)^{\frac{m}{2}} P_{n,3}^{(m)}(t).$$

Here $P_{n,3}^{(m)}(t)$ denotes the mth derivative of the function $P_{n,3}(t)$. Then, an orthonormal basis for $\mathbb{Y}_n^3$ is given by the functions

$$\left[\frac{(2n+1)(n-m)!}{2\pi(n+m)!} \right]^{\frac{1}{2}} (\sin\theta)^m P_{n,3}^{(m)}(\cos\theta) \cos(m\phi), \quad 0 \le m \le n,$$

$$\left[\frac{(2n+1)(n-m)!}{2\pi(n+m)!} \right]^{\frac{1}{2}} (\sin\theta)^m P_{n,3}^{(m)}(\cos\theta) \sin(m\phi), \quad 1 \le m \le n.$$

The basis is usually also written as

$$(-1)^{(m+|m|)/2} \left[\frac{(2n+1)(n-|m|)!}{4\pi(n+|m|)!} \right]^{\frac{1}{2}} (\sin\theta)^m P_{n,3}^{(m)}(\cos\theta)\, e^{im\phi},$$

$$-n \le m \le n.$$

This latter form is more convenient to use in some calculations. $\qquad\qquad\square$

We now use the orthonormal system (2.164) to express the addition theorem. Set

$$\boldsymbol{\xi}_{(d)} = t\, \boldsymbol{e}_d + (1 - t^2)^{\frac{1}{2}} \boldsymbol{\xi}_{(d-1)}, \quad -1 \le t \le 1,$$

$$\boldsymbol{\eta}_{(d)} = s\, \boldsymbol{e}_d + (1 - s^2)^{\frac{1}{2}} \boldsymbol{\eta}_{(d-1)}, \quad -1 \le s \le 1.$$

Then the identity (2.24)

$$\frac{N_{n,d}}{|\mathbb{S}^{d-1}|} P_{n,d}(\boldsymbol{\xi} \cdot \boldsymbol{\eta}) = \sum_{k=1}^{N_{n,d}} Y_{n,k}(\boldsymbol{\xi}) \overline{Y_{n,k}(\boldsymbol{\eta})}$$

is rewritten as

$$\frac{N_{n,d}}{|\mathbb{S}^{d-1}|} P_{n,d}\big(s\,t + (1 - s^2)^{\frac{1}{2}}(1 - t^2)^{\frac{1}{2}} \boldsymbol{\xi}_{(d-1)} \cdot \boldsymbol{\eta}_{(d-1)}\big)$$

$$= \sum_{m=0}^{n} \tilde{P}_{n,d,m}(s) \tilde{P}_{n,d,m}(t) \sum_{k=1}^{N_{m,d-1}} Y_{m,k}(\boldsymbol{\xi}_{(d-1)}) \overline{Y_{m,k}(\boldsymbol{\eta}_{(d-1)})}$$

$$= \sum_{m=0}^{n} \frac{N_{m,d-1}}{|\mathbb{S}^{d-2}|} \tilde{P}_{n,d,m}(s) \tilde{P}_{n,d,m}(t) P_{m,d-1}(\boldsymbol{\xi}_{(d-1)} \cdot \boldsymbol{\eta}_{(d-1)}),$$

where in the last step, the identity (2.24) is applied again. Denote $u = \boldsymbol{\xi}_{(d-1)} \cdot \boldsymbol{\eta}_{(d-1)}$. Then for $d \ge 3$ and $s, t, u \in [-1, 1]$,

$$\sum_{m=0}^{n} N_{m,d-1} \tilde{P}_{n,d,m}(s) \tilde{P}_{n,d,m}(t) P_{m,d-1}(u)$$

$$= \frac{N_{n,d} |\mathbb{S}^{d-2}|}{|\mathbb{S}^{d-1}|} P_{n,d}\big(s\,t + (1 - s^2)^{\frac{1}{2}}(1 - t^2)^{\frac{1}{2}} u\big). \qquad (2.165)$$

Another identity can be derived from (2.165) as follows. Multiply both sides of (2.165) by $P_{k,d-1}(u)\,(1 - u^2)^{\frac{d-4}{2}}$, $0 \le k \le n$, integrate with respect to u from -1 to 1, and use the orthogonality relation (2.79) for the Legendre polynomials,

$$\frac{N_{n,d}}{|\mathbb{S}^{d-1}|} \int_{-1}^{1} P_{n,d}\big(s\,t + (1 - s^2)^{\frac{1}{2}}(1 - t^2)^{\frac{1}{2}} u\big) P_{k,d-1}(u)\,(1 - u^2)^{\frac{d-4}{2}}\, du$$

$$= \frac{1}{|\mathbb{S}^{d-3}|} \tilde{P}_{n,d,k}(s) \tilde{P}_{n,d,k}(t),$$

i.e.,

$$\int_{-1}^{1} P_{n,d}(s\,t + u\,(1-s^2)^{\frac{1}{2}}(1-t^2)^{\frac{1}{2}})P_{k,d-1}(u)\,(1-u^2)^{\frac{d-4}{2}}\,du$$

$$= \frac{2\pi}{(d-2)\,N_{n,d}}\tilde{P}_{n,d,k}(s)\tilde{P}_{n,d,k}(t). \qquad (2.166)$$

In particular, taking $k = 0$ in (2.166) and noting that

$$\tilde{P}_{n,d,0}(t) = \left(\frac{N_{n,d}|\mathbb{S}^{d-2}|}{|\mathbb{S}^{d-1}|}\right)^{\frac{1}{2}} P_{n,d}(t),$$

we arrive at an identity for the Legendre polynomials,

$$\int_{-1}^{1} P_{n,d}(s\,t + (1-s^2)^{\frac{1}{2}}(1-t^2)^{\frac{1}{2}}u)(1-u^2)^{\frac{d-4}{2}}\,du = \frac{|\mathbb{S}^{d-2}|}{|\mathbb{S}^{d-1}|}P_{n,d}(s)P_{n,d}(t)$$

$$(2.167)$$

for $d \geq 3$.

Chapter 3
Differentiation and Integration over the Sphere

In this chapter, we discuss some properties and formulas for differentiation and integration involving spherical harmonics. In Sect. 3.1, we derive representation formulas for the Laplace–Beltrami operator, which is defined to be the restriction of the Laplace operator on the unit sphere. In Sect. 3.2, a concrete formula is shown for the Laplace–Beltrami operator through coordinates and derivatives with respect to the coordinates of the x variable in $\mathbb{R}^d$. It turns out that spherical harmonics are eigenfunctions of the Laplace–Beltrami operator, and this property provides convenience in some calculations involving spherical harmonics; this is the content of Sect. 3.3. Section 3.4 discusses some integration formulas over the unit sphere. In Sect. 3.5, we present some differentiation formulas that are related to the spherical harmonics. Section 3.6 is devoted to some integral identities for spherical harmonics and the section begins with a review of some basic properties of harmonic functions for a general dimension $d \geq 3$. In Sect. 3.7, we show the derivation of some integral identities through a straightforward application of the Funk–Hecke formula. In Sect. 3.8, we introduce Sobolev spaces over the unit sphere through expansions in terms of an orthonormal basis of spherical harmonics. It is possible to study Sobolev spaces over regions of the unit sphere as well as polynomial or spline approximations, see e.g., [37,71,87], or even Sobolev spaces of vector-valued functions [69]. Finally, in Sect. 3.9, we discuss positive definite functions, a concept important in the study of meshless discretization methods for handling discrete data with no associated mesh or grid.

3.1 The Laplace–Beltrami Operator

The Laplace–Beltrami operator is the restriction of the Laplace operator on the unit sphere.

K. Atkinson and W. Han, *Spherical Harmonics and Approximations on the Unit Sphere: An Introduction*, Lecture Notes in Mathematics 2044, DOI 10.1007/978-3-642-25983-8_3, © Springer-Verlag Berlin Heidelberg 2012

For a bijection $\boldsymbol{x} = \boldsymbol{x}(\boldsymbol{u})$ or $\boldsymbol{u} = \boldsymbol{u}(\boldsymbol{x})$ over $\mathbb{R}^d$, define the following quantities

$$g_{ij} := \frac{\partial \boldsymbol{x}}{\partial u_i} \cdot \frac{\partial \boldsymbol{x}}{\partial u_j} = \sum_{k=1}^{d} \frac{\partial x_k}{\partial u_i} \frac{\partial x_k}{\partial u_j},$$

$$g^{ij} := \sum_{k=1}^{d} \frac{\partial u_i}{\partial x_k} \frac{\partial u_j}{\partial x_k}$$

for $1 \le i, j \le d$, and

$$g := \det(g_{ij}).$$

By the chain rule,

$$(g^{ij}) = (g_{ij})^{-1}.$$

Then we have the following transformation formula for the Laplacian operator

$$\Delta_{(d)} := \sum_{j=1}^{d} \left(\frac{\partial}{\partial x_j} \right)^2 = \frac{1}{\sqrt{g}} \sum_{i,j=1}^{d} \frac{\partial}{\partial u_i} \left(g^{ij} \sqrt{g} \frac{\partial}{\partial u_j} \right). \tag{3.1}$$

Now assume $\boldsymbol{\xi}_{(d)} = \boldsymbol{\xi}(u_1, \dots, u_{d-1})$ is a bijection between a set U of $\mathbb{R}^{d-1}$ and $\mathbb{S}^{d-1}$ such that the mapping is C^2 in the interior of U. We use $u_1, \dots, u_{d-1}$ and $u_d = r = |\boldsymbol{x}|$ as the polar coordinates. Then $\boldsymbol{x}_{(d)} = r\boldsymbol{\xi}_{(d)}$. Similar to the quantities g_{ij}, g^{ij} and g, we define

$$\gamma_{ij} := \frac{\partial \boldsymbol{\xi}_{(d)}}{\partial u_i} \cdot \frac{\partial \boldsymbol{\xi}_{(d)}}{\partial u_j} = \sum_{k=1}^{d} \frac{\partial \xi_k}{\partial u_i} \frac{\partial \xi_k}{\partial u_j},$$

$$(\gamma^{ij}) := (\gamma_{ij})^{-1}, \quad 1 \le i, j \le d-1$$

and

$$\gamma := \det(\gamma_{ij}).$$

We have

$$g_{ij} = r^2 \gamma_{ij}, \quad 1 \le i, j \le d-1.$$

Differentiate the identity

$$1 = \boldsymbol{\xi}_{(d)} \cdot \boldsymbol{\xi}_{(d)} = \sum_{k=1}^{d} (\xi_k)^2$$

with respect to u_i, $1 \le i \le d-1$, to obtain

$$\boldsymbol{\xi}_{(d)} \cdot \frac{\partial \boldsymbol{\xi}_{(d)}}{\partial u_i} = \sum_{k=1}^{d} \xi_k \frac{\partial \xi_k}{\partial u_i} = 0, \quad 1 \le i \le d-1.$$

Thus,

$$g_{id} = r\, \boldsymbol{\xi}_{(d)} \cdot \frac{\partial \boldsymbol{\xi}_{(d)}}{\partial u_i} = r \sum_{k=1}^{d} \xi_k \frac{\partial \xi_k}{\partial u_i} = 0, \quad 1 \le i \le d-1.$$

Moreover,

$$g_{dd} = \boldsymbol{\xi}_{(d)} \cdot \boldsymbol{\xi}_{(d)} = 1.$$

Hence,

$$g = r^{2(d-1)} \gamma, \quad \sqrt{g} = r^{d-1} \sqrt{\gamma}.$$

From $(g^{ij}) = (g_{ij})^{-1}$, we obtain

$$g^{ij} = \frac{1}{r^2} \gamma^{ij},\ 1 \le i,j \le d-1,$$

$$g^{di} = g^{id} = 0,\ 1 \le i \le d-1,$$

$$g^{dd} = 1.$$

Using these relations in (3.1), we have

$$\Delta_{(d)} = \frac{1}{r^{d-1}} \frac{\partial}{\partial r} \left(r^{d-1} \frac{\partial}{\partial r} \right) + \frac{1}{r^2} \Delta^*_{(d-1)}, \tag{3.2}$$

where the Laplace–Beltrami operator

$$\Delta^*_{(d-1)} = \frac{1}{\sqrt{\gamma}} \sum_{i,j=1}^{d-1} \frac{\partial}{\partial u_i} \left(\gamma^{ij} \sqrt{\gamma} \frac{\partial}{\partial u_j} \right). \tag{3.3}$$

Recall the extension f^* of f, (1.22). Then, if $f \in C^2(\mathbb{S}^{d-1})$,

$$\Delta_{(d)} f^*(\boldsymbol{x})|_{|\boldsymbol{x}|=1} = \Delta^*_{(d-1)} f(\boldsymbol{\xi}). \tag{3.4}$$

Thus, the value of $\Delta^*_{(d-1)} f(\boldsymbol{\xi})$ does not depend on the coordinate system for $\mathbb{S}^{d-1}$.

In the polar coordinates $u_1, \ldots, u_{d-1}$, $u_d = r = |\boldsymbol{x}|$, the gradient operator can be expressed as

$$\boldsymbol{\nabla} = \boldsymbol{\xi}\frac{\partial}{\partial r} + \frac{1}{r}\boldsymbol{\nabla}^*_{(d-1)}, \tag{3.5}$$

where

$$\boldsymbol{\nabla}^*_{(d-1)} := \sum_{i,j=1}^{d-1} \gamma^{ij}\frac{\partial\boldsymbol{\xi}}{\partial u_i}\frac{\partial}{\partial u_j} \tag{3.6}$$

is the first-order Beltrami operator on $\mathbb{S}^{d-1}$. Observe that the operator

$$\boldsymbol{\nabla}^*_{(d-1)} = r\boldsymbol{\nabla} - \boldsymbol{\xi}r\frac{\partial}{\partial r}$$

does not depend on the coordinate system for $\mathbb{S}^{d-1}$. Similar to (3.4), we have

$$\boldsymbol{\nabla}_{(d)}f^*(\boldsymbol{x})|_{|\boldsymbol{x}|=1} = \boldsymbol{\nabla}^*_{(d-1)}f(\boldsymbol{\xi}) \tag{3.7}$$

for $f \in C^1(\mathbb{S}^{d-1})$.

Example 3.1. Consider the case $d = 2$. We have

$$\boldsymbol{x} = r\boldsymbol{\xi}, \quad \boldsymbol{\xi} = \begin{pmatrix} \cos\theta \\ \sin\theta \end{pmatrix}, \ 0 \le \theta < 2\pi.$$

We let $u_1 = \theta$. Then,

$$\gamma = \gamma_{11} = (-\sin\theta)^2 + (\cos\theta)^2 = 1, \quad \gamma^{11} = 1,$$
$$g = g_{11} = r^2, \quad \sqrt{g} = r.$$

By (3.6) and (3.3), we obtain

$$\boldsymbol{\nabla}^*_{(1)} = \begin{pmatrix} -\sin\theta \\ \cos\theta \end{pmatrix}\frac{\partial}{\partial\theta}, \quad \Delta^*_{(1)} = \frac{\partial^2}{\partial\theta^2}.$$

It can be verified that $\Delta^*_{(1)} = \boldsymbol{\nabla}^*_{(1)} \cdot \boldsymbol{\nabla}^*_{(1)}$. $\square$

Example 3.2. Consider the case $d = 3$. We have

$$\boldsymbol{x} = r\boldsymbol{\xi}, \quad \boldsymbol{\xi} = \begin{pmatrix} \cos\phi\sin\theta \\ \sin\phi\sin\theta \\ \cos\theta \end{pmatrix}, \ 0 \le \theta \le \pi, \ 0 \le \phi < 2\pi.$$

We let $u_1 = \theta$ and $u_2 = \phi$. Then,

$$\gamma_{11} = 1, \quad \gamma_{22} = \sin^2\theta, \quad \gamma_{12} = \gamma_{21} = 0,$$

$$\gamma = \sin^2\theta, \quad \sqrt{\gamma} = \sin\theta,$$

$$\gamma^{11} = 1, \quad \gamma^{22} = \sin^{-2}\theta, \quad \gamma^{12} = \gamma^{21} = 0.$$

By (3.6) and (3.3), we obtain

$$\boldsymbol{\nabla}^*_{(2)} = \begin{pmatrix} \cos\theta\cos\phi \\ \cos\theta\sin\phi \\ -\sin\theta \end{pmatrix} \frac{\partial}{\partial\theta} + \frac{1}{\sin\theta} \begin{pmatrix} -\sin\phi \\ \cos\phi \\ 0 \end{pmatrix} \frac{\partial}{\partial\phi},$$

$$\Delta^*_{(2)} = \frac{1}{\sin\theta} \frac{\partial}{\partial\theta}\left(\sin\theta\frac{\partial}{\partial\theta}\right) + \frac{1}{\sin^2\theta}\frac{\partial^2}{\partial\phi^2}.$$

It can be verified that $\Delta^*_{(2)} = \boldsymbol{\nabla}^*_{(2)} \cdot \boldsymbol{\nabla}^*_{(2)}$.

Another possibility is to use $t = \cos\theta$ and write

$$\boldsymbol{x} = \boldsymbol{\Phi}(r, t, \phi) \equiv r\,\boldsymbol{\xi}, \quad \boldsymbol{\xi} = \begin{pmatrix} \sqrt{1-t^2}\,\cos\phi \\ \sqrt{1-t^2}\,\sin\phi \\ t \end{pmatrix}, \quad -1 \le t \le 1,\ 0 \le \phi < 2\pi.$$

We let $u_1 = t$ and $u_2 = \phi$. Then,

$$\gamma_{11} = \frac{1}{1-t^2}, \quad \gamma_{22} = 1-t^2, \quad \gamma_{12} = \gamma_{21} = 0,$$

$$\gamma = 1,$$

$$\gamma^{11} = 1-t^2, \quad \gamma^{22} = \frac{1}{1-t^2}, \quad \gamma^{12} = \gamma^{21} = 0.$$

By (3.6) and (3.3), we obtain

$$\boldsymbol{\nabla}^*_{(2)} = \boldsymbol{e}_t\sqrt{1-t^2}\frac{\partial}{\partial t} + \boldsymbol{e}_\phi\frac{1}{\sqrt{1-t^2}}\frac{\partial}{\partial\phi},$$

$$\Delta^*_{(2)} = \frac{\partial}{\partial t}\left((1-t^2)\frac{\partial}{\partial t}\right) + \frac{1}{1-t^2}\frac{\partial^2}{\partial\phi^2},$$

where $\boldsymbol{e}_\phi$ and $\boldsymbol{e}_t$ together with $\boldsymbol{e}_r$ form a local vector basis for $\mathbb{R}^3$:

$$\boldsymbol{e}_r = \left|\frac{\partial\boldsymbol{\Phi}(1, t, \phi)}{\partial r}\right|^{-1}\frac{\partial\boldsymbol{\Phi}(1, t, \phi)}{\partial r} = \begin{pmatrix} \sqrt{1-t^2}\,\cos\phi \\ \sqrt{1-t^2}\,\sin\phi \\ t \end{pmatrix},$$

$$\boldsymbol{e}_\phi = \left| \frac{\partial \boldsymbol{\Phi}(1, t, \phi)}{\partial \phi} \right|^{-1} \frac{\partial \boldsymbol{\Phi}(1, t, \phi)}{\partial \phi} = \begin{pmatrix} -\sin \phi \\ \cos \phi \\ 0 \end{pmatrix},$$

$$\boldsymbol{e}_t = \left| \frac{\partial \boldsymbol{\Phi}(1, t, \phi)}{\partial t} \right|^{-1} \frac{\partial \boldsymbol{\Phi}(1, t, \phi)}{\partial t} = \begin{pmatrix} -t \cos \phi \\ -t \sin \phi \\ \sqrt{1 - t^2} \end{pmatrix}.$$

The second possibility is more widely used in applications (cf. [47, 48]). $\square$

We now explore a relation between $\Delta_{(d-1)}^*$ and $\Delta_{(d-2)}^*$. This will allow us to find a representation formula for $\Delta_{(d-1)}^*$ explicitly. For $\boldsymbol{\xi} = \boldsymbol{\xi}_{(d)} \in \mathbb{S}^{d-1}$, we write (cf. (1.15))

$$\boldsymbol{\xi}_{(d)} = t \, \boldsymbol{e}_d + \sqrt{1 - t^2} \, \boldsymbol{\xi}_{(d-1)},$$

where $t \in [-1, 1]$ is identified with u_{d-1}, and $\boldsymbol{\xi}_{(d-1)}$ is a d-dimensional vector $(\xi_1, \ldots, \xi_{d-1}, 0)^T \in \mathbb{S}^{d-1}$ and we will also use $\boldsymbol{\xi}_{(d-1)}$ to represent a $(d-1)$-dimensional vector $(\xi_1, \ldots, \xi_{d-1})^T \in \mathbb{S}^{d-2}$. Then,

$$\frac{\partial \boldsymbol{\xi}_{(d)}}{\partial u_{d-1}} = \boldsymbol{e}_d - \frac{t}{\sqrt{1 - t^2}} \boldsymbol{\xi}_{(d-1)},$$

$$\frac{\partial \boldsymbol{\xi}_{(d)}}{\partial u_i} = \sqrt{1 - t^2} \, \frac{\partial \boldsymbol{\xi}_{(d-1)}}{\partial u_i}, \quad 1 \leq i \leq d - 2.$$

Thus,

$$\gamma_{d-1,d-1} = \frac{\partial \boldsymbol{\xi}_{(d)}}{\partial u_{d-1}} \cdot \frac{\partial \boldsymbol{\xi}_{(d)}}{\partial u_{d-1}} = \frac{1}{1 - t^2},$$

$$\gamma_{i,d-1} = \gamma_{d-1,i} = \frac{\partial \boldsymbol{\xi}_{(d)}}{\partial u_{d-1}} \cdot \frac{\partial \boldsymbol{\xi}_{(d)}}{\partial u_i} = 0, \quad 1 \leq i \leq d - 2,$$

$$\gamma_{ij} = \frac{\partial \boldsymbol{\xi}_{(d)}}{\partial u_i} \cdot \frac{\partial \boldsymbol{\xi}_{(d)}}{\partial u_j} = \left(1 - t^2\right) \frac{\partial \boldsymbol{\xi}_{(d-1)}}{\partial u_i} \cdot \frac{\partial \boldsymbol{\xi}_{(d-1)}}{\partial u_j}, \quad 1 \leq i, j \leq d - 2.$$

Denote

$$\overline{\gamma}_{ij} = \frac{\partial \boldsymbol{\xi}_{(d-1)}}{\partial u_i} \cdot \frac{\partial \boldsymbol{\xi}_{(d-1)}}{\partial u_j}, \quad 1 \leq i, j \leq d - 2$$

and let

$$\left(\overline{\gamma}^{ij}\right) = \left(\overline{\gamma}_{ij}\right)^{-1}.$$

Then

$$\gamma := \det(\gamma_{ij})_{(d-1)\times(d-1)} = (1 - t^2)^{d-3}\overline{\gamma}$$

where

$$\overline{\gamma} := \det(\overline{\gamma}_{ij})_{(d-2)\times(d-2)}$$

and

$$\gamma^{d-1,d-1} = 1 - t^2,$$
$$\gamma^{i,d-1} = \gamma^{d-1,i} = 0, \quad 1 \leq i \leq d - 2,$$
$$\gamma^{ij} = \frac{1}{1 - t^2}\,\overline{\gamma}^{ij}, \quad 1 \leq i, j \leq d - 2.$$

Hence, by the formula (3.3),

$$\Delta^*_{(d-1)} = \frac{1}{\sqrt{\gamma}}\frac{\partial}{\partial u_{d-1}}\left(\gamma^{d-1,d-1}\sqrt{\gamma}\frac{\partial}{\partial u_{d-1}}\right) + \frac{1}{\sqrt{\gamma}}\sum_{i,j=1}^{d-2}\frac{\partial}{\partial u_i}\left(\gamma^{ij}\sqrt{\gamma}\frac{\partial}{\partial u_j}\right)$$

$$= \frac{1}{(1-t^2)^{(d-3)/2}}\frac{\partial}{\partial t}\left((1-t^2)^{(d-1)/2}\frac{\partial}{\partial t}\right)$$

$$+ \frac{1}{(1-t^2)\sqrt{\overline{\gamma}}}\sum_{i,j=1}^{d-2}\frac{\partial}{\partial u_i}\left(\overline{\gamma}^{ij}\sqrt{\overline{\gamma}}\frac{\partial}{\partial u_j}\right).$$

Therefore, we have the following recursion formula,

$$\Delta^*_{(d-1)} = \frac{1}{(1-t^2)^{(d-3)/2}}\frac{\partial}{\partial t}\left((1-t^2)^{(d-1)/2}\frac{\partial}{\partial t}\right) + \frac{1}{(1-t^2)}\Delta^*_{(d-2)}. \quad (3.8)$$

As an example, consider using the d-dimensional spherical coordinates

$$x_1 = r\,\cos\theta_1\,\sin\theta_2\,\sin\theta_3\cdots\sin\theta_{d-1},$$
$$x_2 = r\,\sin\theta_1\,\sin\theta_2\,\sin\theta_3\cdots\sin\theta_{d-1},$$
$$x_3 = r\,\cos\theta_2\,\sin\theta_3\cdots\sin\theta_{d-1},$$

$$\vdots$$

$$x_{d-1} = r\,\cos\theta_{d-2}\,\sin\theta_{d-1},$$
$$x_d = r\,\cos\theta_{d-1},$$
$$r \geq 0,\ 0 \leq \theta_1 < 2\pi,\ 0 \leq \theta_j \leq \pi \text{ for } 2 \leq j \leq d - 1.$$

For $t = \cos\theta$ with $0 \le \theta \le \pi$, it can be verified that

$$\frac{1}{(1-t^2)^{(d-3)/2}} \frac{\partial}{\partial t}\left((1-t^2)^{(d-1)/2}\frac{\partial}{\partial t}\right) = \frac{1}{(\sin\theta)^{d-2}}\frac{\partial}{\partial\theta}\left((\sin\theta)^{d-2}\frac{\partial}{\partial\theta}\right).$$

Therefore, using the results from Example 3.1 and (3.8) with $t = \cos\theta_{d-1}$, we have

$$\Delta^*_{(1)} = \frac{\partial^2}{\partial\theta_1^2},$$

$$\Delta^*_{(d)} = \frac{1}{\sin^{d-1}\theta_d}\frac{\partial}{\partial\theta_d}\left(\sin^{d-1}\theta_d\frac{\partial}{\partial\theta_d}\right) + \frac{1}{\sin^2\theta_d}\Delta^*_{(d-1)}, \quad d \ge 2.$$

We can use (3.4) and (3.7) to prove integral identities over the sphere. One such example is the following.

Proposition 3.3. (Green–Beltrami identity) *For any $f \in C^2(\mathbb{S}^{d-1})$ and any $g \in C^1(\mathbb{S}^{d-1})$,*

$$\int_{\mathbb{S}^{d-1}} g\Delta^*_{(d-1)}f\,dS^{d-1} = -\int_{\mathbb{S}^{d-1}} \nabla^*_{(d-1)}g\cdot\nabla^*_{(d-1)}f\,dS^{d-1}. \qquad (3.9)$$

Proof. For the extensions $f^*(\boldsymbol{x}) = f(\boldsymbol{x}/|\boldsymbol{x}|)$ and $g^*(\boldsymbol{x}) = g(\boldsymbol{x}/|\boldsymbol{x}|)$, we apply the Green's identity

$$\int_{1-\delta \le |\boldsymbol{x}| \le 1+\delta} (g^*\Delta f^* + \nabla g^*\cdot\nabla f^*)\,dx = 0,$$

i.e.,

$$\int_{1-\delta}^{1+\delta} r^{d-3}\left[\int_{\mathbb{S}^{d-1}}\left(g\Delta^*_{(d-1)}f + \nabla^*_{(d-1)}g\cdot\nabla^*_{(d-1)}f\right)dS^{d-1}\right]dr = 0.$$

So (3.9) holds.   □

As a consequence of Proposition 3.3, we have the next integral identity.

Corollary 3.4. *For any $f, g \in C^2(\mathbb{S}^{d-1})$,*

$$\int_{\mathbb{S}^{d-1}} g\,\Delta^*_{(d-1)}f\,dS^{d-1} = \int_{\mathbb{S}^{d-1}} f\,\Delta^*_{(d-1)}g\,dS^{d-1}. \qquad (3.10)$$

3.2 A Formula for the Laplace–Beltrami Operator

In this section, we derive a formula for the Laplace–Beltrami operator written in terms of the variable $\boldsymbol{x}$ and derivatives with respect to the components of $\boldsymbol{x}$.

Consider the operator

$$D_{(d)} := - \sum_{1 \le j < i \le d} \left(x_i \frac{\partial}{\partial x_j} - x_j \frac{\partial}{\partial x_i} \right)^2 = -\frac{1}{2} \sum_{i,j=1}^{d} \left(x_i \frac{\partial}{\partial x_j} - x_j \frac{\partial}{\partial x_i} \right)^2.$$

$$(3.11)$$

For $i \ne j$, we have

$$\left(x_i \frac{\partial}{\partial x_j} - x_j \frac{\partial}{\partial x_i} \right)^2 = x_i \frac{\partial}{\partial x_j} \left(x_i \frac{\partial}{\partial x_j} - x_j \frac{\partial}{\partial x_i} \right)$$

$$- x_j \frac{\partial}{\partial x_i} \left(x_i \frac{\partial}{\partial x_j} - x_j \frac{\partial}{\partial x_i} \right)$$

$$= x_i^2 \frac{\partial^2}{\partial x_j^2} + x_j^2 \frac{\partial^2}{\partial x_i^2} - x_i \frac{\partial}{\partial x_i} - x_j \frac{\partial}{\partial x_j} - 2\, x_i x_j \frac{\partial^2}{\partial x_i \partial x_j}.$$

Then,

$$D_{(d)} = - \sum_{j=1}^{d-1} \left(\sum_{i=j+1}^{d} x_i^2 \right) \frac{\partial^2}{\partial x_j^2} - \sum_{i=2}^{d} \left(\sum_{j=1}^{i-1} x_j^2 \right) \frac{\partial^2}{\partial x_i^2}$$

$$+ \sum_{i \ne j} x_i x_j \frac{\partial^2}{\partial x_i \partial x_j} + (d-1) \sum_{i=1}^{d} x_i \frac{\partial}{\partial x_i},$$

i.e.,

$$D_{(d)} = -r^2 \Delta_{(d)} + \sum_{i,j=1}^{d} x_i x_j \frac{\partial^2}{\partial x_i \partial x_j} + (d-1) \sum_{i=1}^{d} x_i \frac{\partial}{\partial x_i}. \qquad (3.12)$$

We will use more compact expressions for the two sums in (3.12). Consider an arbitrary differentiable function $f(\boldsymbol{x}) = f(r\boldsymbol{\xi})$, $\boldsymbol{\xi} \in \mathbb{S}^{d-1}$. We differentiate the function with respect to r:

$$\frac{\partial f(r\boldsymbol{\xi})}{\partial r} = \sum_{i=1}^{d} \xi_i \frac{\partial f(r\boldsymbol{\xi})}{\partial x_i}. \qquad (3.13)$$

Multiply both sides by r,

$$r \frac{\partial f}{\partial r} = \sum_{i=1}^{d} x_i \frac{\partial f}{\partial x_i}.$$

Since f is arbitrary,

$$\sum_{i=1}^{d} x_i \frac{\partial}{\partial x_i} = r \frac{\partial}{\partial r}. \qquad (3.14)$$

Now assume f is twice differentiable and differentiate (3.13) with respect to r,

$$\frac{\partial^2 f(r\boldsymbol{\xi})}{\partial r^2} = \sum_{i,j=1}^{d} \xi_i \xi_j \frac{\partial^2 f(r\boldsymbol{\xi})}{\partial x_i \partial x_j}.$$

Multiply both sides by r^2,

$$r^2 \frac{\partial^2 f}{\partial r^2} = \sum_{i,j=1}^{d} x_i x_j \frac{\partial^2 f}{\partial x_i \partial x_j}.$$

Since f is arbitrary, we have

$$\sum_{i,j=1}^{d} x_i x_j \frac{\partial^2}{\partial x_i \partial x_j} = r^2 \frac{\partial}{\partial r^2}. \tag{3.15}$$

Using (3.14) and (3.15) in (3.12), we obtain

$$D_{(d)} = -r^2 \Delta_{(d)} + r^2 \frac{\partial}{\partial r^2} + (d-1) r \frac{\partial}{\partial r}. \tag{3.16}$$

This relation can be rewritten in the form

$$\Delta_{(d)} = \frac{1}{r^{d-1}} \frac{\partial}{\partial r} \left(r^{d-1} \frac{\partial}{\partial r} \right) - \frac{1}{r^2} D_{(d)}. \tag{3.17}$$

Comparing (3.17) and (3.2), we see that $D_{(d)}$ is the Laplace–Beltrami operator. Therefore,

$$\Delta^* = - \sum_{1 \le j < i \le d} \left(x_i \frac{\partial}{\partial x_j} - x_j \frac{\partial}{\partial x_i} \right)^2. \tag{3.18}$$

3.3 Spherical Harmonics As Eigenfunctions of the Laplace–Beltrami Operator

Consider any non-zero function $Y_n \in \mathbb{Y}_n^d$, write $Y_n(\boldsymbol{x}) = Y_n(r\boldsymbol{\xi}) = r^n Y_n(\boldsymbol{\xi})$. Then

$$0 = \Delta Y_n(\boldsymbol{x}) = \left(\frac{1}{r^{d-1}} \frac{\partial}{\partial r} \left(r^{d-1} \frac{\partial}{\partial r} \right) + \frac{1}{r^2} \Delta^*_{(d-1)} \right) r^n Y_n(\boldsymbol{\xi}),$$

i.e.,

$$-\Delta^*_{(d-1)} Y_n(\boldsymbol{\xi}) = n\,(n+d-2)\,Y_n(\boldsymbol{\xi}). \tag{3.19}$$

In other words, $Y_n(\boldsymbol{\xi})$ is an eigenfunction of the operator $-\Delta^*_{(d-1)}$ corresponding to the eigenvalue $n\,(n+d-2)$. More precisely, we have the following result.

Proposition 3.5. *Non-zero functions in the space $\mathbb{Y}^d_n$ are eigenfunctions of the Laplace–Beltrami operator $-\Delta^*_{(d-1)}$ on $\mathbb{S}^{d-1}$ corresponding to the eigenvalue $n\,(n+d-2)$. The dimension $N_{n,d} = \dim \mathbb{Y}^d_n$ is the multiplicity of the eigenvalue $n\,(n+d-2)$.*

For $f \in C^1(\mathbb{S}^{d-1})$ and $Y_n \in \mathbb{Y}^d_n$, we apply the Green-Beltrami identity (3.9),

$$\int_{\mathbb{S}^{d-1}} \boldsymbol{\nabla}^* f \cdot \boldsymbol{\nabla}^* Y_n \, dS^{d-1} = -\int_{\mathbb{S}^{d-1}} f \Delta^* Y_n \, dS^{d-1}.$$

By (3.19),

$$\int_{\mathbb{S}^{d-1}} \boldsymbol{\nabla}^* f \cdot \boldsymbol{\nabla}^* Y_n \, dS^{d-1} = n\,(n+d-2) \int_{\mathbb{S}^{d-1}} f\, Y_n \, dS^{d-1}. \tag{3.20}$$

For any fixed $\boldsymbol{\xi} \in \mathbb{S}^{d-1}$, let $Y_n(\boldsymbol{\eta}) = P_{n,d}(\boldsymbol{\xi}\cdot\boldsymbol{\eta})$ in (3.20) to obtain

$$\int_{\mathbb{S}^{d-1}} \boldsymbol{\nabla}^* f(\boldsymbol{\eta}) \cdot \boldsymbol{\nabla}^*_{\boldsymbol{\eta}} P_{n,d}(\boldsymbol{\xi}\cdot\boldsymbol{\eta}) \, dS^{d-1}(\boldsymbol{\eta})$$

$$= n\,(n+d-2) \int_{\mathbb{S}^{d-1}} f(\boldsymbol{\eta})\, P_{n,d}(\boldsymbol{\xi}\cdot\boldsymbol{\eta}) \, dS^{d-1}(\boldsymbol{\eta}). \tag{3.21}$$

The next result can be derived from (3.21).

Proposition 3.6.

$$\int_{\mathbb{S}^{d-1}} |\boldsymbol{\nabla}^*_{\boldsymbol{\eta}} P_{n,d}(\boldsymbol{\xi}\cdot\boldsymbol{\eta})|^2 dS^{d-1}(\boldsymbol{\eta}) = n\,(n+d-2)\frac{|\mathbb{S}^{d-1}|}{N_{n,d}} \quad \forall \boldsymbol{\xi} \in \mathbb{S}^{d-1}. \tag{3.22}$$

Proof. Take $f(\boldsymbol{\eta}) = P_{n,d}(\boldsymbol{\xi}\cdot\boldsymbol{\eta})$ in (3.21),

$$\int_{\mathbb{S}^{d-1}} |\boldsymbol{\nabla}^*_{\boldsymbol{\eta}} P_{n,d}(\boldsymbol{\xi}\cdot\boldsymbol{\eta})|^2 dS^{d-1}(\boldsymbol{\eta}) = n\,(n+d-2) \int_{\mathbb{S}^{d-1}} |P_{n,d}(\boldsymbol{\xi}\cdot\boldsymbol{\eta})|^2 dS^{d-1}(\boldsymbol{\eta}).$$

Note that the integral on the right side equals $|\mathbb{S}^{d-1}|/N_{n,d}$ by (2.40). $\qquad \square$

Now we study relations between the order of growth of the Fourier–Laplace components $\{\mathcal{P}_{n,d}f\}_{n\geq0}$ and the order of differentiability. Recall (2.44) for the definition of the projection $\mathcal{P}_{n,d}f$.

First, assume $f \in C^1(\mathbb{S}^{d-1})$. Then, using (3.21), we can write

$$(\mathcal{P}_{n,d}f)(\boldsymbol{\xi}) = \frac{N_{n,d}}{n\,(n+d-2)\,|\mathbb{S}^{d-1}|} \int_{\mathbb{S}^{d-1}} \boldsymbol{\nabla}^*_{\boldsymbol{\eta}} P_{n,d}(\boldsymbol{\xi}\cdot\boldsymbol{\eta})\cdot\boldsymbol{\nabla}^* f(\boldsymbol{\eta})\, dS^{d-1}(\boldsymbol{\eta}).$$

We bound the integral term first by applying the Cauchy–Schwarz inequality:

$$\left|\int_{\mathbb{S}^{d-1}} \boldsymbol{\nabla}^*_{\boldsymbol{\eta}} P_{n,d}(\boldsymbol{\xi}\cdot\boldsymbol{\eta})\cdot\boldsymbol{\nabla}^* f(\boldsymbol{\eta})\, dS^{d-1}(\boldsymbol{\eta})\right| \leq \left[\int_{\mathbb{S}^{d-1}} |\boldsymbol{\nabla}^*_{\boldsymbol{\eta}} P_{n,d}(\boldsymbol{\xi}\cdot\boldsymbol{\eta})|^2 dS^{d-1}(\boldsymbol{\eta})\right.$$

$$\left.\int_{\mathbb{S}^{d-1}} |\boldsymbol{\nabla}^* f(\boldsymbol{\eta})|^2 dS^{d-1}(\boldsymbol{\eta})\right]^{1/2},$$

and then by applying (3.22),

$$\left|\int_{\mathbb{S}^{d-1}} \boldsymbol{\nabla}^*_{\boldsymbol{\eta}} P_{n,d}(\boldsymbol{\xi}\cdot\boldsymbol{\eta})\cdot\boldsymbol{\nabla}^* f(\boldsymbol{\eta})\, dS^{d-1}(\boldsymbol{\eta})\right| \leq \left[n\,(n+d-2)\,\frac{|\mathbb{S}^{d-1}|}{N_{n,d}}\right]^{1/2}$$

$$\left[\int_{\mathbb{S}^{d-1}} |\boldsymbol{\nabla}^* f(\boldsymbol{\eta})|^2 dS^{d-1}(\boldsymbol{\eta})\right]^{1/2}.$$

So

$$\|\mathcal{P}_{n,d}f\|_{C(\mathbb{S}^{d-1})} \leq \left[\frac{N_{n,d}}{n\,(n+d-2)\,|\mathbb{S}^{d-1}|}\right]^{1/2} \|\boldsymbol{\nabla}^* f\|_{L^2(\mathbb{S}^{d-1})}. \qquad (3.23)$$

Since $N_{n,d} = \mathcal{O}(n^{d-2})$ (cf. (2.12)), we conclude

$$\|\mathcal{P}_{n,d}f\|_{C(\mathbb{S}^{d-1})} = \mathcal{O}(n^{\frac{d}{2}-2}) \text{ if } f \in C^1(\mathbb{S}^{d-1}). \qquad (3.24)$$

Then, assume $f \in C^2(\mathbb{S}^{d-1})$. Note that by the Green-Beltrami identity (3.9),

$$(\mathcal{P}_{n,d}f)(\boldsymbol{\xi}) = -\frac{1}{n\,(n+d-2)}(\mathcal{P}_{n,d}\Delta^* f)(\boldsymbol{\xi})$$

and also recall the bound (2.48). So

$$\|\mathcal{P}_{n,d}f\|_{C(\mathbb{S}^{d-1})} \leq \frac{1}{n\,(n+d-2)}\|\mathcal{P}_{n,d}\Delta^* f\|_{C(\mathbb{S}^{d-1})}$$

$$\leq \frac{1}{n\,(n+d-2)}\left(\frac{N_{n,d}}{|\mathbb{S}^{d-1}|}\right)^{1/2}\|\Delta^* f\|_{L^2(\mathbb{S}^{d-1})}$$

and we obtain

$$\|\mathcal{P}_{n,d}f\|_{C(\mathbb{S}^{d-1})} = \mathcal{O}(n^{\frac{d}{2}-3}) \quad \text{if } f \in C^2(\mathbb{S}^{d-1}). \tag{3.25}$$

In general, we have

$$\|\mathcal{P}_{n,d}f\|_{C(\mathbb{S}^{d-1})} = \mathcal{O}(n^{\frac{d}{2}-k-1}) \quad \text{if } f \in C^k(\mathbb{S}^{d-1}). \tag{3.26}$$

From this we see that the Fourier–Laplace series

$$\sum_{n=0}^{\infty} \mathcal{P}_{n,d}f(\boldsymbol{\xi}) \text{ converges for } f \in C^k(\mathbb{S}^{d-1}) \text{ with } k > \frac{d}{2}.$$

For $f, g \in C^2(\mathbb{S}^{d-1})$, by the integral identity (3.10),

$$(f, \Delta^* g)_{\mathbb{S}^{d-1}} = (\Delta^* f, g)_{\mathbb{S}^{d-1}}.$$

So for $n \neq m$, $Y_n \in \mathbb{Y}_n^d$ and $Y_m \in \mathbb{Y}_m^d$,

$$(Y_n, Y_m)_{\mathbb{S}^{d-1}} = 0, \tag{3.27}$$

i.e., eigenfunctions corresponding to distinct eigenvalues of $-\Delta^*$ are orthogonal. The orthogonality (3.27) also follows from Corollary 2.15.

We deduce from (3.20) that

$$(\boldsymbol{\nabla}^* \mathcal{P}_{n,d}f, \boldsymbol{\nabla}^* \mathcal{P}_{m,d}f)_{\mathbb{S}^{d-1}} = n(n+d-2)\|f\|^2_{L^2(\mathbb{S}^{d-1})}\,\delta_{nm}.$$

If $f \in C^1(\mathbb{S}^{d-1})$, then

$$(\boldsymbol{\nabla}^* f, \boldsymbol{\nabla}^* \mathcal{P}_{n,d}f)_{\mathbb{S}^{d-1}} = n(n+d-2)(f, \mathcal{P}_{n,d}f)_{\mathbb{S}^{d-1}}$$
$$= n(n+d-2)\|\mathcal{P}_{n,d}f\|^2_{L^2(\mathbb{S}^{d-1})}.$$

We use these two properties in the proof of the next result.

Proposition 3.7. Let $\mathbb{L}_n^{\perp}$ be the set of all $f \in C^1(\mathbb{S}^{d-1})$ with $\|f\|_{L^2(\mathbb{S}^{d-1})} = 1$ and $f \perp \mathbb{Y}_k^d$ for $0 \leq k \leq n$. Then

$$\inf\left\{\|\boldsymbol{\nabla}^* f\|^2_{L^2(\mathbb{S}^{d-1})} : f \in \mathbb{L}_n^{\perp}\right\} = (n+1)(n+d-1). \tag{3.28}$$

Proof. First, for any $f \in C^1(\mathbb{S}^{d-1})$ and any $m \in \mathbb{N}$,

$$0 \leq \left\|\boldsymbol{\nabla}^* f - \boldsymbol{\nabla}^* \sum_{k=0}^{m} \mathcal{P}_{k,d}f\right\|^2_{L^2(\mathbb{S}^{d-1})}$$

$$= \|\nabla^* f\|^2_{L^2(\mathbb{S}^{d-1})} - \sum_{k=0}^{m} k\,(k+d-2)\,\|\mathcal{P}_{k,d}f\|^2_{L^2(\mathbb{S}^{d-1})}.$$

Consequently, by letting $m \to \infty$,

$$\|\nabla^* f\|^2_{L^2(\mathbb{S}^{d-1})} \geq \sum_{k=0}^{\infty} k\,(k+d-2)\,\|\mathcal{P}_{k,d}f\|^2_{L^2(\mathbb{S}^{d-1})}.$$

Thus, for $f \in \mathbb{L}_n^{\perp}$,

$$\|\nabla^* f\|^2_{L^2(\mathbb{S}^{d-1})} \geq \sum_{k=n+1}^{\infty} k\,(k+d-2)\,\|\mathcal{P}_{k,d}f\|^2_{L^2(\mathbb{S}^{d-1})}.$$

Also, note that

$$1 = \|f\|^2_{L^2(\mathbb{S}^{d-1})} = \sum_{k=n+1}^{\infty} \|\mathcal{P}_{k,d}f\|^2_{L^2(\mathbb{S}^{d-1})}.$$

So (3.28) holds and the infimum is assumed by elements in $\mathbb{Y}_{n+1}^d$.   □

3.4   Some Integration Formulas

Let $Y_n(\boldsymbol{x})$ be a homogeneous harmonic polynomial of order n. Then $Y_n(\nabla)$ defines a harmonic differential operator of order n.

First, we present an integration formula which can be evaluated through applying a harmonic differential operator to a homogeneous harmonic polynomial.

Proposition 3.8. *For $n \geq k$,*

$$\int_{\mathbb{S}^{d-1}} (\boldsymbol{x} \cdot \boldsymbol{\zeta})^{n-k} Y_k(\boldsymbol{\zeta})\, Y_n(\boldsymbol{\zeta})\, dS^{d-1}(\boldsymbol{\zeta}) = \frac{\pi^{\frac{d}{2}}(n-k)!}{2^{n-1}\Gamma(n+\frac{d}{2})}\, Y_k(\nabla)Y_n(\boldsymbol{x}). \quad (3.29)$$

Proof. Writing $\boldsymbol{x} = |\boldsymbol{x}|\,\boldsymbol{\xi}$, $\boldsymbol{\xi} \in \mathbb{S}^{d-1}$, we have

$$\int_{\mathbb{S}^{d-1}} (\boldsymbol{x} \cdot \boldsymbol{\zeta})^{n} Y_n(\boldsymbol{\zeta})\, dS^{d-1}(\boldsymbol{\zeta}) = |\boldsymbol{x}|^{n} \int_{\mathbb{S}^{d-1}} (\boldsymbol{\xi} \cdot \boldsymbol{\zeta})^{n} Y_n(\boldsymbol{\zeta})\, dS^{d-1}(\boldsymbol{\zeta}).$$

By the Funk–Hecke formula (Theorem 2.22),

$$\int_{\mathbb{S}^{d-1}} (\boldsymbol{x} \cdot \boldsymbol{\zeta})^{n} Y_n(\boldsymbol{\zeta})\, dS^{d-1}(\boldsymbol{\zeta}) = \lambda_{n,d}|\boldsymbol{x}|^{n} Y_n(\boldsymbol{\xi}) = \lambda_{n,d} Y_n(\boldsymbol{x}),$$

where (cf. (2.62)),

$$\lambda_{n,d} = |\mathbb{S}^{d-2}| \int_{-1}^{1} t^n P_{n,d}(t) (1 - t^2)^{\frac{d-3}{2}} dt.$$

We apply Proposition 2.26 to compute the integral in the formula for $\lambda_{n,d}$,

$$\lambda_{n,d} = |\mathbb{S}^{d-2}| n! R_{n,d} \int_{-1}^{1} (1 - t^2)^{n+\frac{d-3}{2}} dt$$

$$= \frac{\pi^{\frac{d}{2}} n!}{2^{n-1} \Gamma(n + \frac{d}{2})}.$$

Thus,

$$Y_n(\boldsymbol{x}) = \frac{1}{\lambda_{n,d}} \int_{\mathbb{S}^{d-1}} (\boldsymbol{x} \cdot \boldsymbol{\zeta})^n Y_n(\boldsymbol{\zeta}) \, dS^{d-1}(\boldsymbol{\zeta}), \tag{3.30}$$

$$Y_k(\boldsymbol{\nabla}) = \frac{1}{\lambda_{k,d}} \int_{\mathbb{S}^{d-1}} Y_k(\boldsymbol{\eta})(\boldsymbol{\eta} \cdot \boldsymbol{\nabla})^k dS^{d-1}(\boldsymbol{\eta}). \tag{3.31}$$

Hence, $Y_k(\boldsymbol{\nabla}) Y_n(\boldsymbol{x})$ equals

$$\frac{1}{\lambda_{k,d} \lambda_{n,d}} \int_{\mathbb{S}^{d-1}} Y_n(\boldsymbol{\zeta}) \left[\int_{\mathbb{S}^{d-1}} Y_k(\boldsymbol{\eta})(\boldsymbol{\eta} \cdot \boldsymbol{\nabla})^k (\boldsymbol{x} \cdot \boldsymbol{\zeta})^n dS^{d-1}(\boldsymbol{\eta}) \right] dS^{d-1}(\boldsymbol{\zeta})$$

$$= \frac{1}{\lambda_{k,d} \lambda_{n.d}} \int_{\mathbb{S}^{d-1}} Y_n(\boldsymbol{\zeta})$$

$$\left[\frac{n!}{(n-k)!} \int_{\mathbb{S}^{d-1}} Y_k(\boldsymbol{\eta})(\boldsymbol{x} \cdot \boldsymbol{\zeta})^{n-k} (\boldsymbol{\eta} \cdot \boldsymbol{\zeta})^k dS^{d-1}(\boldsymbol{\eta}) \right] dS^{d-1}(\boldsymbol{\zeta})$$

$$= \frac{1}{\lambda_{k,d} \lambda_{n,d}} \frac{n!}{(n-k)!} \int_{\mathbb{S}^{d-1}} Y_n(\boldsymbol{\zeta})(\boldsymbol{x} \cdot \boldsymbol{\zeta})^{n-k} \lambda_{k,d} Y_k(\boldsymbol{\zeta}) \, dS^{d-1}(\boldsymbol{\zeta})$$

$$= \frac{n!}{\lambda_{n,d}(n-k)!} \int_{\mathbb{S}^{d-1}} (\boldsymbol{x} \cdot \boldsymbol{\zeta})^{n-k} Y_k(\boldsymbol{\zeta}) Y_n(\boldsymbol{\zeta}) \, dS^{d-1}(\boldsymbol{\zeta}).$$

The formula (3.29) follows from this relation and the formula for the constant $\lambda_{n,d}$. $\qquad\square$

In the rest of this section, we follow [17] and discuss some integration formulas over the unit sphere.

We recall that spherical harmonics of different degrees are orthogonal. In particular, this implies

$$\int_{\mathbb{S}^{d-1}} Y_j(\boldsymbol{\xi}) \, dS^{d-1}(\boldsymbol{\xi}) = 0 \quad \forall Y_j \in \mathbb{Y}_j^d, \; j \geq 1. \tag{3.32}$$

Using this property and (2.52), we see that

$$\int_{\mathbb{S}^{d-1}} H_n(\boldsymbol{\xi})\, dS^{d-1}(\boldsymbol{\xi}) = 0 \quad \forall\, H_n \in \mathbb{H}_n^d, \; n \text{ odd.} \tag{3.33}$$

For n even, by (2.52), (2.59) and (3.32), we have

$$\int_{\mathbb{S}^{d-1}} H_n(\boldsymbol{\xi})\, dS^{d-1}(\boldsymbol{\xi}) = |\mathbb{S}^{d-1}|\, Y_0,$$

i.e.,

$$\int_{\mathbb{S}^{d-1}} H_n(\boldsymbol{\xi})\, dS^{d-1}(\boldsymbol{\xi}) = \frac{(d-2)!!\,|\mathbb{S}^{d-1}|}{n!!\,(d+n-2)!!}\, \Delta^{n/2} H_n(\boldsymbol{x}) \quad \forall\, H_n \in \mathbb{H}_n^d, \; n \text{ even.} \tag{3.34}$$

Note that for $H_n \in \mathbb{H}_n^d$, $\Delta^{n/2} H_n(\boldsymbol{x})$ is a constant.

As some examples of (3.34), we have

$$\int_{\mathbb{S}^{d-1}} \xi_1^2\, dS^{d-1}(\boldsymbol{\xi}) = \frac{(d-2)!!\,|\mathbb{S}^{d-1}|}{2!!\,d!!}\, \Delta(x_1^2) = \frac{|\mathbb{S}^{d-1}|}{d},$$

$$\int_{\mathbb{S}^{d-1}} \xi_1^4\, dS^{d-1}(\boldsymbol{\xi}) = \frac{(d-2)!!\,|\mathbb{S}^{d-1}|}{4!!\,(d+2)!!}\, \Delta^2(x_1^4) = \frac{3\,|\mathbb{S}^{d-1}|}{d\,(d+2)}.$$

An integration formula is given in [17] for an integrand that can be expanded into a power series.

Proposition 3.9. *Assume $f \in L^1(\mathbb{S}^{d-1})$ and*

$$\left\| f - \sum_{j=0}^{n} f_j \right\|_{L^1(\mathbb{S}^{d-1})} \to 0 \quad \text{as } n \to \infty, \tag{3.35}$$

where $f_j \in \mathbb{Y}_j^d$, $j \geq 0$. Then

$$\int_{\mathbb{S}^{d-1}} f(\boldsymbol{\xi})\, dS^{d-1}(\boldsymbol{\xi}) = \frac{2\,\pi^{d/2}(d-2)!!}{\Gamma(d/2)} \sum_{j=0}^{\infty} \frac{1}{(2j)!!\,(2j+d-2)!!}\, \Delta^j f_{2j}(\boldsymbol{x}). \tag{3.36}$$

Proof. By the assumption (3.35), we have, as $n \to \infty$,

$$\left| \int_{\mathbb{S}^{d-1}} f(\boldsymbol{\xi})\, dS^{d-1}(\boldsymbol{\xi}) - \sum_{j=0}^{n} \int_{\mathbb{S}^{d-1}} f_j(\boldsymbol{\xi})\, dS^{d-1}(\boldsymbol{\xi}) \right| \leq \left\| f - \sum_{j=0}^{n} f_j \right\|_{L^1(\mathbb{S}^{d-1})}$$

$$\to 0.$$

Thus,

$$\int_{\mathbb{S}^{d-1}} f(\boldsymbol{\xi}) \, dS^{d-1}(\boldsymbol{\xi}) = \sum_{j=0}^{\infty} \int_{\mathbb{S}^{d-1}} f_j(\boldsymbol{\xi}) \, dS^{d-1}(\boldsymbol{\xi}).$$

We then apply the formulas (3.33) and (3.34) to get (3.36).   □

A sufficient condition for (3.35) is that the sequence of the partial sums $\{\sum_{j=0}^{n} f_j\}_{n \geq 0}$ converges uniformly on $\mathbb{S}^{d-1}$ to the function f.

Corollary 3.10. *Suppose a real variable function g has a Taylor expansion at $t = 0$ with a convergence radius $r_0 > 0$:*

$$g(t) = \sum_{j=0}^{\infty} \frac{g^{(j)}(0)}{j!} t^j, \quad |t| < r_0. \tag{3.37}$$

For a fixed vector $\boldsymbol{k} \in \mathbb{R}^d$, define a vector variable function

$$f(\boldsymbol{x}) := g(\boldsymbol{k}\cdot\boldsymbol{x}), \quad \boldsymbol{x} \in \mathbb{R}^d.$$

Write $\boldsymbol{x} = r\,\boldsymbol{\xi}$, $r > 0$, $\boldsymbol{\xi} \in \mathbb{S}^{d-1}$, and assume $r\,|\boldsymbol{k}| < r_0$. Then,

$$\int_{\mathbb{S}^{d-1}} f(r\boldsymbol{\xi}) \, dS^{d-1}(\boldsymbol{\xi}) = \frac{2\,\pi^{d/2}(d-2)!!}{\Gamma(d/2)} \sum_{j=0}^{\infty} \frac{g^{(2j)}(0)\, r^{2j}|\boldsymbol{k}|^{2j}}{(2j)!!\,(2j+d-2)!!}. \tag{3.38}$$

Proof. Since $r\,|\boldsymbol{k}| < r_0$, the power series in (3.37) with t replaced by $\boldsymbol{k}\cdot\boldsymbol{x}$ converges uniformly with respect to $\boldsymbol{\xi} \in \mathbb{S}^{d-1}$. Hence,

$$\int_{\mathbb{S}^{d-1}} f(r\boldsymbol{\xi}) \, dS^{d-1}(\boldsymbol{\xi}) = \sum_{j=0}^{\infty} \frac{g^{(j)}(0)}{j!} \int_{\mathbb{S}^{d-1}} (r\,\boldsymbol{k}\cdot\boldsymbol{\xi})^j dS^{d-1}(\boldsymbol{\xi}).$$

By a straightforward calculation,

$$\Delta^j (\boldsymbol{k}\cdot\boldsymbol{x})^{2j} = (2j)!\,|\boldsymbol{k}|^{2j}.$$

Therefore, the formula (3.38) follows from an application of (3.34).   □

As some examples of the formula (3.38), we note

$$\int_{\mathbb{S}^{d-1}} \sin(\boldsymbol{k}\cdot\boldsymbol{\xi}) \, dS^{d-1}(\boldsymbol{\xi}) = 0,$$

$$\int_{\mathbb{S}^{d-1}} \cos(\boldsymbol{k}\cdot\boldsymbol{\xi}) \, dS^{d-1}(\boldsymbol{\xi}) = \frac{2\,\pi^{d/2}(d-2)!!}{\Gamma(d/2)} \sum_{j=0}^{\infty} \frac{(-1)^j |\boldsymbol{k}|^{2j}}{(2j)!!\,(2j+d-2)!!}.$$

Recall the Taylor expansion

$$(1+t)^\alpha = \sum_{j=0}^{\infty} \binom{\alpha}{j} t^j, \quad |t| < 1.$$

Here,

$$\binom{\alpha}{j} := \frac{\alpha(\alpha-1)\cdots(\alpha-(j-1))}{j!}, \quad j \geq 1, \quad \binom{\alpha}{0} := 1.$$

Applying the formula (3.38), for $r\,|\boldsymbol{k}| < 1$,

$$\int_{\mathbb{S}^{d-1}} (1+\boldsymbol{k}\!\cdot\!\boldsymbol{\xi})^\alpha \, dS^{d-1}(\boldsymbol{\xi}) = \frac{2\,\pi^{d/2}(d-2)!!}{\Gamma(d/2)} \sum_{j=0}^{\infty} \frac{\prod_{l=0}^{2j-1}(\alpha-l)\,|\boldsymbol{k}|^{2j}}{(2j)!!\,(2j+d-2)!!}.$$

3.5 Some Differentiation Formulas

There are formulas that simplify the calculation of applying the harmonic differential operator to certain functions. Consider applying the harmonic differential operator $Y_n(\boldsymbol{\nabla})$ to a radial function $f(\boldsymbol{x}) = \varphi(|\boldsymbol{x}|^2)$. Suppose f is n times continuously differentiable. Let $\boldsymbol{x}$ be fixed. For $\boldsymbol{h} \in \mathbb{R}^d$, denote $h = |\boldsymbol{h}|$. As $h \to 0$, by the Taylor theorem,

$$f(\boldsymbol{x}+\boldsymbol{h}) = \sum_{j=0}^{n} \frac{1}{j!} (\boldsymbol{h}\cdot\boldsymbol{\nabla}_{\boldsymbol{x}})^j f(\boldsymbol{x}) + o(h^n).$$

With $\boldsymbol{x} = |\boldsymbol{x}|\,\boldsymbol{\xi}$ and $\boldsymbol{h} = h\,\boldsymbol{\eta}$, $\boldsymbol{\xi}, \boldsymbol{\eta} \in \mathbb{S}^{d-1}$, we rewrite the above relation as

$$\varphi(|\boldsymbol{x}|^2 + 2\,|\boldsymbol{x}|\,\boldsymbol{\xi}\!\cdot\!\boldsymbol{\eta}\,h + h^2) = \sum_{j=0}^{n} \frac{h^j}{j!} (\boldsymbol{\eta}\cdot\boldsymbol{\nabla}_{\boldsymbol{x}})^j \varphi(|\boldsymbol{x}|^2) + o(h^n). \qquad (3.39)$$

Use (3.31),

$$\int_{\mathbb{S}^{d-1}} Y_n(\boldsymbol{\eta})(\boldsymbol{\eta}\cdot\boldsymbol{\nabla})^j dS^{d-1}(\boldsymbol{\eta}) = \begin{cases} 0, & 0 \leq j \leq n-1, \\ \dfrac{\pi^{\frac{d}{2}} n!}{2^{n-1}\Gamma(n+\frac{d}{2})} Y_n(\boldsymbol{\nabla}), & j = n. \end{cases}$$

We then derive from (3.39) that

$$\int_{\mathbb{S}^{d-1}} \varphi(|\boldsymbol{x}|^2 + 2\,|\boldsymbol{x}|\,\boldsymbol{\xi}\!\cdot\!\boldsymbol{\eta}\,h + h^2)\, Y_n(\boldsymbol{\eta})\, dS^{d-1}(\boldsymbol{\eta})$$

$$= \frac{\pi^{\frac{d}{2}} h^n}{2^{n-1}\Gamma(n+\frac{d}{2})} Y_n(\boldsymbol{\nabla}_{\boldsymbol{x}})\, \varphi(|\boldsymbol{x}|^2) + o(h^n). \qquad (3.40)$$

By the Funk–Hecke formula (Theorem 2.22), the left side of (3.40) equals $\lambda_{n,d} Y_n(\boldsymbol{\xi})$ with

$$\lambda_{n,d} = |\mathbb{S}^{d-2}| \int_{-1}^{1} \varphi(|\boldsymbol{x}|^2 + 2\,|\boldsymbol{x}|\,h\,t + h^2)\,P_{n,d}(t)\,(1-t^2)^{\frac{d-3}{2}}\,dt.$$

Apply the formula (2.77),

$$\lambda_{n,d} = |\mathbb{S}^{d-2}|\,\frac{\Gamma(\frac{d-1}{2})\,(|\boldsymbol{x}|\,h)^n}{\Gamma(n+\frac{d-1}{2})}$$

$$\cdot \int_{-1}^{1} \varphi^{(n)}(|\boldsymbol{x}|^2 + 2\,|\boldsymbol{x}|\,h\,t + h^2)\,(1-t^2)^{n+\frac{d-3}{2}}\,dt$$

$$= \frac{2\,\pi^{\frac{d}{2}}\,|\boldsymbol{x}|^n h^n}{\Gamma(n+\frac{d}{2})}\,\varphi^{(n)}(|\boldsymbol{x}|^2) + o(h^n).$$

Thus, from (3.40),

$$\frac{2\,\pi^{\frac{d}{2}}\,|\boldsymbol{x}|^n h^n}{\Gamma(n+\frac{d}{2})}\,\varphi^{(n)}(|\boldsymbol{x}|^2)\,Y_n(\boldsymbol{\xi}) = \frac{\pi^{\frac{d}{2}}\,h^n}{2^{n-1}\Gamma(n+\frac{d}{2})}\,Y_n(\boldsymbol{\nabla}_{\boldsymbol{x}})\,\varphi(|\boldsymbol{x}|^2) + o(h^n).$$

Now divide both sides by h^n and take the limit $h \to 0$ to obtain, after some rearrangement

$$Y_n(\boldsymbol{\nabla}_{\boldsymbol{x}})\varphi(|\boldsymbol{x}|^2) = 2^n \varphi^{(n)}(|\boldsymbol{x}|^2) Y_n(\boldsymbol{x}). \tag{3.41}$$

This shows that homogeneous harmonics are reproduced by harmonic differentiations of orthogonally invariant functions. Note that

$$\left(\frac{d}{d(r^2)}\right)^n = \frac{1}{2^n}\left(\frac{1}{r}\frac{d}{dr}\right)^n.$$

We state (3.41) in the form of a theorem.

Theorem 3.11. *Let $f \in C^n(a,b)$. Then for $a < |\boldsymbol{x}| < b$,*

$$Y_n(\boldsymbol{\nabla})f(|\boldsymbol{x}|) = \mu_n(|\boldsymbol{x}|)Y_n(\boldsymbol{x}), \quad \mu_n(r) = \left(\frac{1}{r}\frac{d}{dr}\right)^n f(r). \tag{3.42}$$

With $\boldsymbol{x} = r\boldsymbol{\xi}$,

$$Y_n(\boldsymbol{\nabla})f(r) = \mu_n(r)\,r^n Y_n(\boldsymbol{\xi}), \quad \mu_n(r) = \left(\frac{1}{r}\frac{d}{dr}\right)^n f(r). \tag{3.43}$$

As an application of Theorem 3.11, let there be given a function $f \in C^n[-R,R]$. Denote $|\boldsymbol{x}| = r$. Then for $0 < r < R$,

$$\int_{\mathbb{S}^{d-1}} f(\boldsymbol{x} \cdot \boldsymbol{\eta}) \, dS^{d-1}(\boldsymbol{\eta}) = |\mathbb{S}^{d-2}| \int_{-1}^{1} f(rt) \, (1 - t^2)^{\frac{d-3}{2}} \, dt \equiv F(r).$$

With $\boldsymbol{x} = r\boldsymbol{\xi}$ and $Y_n \in \mathbb{Y}_n^d$,

$$Y_n(\boldsymbol{\nabla}) \int_{\mathbb{S}^{d-1}} f(\boldsymbol{x} \cdot \boldsymbol{\eta}) \, dS^{d-1}(\boldsymbol{\eta}) = \int_{\mathbb{S}^{d-1}} f^{(n)}(\boldsymbol{x} \cdot \boldsymbol{\eta}) \, Y_n(\boldsymbol{\eta}) \, dS^{d-1}(\boldsymbol{\eta}). \quad (3.44)$$

Apply Theorem 3.11,

$$Y_n(\boldsymbol{\nabla}) F(r) = \mu_n(r) \, r^n Y_n(\boldsymbol{\xi}), \quad \mu_n(r) = \left(\frac{1}{r} \frac{d}{dr} \right)^n F(r).$$

Use the Funk–Hecke formula (Theorem 2.22),

$$\int_{\mathbb{S}^{d-1}} f^{(n)}(\boldsymbol{x} \cdot \boldsymbol{\eta}) \, Y_n(\boldsymbol{\eta}) \, dS^{d-1}(\boldsymbol{\eta}) = \lambda_n(r) \, Y_n(\boldsymbol{\xi}),$$

where

$$\lambda_n(r) = |\mathbb{S}^{d-2}| \int_{-1}^{1} f^{(n)}(rt) \, P_{n,d}(t) \, (1 - t^2)^{\frac{d-3}{2}} \, dt.$$

Thus, from (3.44),

$$r^n \mu_n(r) \, Y_n(\boldsymbol{\xi}) = \lambda_n(r) \, Y_n(\boldsymbol{\xi}).$$

Summarizing, we have the following result.

Proposition 3.12. *Let $f \in C^n[-R, R]$. Then*

$$r^n \left(\frac{1}{r} \frac{d}{dr} \right)^n \int_{-1}^{1} f(rt) \, (1 - t^2)^{\frac{d-3}{2}} \, dt$$

$$= \int_{-1}^{1} f^{(n)}(rt) P_{n,d}(t) \, (1 - t^2)^{\frac{d-3}{2}} \, dt, \quad 0 < r < R. \quad (3.45)$$

3.6 Some Integral Identities for Spherical Harmonics

In this section, we first review some basic properties of harmonic functions, as is done in [70] but for a general dimension $d \geq 3$, and then derive an integral identity for spherical harmonics. In the following, we always assume $d \geq 3$. Proofs of the results are straightforward extensions of those for the case $d = 3$. For completeness, we include the proofs.

Proposition 3.13. *For fixed $\boldsymbol{y} \in \mathbb{R}^d$, the function*

$$\boldsymbol{x} \mapsto \Phi(\boldsymbol{x}, \boldsymbol{y}) := \frac{1}{(d-2)\,|\mathbb{S}^{d-1}|}\,|\boldsymbol{x} - \boldsymbol{y}|^{2-d} \tag{3.46}$$

is harmonic in $\mathbb{R}^d \backslash \{\boldsymbol{y}\}$.

Proof. It is convenient to write $|\boldsymbol{x} - \boldsymbol{y}| = \left(|\boldsymbol{x} - \boldsymbol{y}|^2\right)^{1/2}$ when we differentiate $|\boldsymbol{x} - \boldsymbol{y}|$:

$$\frac{\partial}{\partial x_i}|\boldsymbol{x} - \boldsymbol{y}| = \frac{1}{2}\left(|\boldsymbol{x} - \boldsymbol{y}|^2\right)^{-1/2}\frac{\partial}{\partial x_i}|\boldsymbol{x} - \boldsymbol{y}|^2 = \frac{1}{2}|\boldsymbol{x} - \boldsymbol{y}|^{-1}2\,(x_i - y_i).$$

Thus,

$$\frac{\partial}{\partial x_i}|\boldsymbol{x} - \boldsymbol{y}| = (x_i - y_i)|\boldsymbol{x} - \boldsymbol{y}|^{-1}, \quad 1 \le i \le d. \tag{3.47}$$

Use this formula,

$$\frac{\partial}{\partial x_i}\Phi(\boldsymbol{x}, \boldsymbol{y}) = \frac{1}{(d-2)\,|\mathbb{S}^{d-1}|}\,(2-d)\,|\boldsymbol{x} - \boldsymbol{y}|^{1-d}\frac{\partial}{\partial x_i}|\boldsymbol{x} - \boldsymbol{y}|$$

$$= -\frac{1}{|\mathbb{S}^{d-1}|}\,(x_i - y_i)\,|\boldsymbol{x} - \boldsymbol{y}|^{-d}.$$

We note in passing that

$$\nabla_{\boldsymbol{x}}\Phi(\boldsymbol{x}, \boldsymbol{y}) = -\frac{1}{|\mathbb{S}^{d-1}|}|\boldsymbol{x} - \boldsymbol{y}|^{-d}(\boldsymbol{x} - \boldsymbol{y}). \tag{3.48}$$

Differentiate with respect to x_i another time and use (3.47) again,

$$\frac{\partial^2}{\partial x_i^2}\Phi(\boldsymbol{x}, \boldsymbol{y}) = -\frac{1}{|\mathbb{S}^{d-1}|}\left[|\boldsymbol{x} - \boldsymbol{y}|^{-d} - d\,(x_i - y_i)^2|\boldsymbol{x} - \boldsymbol{y}|^{-d-2}\right].$$

Hence, for fixed $\boldsymbol{y}$,

$$\Delta\Phi(\boldsymbol{x}, \boldsymbol{y}) = \sum_{i=1}^d \frac{\partial^2}{\partial x_i^2}\Phi(\boldsymbol{x}, \boldsymbol{y}) = 0$$

and $\boldsymbol{x} \mapsto \Phi(\boldsymbol{x}, \boldsymbol{y})$ is harmonic for $\boldsymbol{x} \ne \boldsymbol{y}$. $\qquad\square$

The function $\Phi(\boldsymbol{x}, \boldsymbol{y})$ defined by (3.46) is usually called the fundamental solution of the Laplace equation.

Next we review some integration by parts formulas.

Theorem 3.14. *Let $\Omega \subset \mathbb{R}^d$ be a bounded domain with a C^1 boundary $\partial\Omega$. Denote by $\boldsymbol{\nu} = (\nu_1, \ldots, \nu_d)^T$ the unit outward normal vector on $\partial\Omega$. Then we have the formulas*

$$\int_\Omega (u\,\Delta v + \boldsymbol{\nabla}u\cdot\boldsymbol{\nabla}v)\,d\boldsymbol{x} = \int_{\partial\Omega} u\,\frac{\partial v}{\partial \boldsymbol{\nu}}\,d\sigma \quad \forall\, u \in C^1(\overline{\Omega}),\ v \in C^2(\overline{\Omega}), \qquad (3.49)$$

$$\int_\Omega (u\,\Delta v - v\,\Delta u)\,d\boldsymbol{x} = \int_{\partial\Omega} \left(u\,\frac{\partial v}{\partial \boldsymbol{\nu}} - v\,\frac{\partial u}{\partial \boldsymbol{\nu}} \right)\,d\sigma \quad \forall\, u, v \in C^2(\overline{\Omega}). \quad (3.50)$$

Here,

$$\frac{\partial v}{\partial \boldsymbol{\nu}} := \boldsymbol{\nabla}v \cdot \boldsymbol{\nu}$$

is the normal derivative of v on $\partial\Omega$.

Proof. Recall the following integration by parts formula

$$\int_\Omega u\,\frac{\partial v}{\partial x_i}\,d\boldsymbol{x} = \int_{\partial\Omega} u\,v\,\nu_i\,d\sigma - \int_\Omega v\,\frac{\partial u}{\partial x_i}\,d\boldsymbol{x} \quad \forall\, u, v \in C^1(\overline{\Omega}). \qquad (3.51)$$

For $u \in C^1(\overline{\Omega})$ and $v \in C^2(\overline{\Omega})$, we apply (3.51),

$$\int_\Omega u\,\Delta v\,d\boldsymbol{x} = \int_\Omega u \sum_{i=1}^d \frac{\partial^2 v}{\partial x_i^2}\,d\boldsymbol{x} = \sum_{i=1}^d \int_\Omega u\,\frac{\partial^2 v}{\partial x_i^2}\,d\boldsymbol{x}$$

$$= \sum_{i=1}^d \left(\int_{\partial\Omega} u\,\frac{\partial v}{\partial x_i}\,\nu_i\,d\sigma - \int_\Omega \frac{\partial u}{\partial x_i}\,\frac{\partial v}{\partial x_i}\,d\boldsymbol{x} \right)$$

$$= \int_{\partial\Omega} u\,\frac{\partial v}{\partial \boldsymbol{\nu}}\,d\sigma - \int_\Omega \boldsymbol{\nabla}u\cdot\boldsymbol{\nabla}v\,d\boldsymbol{x}.$$

Thus, the formula (3.49) holds.

Interchanging u and v in (3.49) and subtracting the resulting formula from (3.49), we obtain (3.50). $\square$

The smoothness assumption that $\partial\Omega \in C^1$ in Theorem 3.14 can be weakened. Nevertheless, we will mainly apply the theorem to the case where Ω is an open ball which has an infinitely smooth boundary.

Corollary 3.15. *Let Ω be as in Theorem 3.14 and let $u \in C^1(\overline{\Omega})$ be harmonic in Ω. Then*

$$\int_{\partial\Omega} \frac{\partial u}{\partial \boldsymbol{\nu}}\,d\sigma = 0.$$

Proof. Consider a sequence of domains that approximate Ω:

$$\Omega_n = \{x \in \Omega : \mathrm{dist}(x, \partial\Omega) < 1/n\}, \quad n \in \mathbb{N}.$$

Since $\partial\Omega \in C^1$, for n sufficiently large, $\partial\Omega_n \in C^1$. Note that u is harmonic in Ω, implying $u \in C^2(\overline{\Omega_n})$. We apply (3.50) with $v = 1$ on Ω_n with n sufficiently large:

$$\int_{\partial\Omega_n} \frac{\partial u}{\partial \nu} \, d\sigma = 0.$$

Then take the limit $n \to \infty$ to obtain the stated result. $\qquad\square$

Theorem 3.16. *Let Ω be as in Theorem 3.14 and let $u \in C^1(\overline{\Omega})$ be harmonic in Ω. Then*

$$u(x) = \int_{\partial\Omega} \left[\frac{\partial u}{\partial \nu}(y) \, \Phi(x,y) - u(y) \frac{\partial \Phi(x,y)}{\partial \nu_y} \right] d\sigma(y), \quad x \in \Omega. \qquad (3.52)$$

Proof. As in the proof of Corollary 3.15, it is sufficient to show the result under the assumption $u \in C^2(\overline{\Omega})$. Fix $x \in \Omega$. For $\varepsilon > 0$, denote the sphere

$$S(x; \varepsilon) := \left\{ y \in \mathbb{R}^d : |y - x| = \varepsilon \right\}$$

and the ball

$$B(x; \varepsilon) := \left\{ y \in \mathbb{R}^d : |y - x| < \varepsilon \right\}.$$

If ε is small enough, which we assume is the case, then $B(x; \varepsilon) \subset \Omega$. We apply the formula (3.50) on the set

$$\Omega_\varepsilon := \Omega \backslash B(x; \varepsilon)$$

to obtain

$$\int_{\partial\Omega \cup S(x;\varepsilon)} \left[\frac{\partial u}{\partial \nu}(y) \, \Phi(x,y) - u(y) \frac{\partial \Phi(x,y)}{\partial \nu_y} \right] d\sigma(y) = 0, \qquad (3.53)$$

where ν is the unit outward normal vector on $\partial\Omega_\varepsilon$.

From (3.48),

$$\nabla_y \Phi(x,y) = \frac{1}{|\mathbb{S}^{d-1}|} |x - y|^{-d}(x - y).$$

Note that on $S(\boldsymbol{x}; \varepsilon)$,

$$\boldsymbol{\nu_y} = \frac{\boldsymbol{x} - \boldsymbol{y}}{|\boldsymbol{x} - \boldsymbol{y}|}$$

and so

$$\frac{\partial \Phi(\boldsymbol{x}, \boldsymbol{y})}{\partial \boldsymbol{\nu_y}} = \boldsymbol{\nabla_y} \Phi(\boldsymbol{x}, \boldsymbol{y}) \cdot \boldsymbol{\nu_y} = \frac{1}{|\mathbb{S}^{d-1}|} |\boldsymbol{x} - \boldsymbol{y}|^{1-d} = \frac{\varepsilon^{1-d}}{|\mathbb{S}^{d-1}|}. \tag{3.54}$$

Hence,

$$\lim_{\varepsilon \to 0+} \int_{S(\boldsymbol{x}; \varepsilon)} u(\boldsymbol{y}) \frac{\partial \Phi(\boldsymbol{x}, \boldsymbol{y})}{\partial \boldsymbol{\nu_y}} \, d\sigma(\boldsymbol{y}) = u(\boldsymbol{x}).$$

Since

$$\Phi(\boldsymbol{x}, \boldsymbol{y})|_{S(\boldsymbol{x}; \varepsilon)} = \frac{\varepsilon^{2-d}}{(d-2)\,|\mathbb{S}^{d-1}|},$$

we have

$$\lim_{\varepsilon \to 0+} \int_{S(\boldsymbol{x}; \varepsilon)} \frac{\partial u}{\partial \boldsymbol{\nu}}(\boldsymbol{y}) \, \Phi(\boldsymbol{x}, \boldsymbol{y}) \, d\sigma(\boldsymbol{y}) = 0.$$

By taking the limit $\varepsilon \to 0+$ in (3.53), we obtain the formula (3.52). $\square$

It is easily seen from (3.52) that a harmonic function u is infinitely smooth in Ω. Now we can state and prove the mean value property of harmonic functions.

Theorem 3.17. *Let u be harmonic in an open ball $B(\boldsymbol{x}; r)$ and be continuous in the closed ball $\overline{B(\boldsymbol{x}; r)}$. Then*

$$u(\boldsymbol{x}) = \frac{1}{|\mathbb{S}^{d-1}|\, r^{d-1}} \int_{S(\boldsymbol{x}; r)} u(\boldsymbol{y}) \, d\sigma(\boldsymbol{y}) = \frac{d}{|\mathbb{S}^{d-1}|\, r^d} \int_{B(\boldsymbol{x}; r)} u(\boldsymbol{y}) \, d\boldsymbol{y}. \tag{3.55}$$

Proof. For any $s \in (0, r)$, $u \in C^2(\overline{B(\boldsymbol{x}; s)})$. We apply (3.52) and Corollary 3.15 on $\Omega = B(\boldsymbol{x}; s)$, noting the formula (3.54),

$$u(\boldsymbol{x}) = \frac{1}{|\mathbb{S}^{d-1}|\, s^{d-1}} \int_{S(\boldsymbol{x}; s)} u(\boldsymbol{y}) \, d\sigma(\boldsymbol{y}). \tag{3.56}$$

Take the limit $s \to r-$ and use the continuity of u in $\overline{B(\boldsymbol{x}; r)}$ to obtain the first formula of (3.55).

We multiply (3.56) by s^{d-1} and integrate the equality with respect to s from 0 to r:

$$u(\boldsymbol{x}) \int_0^r s^{d-1} ds = \frac{1}{|\mathbb{S}^{d-1}|} \int_0^r ds \int_{S(\boldsymbol{x};s)} u(\boldsymbol{y}) \, d\sigma(\boldsymbol{y}).$$

Rearranging the terms in the above equality, we obtain the second formula of (3.55). □

The mean value property leads to a maximum–minimum principle for a harmonic function.

Theorem 3.18. *A harmonic function on a domain cannot have its maximum or minimum unless it is a constant function.*

Proof. Let u be harmonic on the domain Ω. It is sufficient for us to show that if u achieves its maximum value M in Ω, then $u(\boldsymbol{x}) = M$ for $\boldsymbol{x} \in \Omega$. Introduce the set

$$\Omega_M := \{\boldsymbol{x} \in \Omega : u(\boldsymbol{x}) = M\}.$$

This set is non-empty and is relatively closed in Ω. Let $\boldsymbol{x} \in \Omega_M$ be such that $B(\boldsymbol{x};r) \subset \Omega$ for some $r > 0$. Apply Theorem 3.17 to the harmonic function $M - u(\boldsymbol{x})$ in $B(\boldsymbol{x};r)$,

$$0 = M - u(\boldsymbol{x}) = \frac{d}{|\mathbb{S}^{d-1}| \, r^d} \int_{B(\boldsymbol{x};r)} [M - u(\boldsymbol{y})] \, d\boldsymbol{y}.$$

Thus, $u(\boldsymbol{y}) = M$ for $\boldsymbol{y} \in B(\boldsymbol{x};r)$ and then $B(\boldsymbol{x};r) \subset \Omega_M$. Hence, Ω_M is also relatively open in Ω. Therefore, $\Omega_M = \Omega$ and u is a constant in Ω. □

Corollary 3.19. *Let Ω be a bounded domain. Then a harmonic function in Ω that is continuous on $\overline{\Omega}$ has both its maximum and minimum values on $\partial\Omega$.*

Returning to (3.52), let $\boldsymbol{x}_0$ be a point on $\partial\Omega$. From Theorem 18.5.1 in [81],

$$\lim_{\substack{\boldsymbol{x} \in \Omega \\ \boldsymbol{x} \to \boldsymbol{x}_0}} \int_{\partial\Omega} u(\boldsymbol{y}) \frac{\partial\Phi(\boldsymbol{x},\boldsymbol{y})}{\partial\boldsymbol{\nu}_{\boldsymbol{y}}} \, d\sigma(\boldsymbol{y}) = -\frac{1}{2} u(\boldsymbol{x}_0) + \int_{\partial\Omega} u(\boldsymbol{y}) \frac{\partial\Phi(\boldsymbol{x}_0,\boldsymbol{y})}{\partial\boldsymbol{\nu}_{\boldsymbol{y}}} \, d\sigma(\boldsymbol{y}).$$

From the proof of Theorem 18.6.1 in [81],

$$\lim_{\substack{\boldsymbol{x} \in \Omega \\ \boldsymbol{x} \to \boldsymbol{x}_0}} \int_{\partial\Omega} \frac{\partial u}{\partial\boldsymbol{\nu}}(\boldsymbol{y}) \, \Phi(\boldsymbol{x},\boldsymbol{y}) \, d\sigma(\boldsymbol{y}) = \int_{\partial\Omega} \frac{\partial u}{\partial\boldsymbol{\nu}}(\boldsymbol{y}) \, \Phi(\boldsymbol{x}_0,\boldsymbol{y}) \, d\sigma(\boldsymbol{y}).$$

Thus, it can be deduced from (3.52) that

$$u(\boldsymbol{x}_0) = \int_{\partial\Omega} \left[\frac{\partial u}{\partial\boldsymbol{\nu}}(\boldsymbol{y}) \, \Phi(\boldsymbol{x}_0,\boldsymbol{y}) - u(\boldsymbol{y}) \frac{\partial\Phi(\boldsymbol{x}_0,\boldsymbol{y})}{\partial\boldsymbol{\nu}_{\boldsymbol{y}}} \right] d\sigma(\boldsymbol{y}) + \frac{1}{2} u(\boldsymbol{x}_0).$$

We state the result as follows.

Theorem 3.20. *Let Ω be a bounded domain with a C^1 boundary. Assume u is harmonic in Ω such that $u \in C^1(\overline{\Omega})$. Then*

$$\frac{1}{2} u(\boldsymbol{x}) = \int_{\partial\Omega} \left[\frac{\partial u}{\partial \boldsymbol{\nu}}(\boldsymbol{y}) \, \Phi(\boldsymbol{x}, \boldsymbol{y}) - u(\boldsymbol{y}) \frac{\partial \Phi(\boldsymbol{x}, \boldsymbol{y})}{\partial \boldsymbol{\nu}_{\boldsymbol{y}}} \right] d\sigma(\boldsymbol{y}), \quad \boldsymbol{x} \in \partial\Omega. \quad (3.57)$$

Now we are ready to derive an integral identity for spherical harmonics. Apply (3.57) with $\partial\Omega = \mathbb{S}^{d-1}$ and

$$u(\boldsymbol{x}) = r^n Y_n(\boldsymbol{\xi}), \quad \boldsymbol{x} = r\boldsymbol{\xi}, \, \boldsymbol{\xi} \in \mathbb{S}^{d-1},$$

where $Y_n \in \mathbb{Y}_n^d$, $n \in \mathbb{N}_0$. Write $\boldsymbol{y} = |\boldsymbol{y}| \, \boldsymbol{\eta}, \, \boldsymbol{\eta} \in \mathbb{S}^{d-1}$. Observe the following relations

$$u(\boldsymbol{y})|_{\mathbb{S}^{d-1}} = Y_n(\boldsymbol{\eta}),$$

$$\frac{\partial u}{\partial \boldsymbol{\nu}}(\boldsymbol{y}) \bigg|_{\mathbb{S}^{d-1}} = n \, |\boldsymbol{y}|^{n-1} Y_n(\boldsymbol{\eta})|_{|\boldsymbol{y}|=1} = n \, Y_n(\boldsymbol{\eta}),$$

$$\Phi(\boldsymbol{x}, \boldsymbol{y})|_{\mathbb{S}^{d-1}} = \frac{1}{(d-2) \, |\mathbb{S}^{d-1}|} \, |\boldsymbol{\xi} - \boldsymbol{\eta}|^{2-d}.$$

Moreover, recalling the formula (3.48) useful in computing $\boldsymbol{\nabla}_{\boldsymbol{y}} \Phi(\boldsymbol{x}, \boldsymbol{y})$ and noticing that

$$\boldsymbol{\nu}(\boldsymbol{y}) = \boldsymbol{\eta},$$

we have

$$\frac{\partial \Phi(\boldsymbol{x}, \boldsymbol{y})}{\partial \boldsymbol{\nu}_{\boldsymbol{y}}} \bigg|_{\mathbb{S}^{d-1}} = \boldsymbol{\nabla}_{\boldsymbol{y}} \Phi(\boldsymbol{x}, \boldsymbol{y}) \cdot \boldsymbol{\nu}_{\boldsymbol{y}}|_{\mathbb{S}^{d-1}} = \frac{1}{|\mathbb{S}^{d-1}|} \, |\boldsymbol{\xi} - \boldsymbol{\eta}|^{-d} (\boldsymbol{\xi} - \boldsymbol{\eta}) \cdot \boldsymbol{\eta}.$$

Now

$$(\boldsymbol{\xi} - \boldsymbol{\eta}) \cdot \boldsymbol{\eta} = \boldsymbol{\xi} \cdot \boldsymbol{\eta} - 1 = -\frac{1}{2} \, |\boldsymbol{\xi} - \boldsymbol{\eta}|^2.$$

Thus,

$$\frac{\partial \Phi(\boldsymbol{x}, \boldsymbol{y})}{\partial \boldsymbol{\nu}_{\boldsymbol{y}}} \bigg|_{\mathbb{S}^{d-1}} = -\frac{1}{2 \, |\mathbb{S}^{d-1}|} \, |\boldsymbol{\xi} - \boldsymbol{\eta}|^{2-d}.$$

Using these relations in (3.57), after some simplifications, we get the following integral identity

$$\int_{\mathbb{S}^{d-1}} \frac{Y_n(\boldsymbol{\eta})}{|\boldsymbol{\xi} - \boldsymbol{\eta}|^{d-2}} \, dS^{d-1}(\boldsymbol{\eta}) = \frac{(d-2) \, |\mathbb{S}^{d-1}|}{2n + d - 2} \, Y_n(\boldsymbol{\xi}), \quad \boldsymbol{\xi} \in \mathbb{S}^{d-1}. \quad (3.58)$$

We turn to another integral identity for spherical harmonics on $\mathbb{S}^2$. First, we recall the Rayleigh–Green identity:

$$\int_\Omega \left[\varphi(\boldsymbol{y}) \, \Delta^2 \psi(\boldsymbol{y}) - \psi(\boldsymbol{y}) \, \Delta^2 \varphi(\boldsymbol{y}) \right] d\boldsymbol{y}$$

$$= \int_{\partial\Omega} \left[\varphi(\boldsymbol{y}) \frac{\partial \left(\Delta\psi(\boldsymbol{y}) \right)}{\partial \boldsymbol{\nu}} - \Delta\varphi(\boldsymbol{y}) \frac{\partial \psi(\boldsymbol{y})}{\partial \boldsymbol{\nu}} \right] d\sigma(\boldsymbol{y})$$

$$- \int_{\partial\Omega} \left[\psi(\boldsymbol{y}) \frac{\partial \left(\Delta\varphi(\boldsymbol{y}) \right)}{\partial \boldsymbol{\nu}} - \Delta\psi(\boldsymbol{y}) \frac{\partial \varphi(\boldsymbol{y})}{\partial \boldsymbol{\nu}} \right] d\sigma(\boldsymbol{y}). \qquad (3.59)$$

It is obtained from the divergence theorem in much the same manner as (3.50) in Theorem 3.14. Restrict Ω to be a subset of $\mathbb{R}^3$. Let φ be biharmonic over Ω, meaning $\Delta^2\varphi = 0$. Also, let $\psi(\boldsymbol{y}) = |\boldsymbol{y} - \boldsymbol{x}|$, $\boldsymbol{y} \neq \boldsymbol{x}$. Much as in Theorem 3.16, this leads to the following identity.

$$2 \int_{\partial\Omega} \left[\varphi(\boldsymbol{y}) \frac{\partial}{\partial \boldsymbol{\nu}_{\boldsymbol{y}}} \left(\frac{1}{|\boldsymbol{y} - \boldsymbol{x}|} \right) - \frac{1}{|\boldsymbol{y} - \boldsymbol{x}|} \frac{\partial \varphi(\boldsymbol{y})}{\partial \boldsymbol{\nu}} \right] d\sigma(\boldsymbol{y})$$

$$- \int_{\partial\Omega} \left[|\boldsymbol{y} - \boldsymbol{x}| \frac{\partial \left(\Delta\varphi(\boldsymbol{y}) \right)}{\partial \boldsymbol{\nu}} - \Delta\varphi(\boldsymbol{y}) \frac{\partial}{\partial \boldsymbol{\nu}_{\boldsymbol{y}}} |\boldsymbol{y} - \boldsymbol{x}| \right] d\sigma(\boldsymbol{y})$$

$$= -4\pi\varphi(\boldsymbol{x}), \qquad \boldsymbol{x} \in \partial\Omega. \qquad (3.60)$$

For details of this argument, see [66, Appendix 9].

The formula (3.60) was applied in [31, Theorem 5.2.1] to show that spherical harmonics on $\mathbb{S}^2$ are eigenfunctions of the single layer biharmonic potential. Letting $\varphi = Y_n$ be a spherical harmonic of order n on $\mathbb{S}^2$, then

$$\int_{\mathbb{S}^2} |\boldsymbol{\xi} - \boldsymbol{\eta}| \, Y_n(\boldsymbol{\eta}) \, dS^2(\boldsymbol{\eta}) = \frac{-16\pi}{(2n+1)(2n-1)(2n+3)} Y_n(\boldsymbol{\xi}) \quad \forall \boldsymbol{\xi} \in \mathbb{S}^2. \tag{3.61}$$

In Sect. 3.7, we will apply the Funk–Hecke formula to derive families of integral identities that include (3.58) and (3.61) as special cases.

3.7 Integral Identities Through the Funk–Hecke Formula

In this section, we show the derivation of some integral identities through a straightforward application of the Funk–Hecke formula. As we will see, such integral identities as (3.58) and (3.61) can be obtained easily. The material of this section follows a recent paper [55].

3.7.1   A Family of Integral Identities for Spherical Harmonics

Let $Y_n \in \mathbb{Y}_n^d$ be an arbitrary spherical harmonic of order n in d dimensions. Consider an integral of the form

$$I(g)(\boldsymbol{\xi}) := \int_{\mathbb{S}^{d-1}} g(|\boldsymbol{\xi} - \boldsymbol{\eta}|)\, Y_n(\boldsymbol{\eta})\, dS^{d-1}(\boldsymbol{\eta}). \tag{3.62}$$

Recall the weighted L^1 space $L^1_{(d-3)/2}(-1,1)$ defined in (2.60). We have the following result.

Proposition 3.21. *Assume*

$$g(2^{1/2}(1-t)^{1/2}) \in L^1_{(d-3)/2}(-1,1). \tag{3.63}$$

Then

$$I(g)(\boldsymbol{\xi}) = \mu_n Y_n(\boldsymbol{\xi}), \tag{3.64}$$

where

$$\mu_n = |\mathbb{S}^{d-2}| \int_{-1}^{1} g(2^{1/2}(1-t)^{1/2})\, P_{n,d}(t)\, (1-t^2)^{\frac{d-3}{2}}\, dt. \tag{3.65}$$

Proof. Since

$$|\boldsymbol{\xi} - \boldsymbol{\eta}| = [2\,(1 - \boldsymbol{\xi} \cdot \boldsymbol{\eta})]^{1/2}, \quad \boldsymbol{\xi}, \boldsymbol{\eta} \in \mathbb{S}^{d-1}, \tag{3.66}$$

we can write

$$I(g) = \int_{\mathbb{S}^{d-1}} g(2^{1/2}(1 - \boldsymbol{\xi} \cdot \boldsymbol{\eta})^{1/2})\, Y_n(\boldsymbol{\eta})\, dS^{d-1}(\boldsymbol{\eta}).$$

Applying Theorem 2.22, we obtain the formula (3.64) with the coefficient μ_n given by (3.65). □

Using the Rodrigues representation formula (2.70) for the Legendre polynomial $P_{n,d}(t)$, we can express μ_n from (3.65) in the form of

$$\mu_n = (-1)^n |\mathbb{S}^{d-2}|\, R_{n,d} \int_{-1}^{1} g(2^{1/2}(1-t)^{1/2})\, \left(\frac{d}{dt}\right)^n (1-t^2)^{n+\frac{d-3}{2}}\, dt. \tag{3.67}$$

Let us apply Proposition 3.21 to the following function

$$g(t) = t^{\nu}. \tag{3.68}$$

The condition (3.63) requires

$$\nu > 1 - d. \tag{3.69}$$

In the following we assume (3.69) is satisfied. Then from (3.64) and (3.67), we have the integral identity

$$\int_{\mathbb{S}^{d-1}} |\boldsymbol{\xi} - \boldsymbol{\eta}|^\nu Y_n(\boldsymbol{\eta}) \, dS^{d-1}(\boldsymbol{\eta}) = \mu_n Y_n(\boldsymbol{\xi}), \quad \boldsymbol{\xi} \in \mathbb{S}^{d-1}, \tag{3.70}$$

where

$$\mu_n = (-1)^n 2^{\nu/2} |\mathbb{S}^{d-2}| R_{n,d} \int_{-1}^1 (1-t)^{\nu/2} \left(\frac{d}{dt}\right)^n (1-t^2)^{n+\frac{d-3}{2}} dt. \tag{3.71}$$

We can simplify the formula for μ_n through computing the integral

$$I(\nu) = \int_{-1}^1 (1-t)^{\nu/2} \left(\frac{d}{dt}\right)^n (1-t^2)^{n+\frac{d-3}{2}} dt. \tag{3.72}$$

Under the condition (3.69), we can perform integration by parts repeatedly on $I(\nu)$ and all the boundary value terms at $t = \pm 1$ vanish. After integrating by parts n times, we have

$$I(\nu) = \frac{\nu}{2} \left(\frac{\nu}{2} - 1\right) \cdots \left(\frac{\nu}{2} - (n-1)\right) J(\nu) = (-1)^n \left(-\frac{\nu}{2}\right)_n J(\nu), \tag{3.73}$$

where

$$J(\nu) := \int_{-1}^1 (1-t)^{\nu/2 - n} (1-t^2)^{n+\frac{d-3}{2}} dt.$$

Write

$$J(\nu) = \int_{-1}^1 (1-t)^{(\nu+d-3)/2} (1+t)^{n+\frac{d-3}{2}} dt$$

and introduce the change of variables $t = 2s - 1$. Then

$$J(\nu) = 2^{n+\frac{\nu}{2}+d-2} \int_0^1 (1-s)^{\frac{\nu+d-1}{2}-1} s^{n+\frac{d-1}{2}-1} ds$$

$$= 2^{n+\frac{\nu}{2}+d-2} \frac{\Gamma(\frac{\nu+d-1}{2}) \Gamma(n + \frac{d-1}{2})}{\Gamma(n + \frac{\nu}{2} + d - 1)}.$$

Therefore, for μ_n of (3.71),

$$\mu_n = 2^{\nu+d-1} \pi^{\frac{d-1}{2}} \left(-\frac{\nu}{2}\right)_n \frac{\Gamma(\frac{\nu+d-1}{2})}{\Gamma(n + \frac{\nu}{2} + d - 1)}. \tag{3.74}$$

From the formula (3.74), we see that

$$\mu_n = 0 \quad \text{if } \nu = 0, 2, 4, \ldots, 2(n-1).$$

Now consider some special cases for the formula (3.70) with (3.74).

Special case 1: $\nu = 2 - d$. Applying (1.12),

$$\left(-\frac{\nu}{2}\right)_n = \left(\frac{d-2}{2}\right)_n = \frac{\Gamma(n + \frac{d-2}{2})}{\Gamma(\frac{d-2}{2})}.$$

Then,

$$\mu_n = 2\,\pi^{\frac{d-1}{2}} \frac{\Gamma(n + \frac{d-2}{2})}{\Gamma(\frac{d-2}{2})} \frac{\Gamma(\frac{1}{2})}{\Gamma(n + \frac{d}{2})}$$

$$= \frac{2\,\pi^{\frac{d}{2}}}{\Gamma(\frac{d}{2})} \frac{d-2}{2\,n + d - 2}.$$

Hence,

$$\mu_n = \frac{(d-2)\,|\mathbb{S}^{d-1}|}{2\,n + d - 2}. \tag{3.75}$$

So we have the integral identity

$$\int_{\mathbb{S}^{d-1}} \frac{Y_n(\boldsymbol{\eta})}{|\boldsymbol{\xi} - \boldsymbol{\eta}|^{d-2}}\, dS^{d-1}(\boldsymbol{\eta}) = \mu_n Y_n(\boldsymbol{\xi}), \quad \boldsymbol{\xi} \in \mathbb{S}^{d-1}, \tag{3.76}$$

where μ_n is given by (3.75). Notice that this is formula (3.58) in Sect. 3.6.
Special case 2: $\nu = -1$. Then,

$$\mu_n = 2^{d-2}\pi^{\frac{d-1}{2}} \frac{1}{2} \cdot \frac{3}{2} \cdots \frac{2n-1}{2} \frac{\Gamma(\frac{d-2}{2})}{\Gamma(n + d - \frac{3}{2})}.$$

After some simplification,

$$\mu_n = 2^{d-2}\pi^{\frac{d-1}{2}} \frac{(2n)!}{2^{2n}n!} \frac{\Gamma(\frac{d-2}{2})}{\Gamma(n + d - \frac{3}{2})}. \tag{3.77}$$

So we have the integral identity

$$\int_{\mathbb{S}^{d-1}} \frac{Y_n(\boldsymbol{\eta})}{|\boldsymbol{\xi} - \boldsymbol{\eta}|}\, dS^{d-1}(\boldsymbol{\eta}) = \mu_n Y_n(\boldsymbol{\xi}), \quad \boldsymbol{\xi} \in \mathbb{S}^{d-1}, \tag{3.78}$$

where μ_n is given by (3.77). Note that for $d = 3$,

$$\mu_n = \frac{4\pi}{2n+1}. \tag{3.79}$$

Special case 3: $\nu = 1$. Then,

$$\mu_n = (-1)^n 2^d \pi^{\frac{d-1}{2}} \frac{1}{2} \left(-\frac{1}{2}\right) \left(-\frac{3}{2}\right) \cdots \left(-\left(n - \frac{3}{2}\right)\right) \frac{\Gamma(\frac{d}{2})}{\Gamma(n + d - \frac{1}{2})}.$$

Since

$$\Gamma\left(n + d - \frac{1}{2}\right) = \left(n + d - \frac{3}{2}\right)\left(n + d - \frac{5}{2}\right) \cdots \frac{1}{2}\Gamma\left(\frac{1}{2}\right),$$

we have

$$\mu_n = -2^{2d-1} \pi^{\frac{d-2}{2}} \frac{\Gamma(\frac{d}{2})}{(2n-1)(2n+1)\cdots(2n+2d-3)}. \tag{3.80}$$

So we have the integral identity

$$\int_{\mathbb{S}^{d-1}} |\boldsymbol{\xi} - \boldsymbol{\eta}| \, Y_n(\boldsymbol{\eta}) \, dS^{d-1}(\boldsymbol{\eta}) = \mu_n Y_n(\boldsymbol{\xi}), \quad \boldsymbol{\xi} \in \mathbb{S}^{d-1}, \tag{3.81}$$

where μ_n is given by (3.80). Note that for $d = 3$,

$$\mu_n = -\frac{16\pi}{(2n-1)(2n+1)(2n+3)}, \tag{3.82}$$

and we recover formula (3.61) derived in Sect. 3.6.

We may also choose g as a log function in applying Proposition 3.21:

$$g(t) = \log t.$$

Note that this function satisfies the condition (3.63). Then we obtain the formula

$$\int_{\mathbb{S}^{d-1}} \log |\boldsymbol{\xi} - \boldsymbol{\eta}| \, Y_n(\boldsymbol{\eta}) \, dS^{d-1}(\boldsymbol{\eta}) = \mu_n Y_n(\boldsymbol{\xi}), \quad \boldsymbol{\xi} \in \mathbb{S}^{d-1}, \tag{3.83}$$

where

$$\mu_n = \frac{|\mathbb{S}^{d-2}|}{2} \int_{-1}^1 \log(2(1-t)) \, P_{n,d}(t) (1 - t^2)^{\frac{d-3}{2}} \, dt. \tag{3.84}$$

Using the orthogonality of the Legendre polynomials, for $n \geq 1$, we can simplify (3.84) to

$$\mu_n = \frac{|\mathbb{S}^{d-2}|}{2} \int_{-1}^{1} \log(1-t)\, P_{n,d}(t)\, (1-t^2)^{\frac{d-3}{2}}\, dt, \quad n \geq 1. \qquad (3.85)$$

3.7.2 Some Extensions

Proposition 3.21 can be extended straightforward to some other similar integrals.

Let a and b be non-zero real numbers. Similar to (3.66),

$$|a\,\boldsymbol{\xi} + b\,\boldsymbol{\eta}| = (a^2 + b^2 + 2\,a\,b\,\boldsymbol{\xi}\cdot\boldsymbol{\eta})^{1/2}, \quad \boldsymbol{\xi}, \boldsymbol{\eta} \in \mathbb{S}^{d-1}. \qquad (3.86)$$

Then for a function g satisfying

$$g\big((a^2 + b^2 + 2\,a\,b\,t)^{1/2}\big) \in L^1_{(d-3)/2}(-1,1),$$

we can apply Theorem 2.22 to get

$$\int_{\mathbb{S}^{d-1}} g(|a\,\boldsymbol{\xi} + b\,\boldsymbol{\eta}|)\, Y_n(\boldsymbol{\eta})\, dS^{d-1}(\boldsymbol{\eta}) = \mu_n Y_n(\boldsymbol{\xi}), \quad \boldsymbol{\xi} \in \mathbb{S}^{d-1}, \qquad (3.87)$$

where

$$\mu_n = (-1)^n |\mathbb{S}^{d-2}|\, R_{n,d} \int_{-1}^{1} g\big((a^2 + b^2 + 2\,a\,b\,t)^{1/2}\big) \left(\frac{d}{dt}\right)^n (1-t^2)^{n+\frac{d-3}{2}}\, dt. \qquad (3.88)$$

This formula includes Proposition 3.21 as a special case with $a = 1$ and $b = -1$. Choosing $a = b = 1$, we obtain another special case formula:

$$\int_{\mathbb{S}^{d-1}} g(|\boldsymbol{\xi} + \boldsymbol{\eta}|)\, Y_n(\boldsymbol{\eta})\, dS^{d-1}(\boldsymbol{\eta}) = \mu_n Y_n(\boldsymbol{\xi}), \quad \boldsymbol{\xi} \in \mathbb{S}^{d-1}, \qquad (3.89)$$

where

$$\mu_n = (-1)^n |\mathbb{S}^{d-2}|\, R_{n,d} \int_{-1}^{1} g(2^{1/2}(1+t)^{1/2}) \left(\frac{d}{dt}\right)^n (1-t^2)^{n+\frac{d-3}{2}}\, dt. \qquad (3.90)$$

More generally, let $g(t_1, \ldots, t_L)$ be a function of L real variables, and let a_l, b_l, $1 \leq l \leq L$, be $2L$ non-zero real numbers. Assume

$$g\big((a_1^2 + b_1^2 + 2\,a_1 b_1 t)^{1/2}, \ldots, (a_L^2 + b_L^2 + 2\,a_L b_L t)^{1/2}\big) \in L^1_{(d-3)/2}(-1,1).$$

Then

$$\int_{\mathbb{S}^{d-1}} g(|a_1\boldsymbol{\xi} + b_1\boldsymbol{\eta}|, \ldots, |a_L\boldsymbol{\xi} + b_L\boldsymbol{\eta}|)\, Y_n(\boldsymbol{\eta})\, dS^{d-1}(\boldsymbol{\eta}) = \mu_n Y_n(\boldsymbol{\xi}), \quad \boldsymbol{\xi} \in \mathbb{S}^{d-1},$$

where

$$\mu_n = (-1)^n |\mathbb{S}^{d-2}|\, R_{n,d} \int_{-1}^{1} \tilde{g}(t)\, dt$$

and

$$\tilde{g}(t) = g\Big((a_1^2 + b_1^2 + 2\,a_1 b_1 t)^{1/2}, \ldots, (a_L^2 + b_L^2 + 2\,a_L b_L t)^{1/2}\Big)$$
$$\cdot \left(\frac{d}{dt}\right)^n (1 - t^2)^{n + \frac{d-3}{2}}.$$

As a particular example, for $\nu_1 > 1 - d$ and $\nu_2 > 1 - d$,

$$\int_{\mathbb{S}^{d-1}} |\boldsymbol{\xi} - \boldsymbol{\eta}|^{\nu_1} |\boldsymbol{\xi} + \boldsymbol{\eta}|^{\nu_2} Y_n(\boldsymbol{\eta})\, dS^{d-1}(\boldsymbol{\eta}) = \mu_n Y_n(\boldsymbol{\xi}), \quad \boldsymbol{\xi} \in \mathbb{S}^{d-1},$$

where μ_n equals

$$(-1)^n 2^{(\nu_1 + \nu_2)/2} |\mathbb{S}^{d-2}|\, R_{n,d} \int_{-1}^{1} (1 - t)^{\nu_1/2} (1 + t)^{\nu_2/2} \left(\frac{d}{dt}\right)^n (1 - t^2)^{n + \frac{d-3}{2}}\, dt.$$

Applying the Funk–Hecke formula, we have derived identities for some integrals involving spherical harmonics over the unit sphere in an arbitrary dimension. Integral identities of the forms (3.70) and (3.83) are useful in numerical approximations of boundary integral equations [10]. Note that direct derivation of such identities as (3.76), (3.78), and (3.81) are quite involved, often using some form of Green's integral identities; see Sect. 3.6 or [81].

3.8   Sobolev Spaces on the Unit Sphere

In this section, we discuss Sobolev spaces over the sphere. The spaces are defined through expansions in terms of an orthonormal basis of spherical harmonics, following [47].

Let $\{Y_{n,j} : 1 \leq j \leq N_{n,d},\ n \geq 0\}$ be an orthonormal basis of spherical harmonics over $\mathbb{S}^{d-1}$. For a function v, introduce a sequence of numbers

$$v_{n,j} := (v, Y_{n,j})_{L^2(\mathbb{S}^{d-1})}, \quad 1 \leq j \leq N_{n,d},\ n \geq 0, \tag{3.91}$$

wherever the integrals are defined.

First, we consider general Sobolev spaces corresponding to a given sequence of numbers $\{a_n : n \geq 0\}$. Introduce an inner product

$$(u, v)_{H(\{a_n\};\mathbb{S}^{d-1})} := \sum_{n=0}^{\infty} |a_n|^2 \sum_{j=1}^{N_{n,d}} u_{n,j}\overline{v_{n,j}}, \qquad (3.92)$$

wherever the right side is defined. We note that this definition does not depend on the choice of the basis functions $\{Y_{n,j} : 1 \leq j \leq N_{n,d}\}$ in $\mathbb{Y}_n^d$, since by the addition formula (2.24),

$$\sum_{j=1}^{N_{n,d}} u_{n,j}\overline{v_{n,j}} = \sum_{j=1}^{N_{n,d}} \int_{\mathbb{S}^{d-1}} u(\boldsymbol{\xi})\, Y_{n,j}(\boldsymbol{\xi})\, dS^{d-1}(\boldsymbol{\xi}) \int_{\mathbb{S}^{d-1}} \overline{v(\boldsymbol{\eta})\, Y_{n,j}(\boldsymbol{\eta})}\, dS^{d-1}(\boldsymbol{\eta})$$

$$= \int_{\mathbb{S}^{d-1}} \int_{\mathbb{S}^{d-1}} u(\boldsymbol{\xi})\, \overline{v(\boldsymbol{\eta})} \sum_{j=1}^{N_{n,d}} Y_{n,j}(\boldsymbol{\xi})\overline{Y_{n,j}(\boldsymbol{\eta})}\, dS^{d-1}(\boldsymbol{\xi})\, dS^{d-1}(\boldsymbol{\eta})$$

$$= \frac{N_{n,d}}{|\mathbb{S}^{d-1}|} \int_{\mathbb{S}^{d-1}} \int_{\mathbb{S}^{d-1}} u(\boldsymbol{\xi})\, \overline{v(\boldsymbol{\eta})}\, P_{n,d}(\boldsymbol{\xi}\cdot\boldsymbol{\eta})\, dS^{d-1}(\boldsymbol{\xi})\, dS^{d-1}(\boldsymbol{\eta}).$$

The norm associated with the inner product (3.92) is

$$\|v\|_{H(\{a_n\};\mathbb{S}^{d-1})} := (v, v)_{H(\{a_n\};\mathbb{S}^{d-1})}^{1/2}. \qquad (3.93)$$

Let $C^{\infty}(\{a_n\};\mathbb{S}^{d-1})$ be the space of all infinitely differentiable functions with a finite $H(\{a_n\};\mathbb{S}^{d-1})$-norm.

Definition 3.22. The Sobolev space $H(\{a_n\};\mathbb{S}^{d-1})$ is the completion of the smooth function space $C^{\infty}(\{a_n\};\mathbb{S}^{d-1})$ with respect to the norm (3.93).

This defines $H(\{a_n\};\mathbb{S}^{d-1})$ as a Hilbert space with the inner product (3.92).

Now we consider Sobolev spaces with particular choices of the numbers a_n, $n \geq 0$. Since

$$-\Delta^* Y_{n,j} = n\,(n + d - 2)\, Y_{n,j},$$

we have

$$\left(-\Delta^* + \delta_d^2\right) Y_{n,j} = (n + \delta_d)^2\, Y_{n,j}, \qquad (3.94)$$

where

$$\delta_d := \frac{d - 2}{2}. \qquad (3.95)$$

For any $s \in \mathbb{R}$, we formally write

$$\left(-\Delta^* + \delta_d^2\right)^{s/2} Y_{n,j} = (n + \delta_d)^s Y_{n,j}. \tag{3.96}$$

For a function $v \in L^2(\mathbb{S}^{d-1})$, we use the expansion

$$v(\boldsymbol{\xi}) = \sum_{n=0}^{\infty} \sum_{j=1}^{N_{n,d}} v_{n,j} Y_{n,j}(\boldsymbol{\xi}). \tag{3.97}$$

Then formally,

$$\left(-\Delta^* + \delta_d^2\right)^{s/2} v(\boldsymbol{\xi}) = \sum_{n=0}^{\infty} \sum_{j=1}^{N_{n,d}} v_{n,j} (n + \delta_d)^s Y_{n,j}(\boldsymbol{\xi})$$

as long as the right side is defined. Thus,

$$\left\|\left(-\Delta^* + \delta_d^2\right)^{s/2} v\right\|_{L^2(\mathbb{S}^{d-1})}^2 = \sum_{n=0}^{\infty} \sum_{j=1}^{N_{n,d}} (n + \delta_d)^{2s} |v_{n,j}|^2.$$

Definition 3.23. The Sobolev space $H^s(\mathbb{S}^{d-1})$ is the completion of the smooth function space $C^\infty(\mathbb{S}^{d-1})$ with respect to the norm

$$\|v\|_{H^s(\mathbb{S}^{d-1})} := \left\|\left(-\Delta^* + \delta_d^2\right)^{s/2} v\right\|_{L^2(\mathbb{S}^{d-1})}$$

$$= \left[\sum_{n=0}^{\infty} \sum_{j=1}^{N_{n,d}} (n + \delta_d)^{2s} |v_{n,j}|^2\right]^{1/2} \tag{3.98}$$

which is induced by the inner product

$$(u, v)_{H^s(\mathbb{S}^{d-1})} := \int_{\mathbb{S}^{d-1}} \left(-\Delta^* + \delta_d^2\right)^{s/2} u(\boldsymbol{\xi}) \left(-\Delta^* + \delta_d^2\right)^{s/2} \overline{v(\boldsymbol{\xi})} \, d\sigma(\boldsymbol{\xi})$$

$$= \sum_{n=0}^{\infty} \sum_{j=1}^{N_{n,d}} (n + \delta_d)^{2s} u_{n,j} \overline{v_{n,j}}. \tag{3.99}$$

We list a few properties of the Sobolev spaces $H^s(\mathbb{S}^{d-1})$ below.

From Definition 3.23, the smooth function space $C^\infty(\mathbb{S}^{d-1})$ is dense in $H^s(\mathbb{S}^{d-1})$ for any $s \in \mathbb{R}$.

If $t < s$, then we have the embedding

$$H^s(\mathbb{S}^{d-1}) \hookrightarrow H^t(\mathbb{S}^{d-1}) \tag{3.100}$$

as well as the inequality

$$\|v\|_{H^t(\mathbb{S}^{d-1})} \le \max\{1, \delta_d^{s-t}\} \|v\|_{H^s(\mathbb{S}^{d-1})}. \tag{3.101}$$

Since

$$\left\|\left(-\Delta^* + \delta_d^2\right)^{t/2} v\right\|_{H^s(\mathbb{S}^{d-1})} = \|v\|_{H^{s+t}(\mathbb{S}^{d-1})} \quad \forall v \in H^{s+t}(\mathbb{S}^{d-1}),$$

we see that the operator $\left(-\Delta^* + \delta_d^2\right)^{t/2}$ is bounded from $H^{s+t}(\mathbb{S}^{d-1})$ to $H^s(\mathbb{S}^{d-1})$.

Let us show the embedding

$$H^s(\mathbb{S}^{d-1}) \hookrightarrow C(\mathbb{S}^{d-1}) \quad \forall s > \frac{d-1}{2}. \tag{3.102}$$

For this purpose, we only need to show the inequality

$$\|v\|_{C(\mathbb{S}^{d-1})} \le c \|v\|_{H^s(\mathbb{S}^{d-1})} \quad \forall v \in H^s(\mathbb{S}^{d-1}), \ s > \frac{d-1}{2}. \tag{3.103}$$

We start with the expansion (3.97) and obtain

$$|v(\boldsymbol{\xi})| \le \sum_{n=0}^{\infty} \sum_{j=1}^{N_{n,d}} |v_{n,j}| \, |Y_{n,j}(\boldsymbol{\xi})|.$$

Apply the Cauchy–Schwarz inequality and then (2.35),

$$|v(\boldsymbol{\xi})| \le \sum_{n=0}^{\infty} \left(\sum_{j=1}^{N_{n,d}} |v_{n,j}|^2\right)^{1/2} \left[\sum_{j=1}^{N_{n,d}} |Y_{n,j}(\boldsymbol{\xi})|^2\right]^{1/2}$$

$$= \sum_{n=0}^{\infty} \left(\frac{N_{n,d}}{|\mathbb{S}^{d-1}|}\right)^{1/2} \left(\sum_{j=1}^{N_{n,d}} |v_{n,j}|^2\right)^{1/2}.$$

Recall from (2.12), $N_{n,d} = \mathcal{O}(n^{d-2})$. Thus,

$$|v(\boldsymbol{\xi})| \le c \sum_{n=0}^{\infty} \left((n+1)^{d-2} \sum_{j=1}^{N_{n,d}} |v_{n,j}|^2\right)^{1/2}.$$

Define $t = 2s - d + 2$. Since $s > (d-1)/2$, $t > 1$ and so

$$\sum_{n=1}^{\infty} n^{-t} < \infty.$$

Thus,

$$|v(\boldsymbol{\xi})| \leq c \sum_{n=0}^{\infty} (n+1)^{-t/2} \left((n+1)^{2s} \sum_{j=1}^{N_{n,d}} |v_{n,j}|^2 \right)^{1/2}$$

$$\leq c \left[\sum_{n=0}^{\infty} (n+1)^{-t} \right]^{1/2} \left[\sum_{n=0}^{\infty} (n+1)^{2s} \sum_{j=1}^{N_{n,d}} |v_{n,j}|^2 \right]^{1/2}$$

$$\leq c \|v\|_{H^s(\mathbb{S}^{d-1})}.$$

So the inequality (3.103), and hence the embedding (3.102), holds.

We note that the condition $s > (d-1)/2$ for the embedding (3.102) is natural. This is explained as follows. For the unit ball $\mathbb{B}^d$ in $\mathbb{R}^d$, the Sobolev embedding

$$H^t(\mathbb{B}^d) \hookrightarrow C(\overline{\mathbb{B}^d})$$

holds for $t > d/2$. Traces on $\mathbb{S}^{d-1} = \partial\mathbb{B}^d$ of $H^t(\mathbb{B}^d)$ functions form the space $H^{t-1/2}(\mathbb{S}^{d-1})$. Thus, a condition for the embedding

$$H^{t-1/2}(\mathbb{S}^{d-1}) \hookrightarrow C(\mathbb{S}^{d-1})$$

is expected to be $t - 1/2 > (d-1)/2$. Then equate $t - 1/2$ with s.

Finally, we mention an approximation result. Denote

$$\mathbb{P}_n(\mathbb{S}^{d-1}) := \operatorname{span}\{Y_{k,j} : 0 \leq j \leq N_{k,d},\ 0 \leq k \leq n\}. \tag{3.104}$$

Assume $v \in H^s(\mathbb{S}^{d-1})$ for some $s > 0$. Then there exists a function $v_n \in \mathbb{P}_n(\mathbb{S}^{d-1})$ such that

$$\|v - v_n\|_{H^t(\mathbb{S}^{d-1})} \leq (n+1+\delta_d)^{-(s-t)} \|v\|_{H^s(\mathbb{S}^{d-1})} \quad \forall t \in [0, s]. \tag{3.105}$$

Indeed, writing

$$v(\boldsymbol{\xi}) = \sum_{k=0}^{\infty} \sum_{j=1}^{N_{k,d}} v_{k,j} Y_{k,j}(\boldsymbol{\xi}),$$

we define

$$v_n(\boldsymbol{\xi}) = \sum_{k=0}^{n} \sum_{j=1}^{N_{k,d}} v_{k,j} Y_{k,j}(\boldsymbol{\xi}).$$

Then $v_n \in \mathbb{P}_n(\mathbb{S}^{d-1})$ and

$$\|v - v_n\|^2_{H^t(\mathbb{S}^{d-1})} = \sum_{k=n+1}^{\infty} \sum_{j=1}^{N_{k,d}} (k + \delta_d)^{2t} |v_{k,j}|^2.$$

Since $v \in H^s(\mathbb{S}^{d-1})$,

$$\|v\|^2_{H^s(\mathbb{S}^{d-1})} = \sum_{k=0}^{\infty} \sum_{j=1}^{N_{k,d}} (k + \delta_d)^{2s} |v_{k,j}|^2 < \infty.$$

Hence,

$$\|v - v_n\|^2_{H^t(\mathbb{S}^{d-1})} = \sum_{k=n+1}^{\infty} \sum_{j=1}^{N_{k,d}} (k + \delta_d)^{2t} |v_{k,j}|^2$$

$$\leq (n + 1 + \delta_d)^{2(t-s)} \sum_{k=n+1}^{\infty} \sum_{j=1}^{N_{k,d}} (k + \delta_d)^{2s} |v_{k,j}|^2$$

$$\leq (n + 1 + \delta_d)^{2(t-s)} \|v\|^2_{H^s(\mathbb{S}^{d-1})}.$$

Thus, (3.105) holds.

The Sobolev spaces discussed here will be needed in Sect. 6.3.

3.9 Positive Definite Functions

An important area of approximation theory is "meshless discretization methods", developed to handle discrete data with no associated mesh or grid. An important topic in the foundations of this subject is that of "positive definite functions"; and in the case of data distributed over the sphere $\mathbb{S}^{d-1}$, spherical harmonics can be used to analyze behaviour of such functions. For this section, we follow closely the development in Wendland [118, Chap. 17]. A number of these ideas trace back to Schoenberg [101].

Definition 3.24. Let $\Phi : \mathbb{S}^{d-1} \times \mathbb{S}^{d-1} \to \mathbb{R}$ be a continuous and symmetric function, symmetry referring to the property that $\Phi(\boldsymbol{\xi}, \boldsymbol{\eta}) \equiv \Phi(\boldsymbol{\eta}, \boldsymbol{\xi}) \, \forall \boldsymbol{\xi}, \boldsymbol{\eta} \in \mathbb{S}^{d-1}$. The function Φ is called *positive semi-definite* if for any $q \geq 1$ and any set of distinct points $\{\boldsymbol{\xi}_1, \ldots, \boldsymbol{\xi}_q\} \subset \mathbb{S}^{d-1}$,

$$\sum_{i,j=1}^{q} \Phi(\boldsymbol{\xi}_i, \boldsymbol{\xi}_j) \zeta_i \zeta_j \geq 0 \quad \forall \boldsymbol{\zeta} \in \mathbb{R}^q. \tag{3.106}$$

If equality holds only when $\boldsymbol{\zeta} = \mathbf{0}$, then Φ is called *positive definite*.

We comment that the above definition is not universal in the literature, and the reader should check the definition used in each source.

The functions $\{\Phi(\cdot, \boldsymbol{\xi}_i) : i = 1, \ldots, q\}$ are used as a basis for interpolation of the data $\{f_1, \ldots, f_q\}$ given at the points $\{\boldsymbol{\xi}_1, \ldots, \boldsymbol{\xi}_q\}$: choose the interpolatory mapping

$$\mathcal{I}_q \boldsymbol{f}(\boldsymbol{\xi}) = \sum_{j=1}^{q} \alpha_j \Phi(\boldsymbol{\xi}, \boldsymbol{\xi}_j)$$

to satisfy

$$\sum_{j=1}^{q} \alpha_j \Phi(\boldsymbol{\xi}_i, \boldsymbol{\xi}_j) = f_i, \qquad i = 1, \ldots, q.$$

When Φ is positive definite, this is a non-singular system, and $\mathcal{I}_q \boldsymbol{f}$ is well-defined. However, this linear system may be quite ill-conditioned.

An important source of positive definite functions are those of the form

$$\Phi(\boldsymbol{\xi}, \boldsymbol{\eta}) = \varphi(\theta(\boldsymbol{\xi}, \boldsymbol{\eta})) \,. \tag{3.107}$$

Recall that $\theta(\boldsymbol{\xi}, \boldsymbol{\eta}) = \cos^{-1}(\boldsymbol{\xi} \cdot \boldsymbol{\eta}) \in [0, \pi]$ is the geodesic distance between $\boldsymbol{\xi}$ and $\boldsymbol{\eta}$ on $\mathbb{S}^{d-1}$ and is equal to the arc-length θ of the shortest path connecting $\boldsymbol{\xi}$ and $\boldsymbol{\eta}$. Equivalently, let $\Phi(\boldsymbol{\xi}, \boldsymbol{\eta}) = \widehat{\varphi}(\|\boldsymbol{\xi} - \boldsymbol{\eta}\|_2)$. Noting that

$$\begin{aligned}
\|\boldsymbol{\xi} - \boldsymbol{\eta}\|_2^2 &= (\boldsymbol{\xi} - \boldsymbol{\eta}) \cdot (\boldsymbol{\xi} - \boldsymbol{\eta}) \\
&= 2 - 2\, \boldsymbol{\xi} \cdot \boldsymbol{\eta} \\
&= 2\,(1 - \cos\theta) \,,
\end{aligned} \tag{3.108}$$

we can write

$$\varphi(\theta) = \widehat{\varphi}\Big(\sqrt{2\,(1 - \cos\theta)}\Big) \,.$$

A function of the variable θ is called a *radial function*. Many important examples of such radial functions $\widehat{\varphi}$ over regions in $\mathbb{R}^d$ are given in Fasshauer [45, Chap. 4].

In the special case of the unit sphere $\mathbb{S}^{d-1}$, $d \geq 3$, we study positive definite functions of the form

$$\Phi(\boldsymbol{\xi}, \boldsymbol{\eta}) = \sum_{n=0}^{\infty} \sum_{j=1}^{N_{n,d}} a_{n,j} Y_{n,j}(\boldsymbol{\xi}) Y_{n,j}(\boldsymbol{\eta}), \qquad \boldsymbol{\xi}, \boldsymbol{\eta} \in \mathbb{S}^{d-1}. \tag{3.109}$$

This uses the notation of Sect. 3.8, with $\{Y_{n,1}, \ldots, Y_{n,N_{n,d}}\}$ an orthonormal basis of $\mathbb{Y}_n^d$. Without loss of generality, the basis functions are chosen to be real-valued.

Theorem 3.25. *The function Φ of (3.109) is a radial function if and only if $\alpha_{n,j}$ is a constant function of j for all $n \geq 0$.*

Proof. Assume that $\alpha_{n,j} = a_n$, $1 \leq j \leq N_{n,d}$, for $n \geq 0$. Then

$$\Phi(\boldsymbol{\xi},\boldsymbol{\eta}) = \sum_{n=0}^{\infty} a_n \sum_{j=1}^{N_{n,d}} Y_{n,j}(\boldsymbol{\xi})Y_{n,j}(\boldsymbol{\eta}), \qquad \boldsymbol{\xi},\boldsymbol{\eta} \in \mathbb{S}^{d-1}.$$

Using Theorem 2.9 (the addition theorem),

$$\sum_{j=1}^{N_{n,d}} Y_{n,j}(\boldsymbol{\xi})Y_{n,j}(\boldsymbol{\eta}) = \frac{N_{n,d}}{|\mathbb{S}^{d-1}|} P_{n,d}(\boldsymbol{\xi}\cdot\boldsymbol{\eta}) \tag{3.110}$$

$$= \frac{N_{n,d}}{|\mathbb{S}^{d-1}|} P_{n,d}(\cos\theta).$$

Thus

$$\Phi(\boldsymbol{\xi},\boldsymbol{\eta}) = \varphi(\theta) = \frac{1}{|\mathbb{S}^{d-1}|} \sum_{n=0}^{\infty} a_n N_{n,d} P_{n,d}(\cos\theta), \tag{3.111}$$

proving Φ is a radial function.

    Conversely, let

$$\Phi(\boldsymbol{\xi},\boldsymbol{\eta}) = \varphi(\theta) \equiv \varphi^*(\boldsymbol{\xi}\cdot\boldsymbol{\eta}).$$

Using the completeness of the Legendre polynomials $\{P_{n,d} : n \geq 0\}$, expand the function $\varphi^*(t)$:

$$\varphi^*(t) = \sum_{n=0}^{\infty} b_n P_{n,d}(t), \qquad -1 \leq t \leq 1.$$

Use Theorem 2.9 to write

$$\varphi^*(\boldsymbol{\xi}\cdot\boldsymbol{\eta}) = |\mathbb{S}^{d-1}| \sum_{n=0}^{\infty} \frac{b_n}{N_{n,d}} \sum_{j=1}^{N_{n,d}} Y_{n,j}(\boldsymbol{\xi})Y_{n,j}(\boldsymbol{\eta}).$$

Using the uniqueness of an orthonormal expansion in $L^2(\mathbb{S}^{d-1})$ with the orthonormal basis

$$\{Y_{n,j} : 1 \leq j \leq N_{d,n}, \ n \geq 0\}$$

concludes the proof that φ^* has the form (3.109). $\qquad\qquad\qquad\square$

What are conditions on the coefficients in the expansion (3.109) that will imply Φ is a continuous function on $\mathbb{S}^{d-1} \times \mathbb{S}^{d-1}$? We begin with the bound (2.116),

$$|P_{n,d}(t)| \le 1, \qquad -1 \le t \le 1.$$

For the special case of a radial function $\varphi(\theta)$ as in (3.111), we have

$$|\Phi(\boldsymbol{\xi}, \boldsymbol{\eta})| \le \frac{1}{|\mathbb{S}^{d-1}|} \sum_{n=0}^{\infty} |a_n| N_{n,d}, \quad \boldsymbol{\xi}, \boldsymbol{\eta} \in \mathbb{S}^{d-1}.$$

If we assume

$$\sum_{n=0}^{\infty} |a_n| N_{n,d} < \infty, \tag{3.112}$$

then the Weierstrass comparison test implies that the series (3.111) is uniformly convergent and the limit is continuous.

For a function $\Phi(\boldsymbol{\xi}, \boldsymbol{\eta})$ of the general form (3.109), let

$$a_n^* = \max_{1 \le j \le N} |\alpha_{n,j}|, \quad n \ge 0.$$

Theorem 3.26. *Assume*

$$\sum_{n=0}^{\infty} a_n^* N_{n,d} < \infty. \tag{3.113}$$

Then the series on the right side of (3.109) is uniformly convergent and defines a continuous function Φ on $\mathbb{S}^{d-1} \times \mathbb{S}^{d-1}$.

Proof. Introduce

$$S_n(\boldsymbol{\xi}; \boldsymbol{\eta}) = \sum_{j=1}^{N_{n,d}} \alpha_{n,j} Y_{n,j}(\boldsymbol{\xi}) Y_{n,j}(\boldsymbol{\eta}).$$

Fix an arbitrary $\boldsymbol{\eta}$ and consider S_n as a function of $\boldsymbol{\xi}$. From the orthonormality of $\{Y_{n,j} : 1 \le j \le N_{n,d}\}$ in $L^2(\mathbb{S}^{d-1})$, we have

$$\|S_n(\cdot; \boldsymbol{\eta})\|_{L^2} = \sqrt{\sum_{j=1}^{N_{n,d}} |\alpha_{n,j}|^2 |Y_{n,j}(\boldsymbol{\eta})|^2} \le a_n^* \sqrt{\sum_{j=1}^{N_{n,d}} |Y_{n,j}(\boldsymbol{\eta})|^2}.$$

Apply (3.110) with $\boldsymbol{\xi} = \boldsymbol{\eta}$ and the normalization condition $P_{n,d}(1) = 1$ to obtain

$$\|S_n(\cdot;\boldsymbol{\eta})\|_{L^2} \leq a_n^* \sqrt{\frac{N_{n,d}}{|\mathbb{S}^{d-1}|}} \quad \forall \boldsymbol{\eta} \in \mathbb{S}^{d-1}.$$

Using (2.48) to obtain

$$|S_n(\boldsymbol{\xi};\boldsymbol{\eta})| \leq \sqrt{\frac{N_{n,d}}{|\mathbb{S}^{d-1}|}} \, \|S_n(\cdot;\boldsymbol{\eta})\|_{L^2} \leq \frac{N_{n,d}}{|\mathbb{S}^{d-1}|} a_n^*.$$

For the full series in (3.109),

$$|\Phi(\boldsymbol{\xi},\boldsymbol{\eta})| \leq \sum_{n=0}^{\infty} |S_n(\boldsymbol{\xi};\boldsymbol{\eta})| \leq \frac{1}{|\mathbb{S}^{d-1}|} \sum_{n=0}^{\infty} N_{n,d} a_n^*.$$

The conclusion of the theorem then follows from (3.113) and the Weierstrass comparison test. □

The condition (3.113) is satisfied if $a_n^* N_{n,d} = \mathcal{O}\big(n^{-(1+\varepsilon)}\big)$ for some $\varepsilon > 0$. Since $N_{n,d} = \mathcal{O}\big(n^{d-2}\big)$, we have convergence of (3.113) if

$$a_n^* = \mathcal{O}\Big(n^{-(d-1+\varepsilon)}\Big).$$

When are functions Φ of the form (3.109) positive definite? We begin with a needed lemma.

Lemma 3.27. *Let $\{\boldsymbol{\xi}_1,\ldots,\boldsymbol{\xi}_q\}$ be q distinct points in $\mathbb{S}^{d-1}$. Then there exists a spherical polynomial p for which*

$$p(\boldsymbol{\xi}_1) = 1,$$
$$p(\boldsymbol{\xi}_j) = 0, \quad j = 2,\ldots,q.$$

Proof. Define

$$p(\boldsymbol{x}) = \prod_{j=2}^{q} \frac{\|\boldsymbol{x} - \boldsymbol{\xi}_j\|_2^2}{\|\boldsymbol{\xi}_1 - \boldsymbol{\xi}_j\|_2^2}, \quad \boldsymbol{x} \in \mathbb{R}^d,$$

a polynomial of degree $2\,(q-1)$. When restricted to $\mathbb{S}^{d-1}$, it is a spherical polynomial, and it satisfies the requirements of the lemma. □

Theorem 3.28. *The function Φ of (3.109) is positive semi-definite if and only if all coefficients $\alpha_{n,k}$ are non-negative, $1 \leq k \leq N_n$, $n \geq 0$. It is positive definite if all the coefficients are positive.*

Proof. Begin with a set of points $\{\boldsymbol{\xi}_1, \ldots, \boldsymbol{\xi}_q\} \subset \mathbb{S}^{d-1}$. When the quadratic form in (3.106) is combined with the formula (3.109),

$$
\sum_{i,j=1}^{q} \Phi(\boldsymbol{\xi}_i, \boldsymbol{\xi}_j) \zeta_i \zeta_j = \sum_{n=0}^{\infty} \sum_{k=1}^{N_{n,d}} \alpha_{n,k} \sum_{i,j=1}^{q} \zeta_i \zeta_j Y_{n,k}(\boldsymbol{\xi}_i) Y_{n,k}(\boldsymbol{\xi}_j)
$$

$$
= \sum_{n=0}^{\infty} \sum_{k=1}^{N_{n,d}} \alpha_{n,k} \left[\sum_{i=1}^{q} \zeta_i Y_{n,k}(\boldsymbol{\xi}_i) \right] \left[\sum_{j=1}^{q} \zeta_j Y_{n,k}(\boldsymbol{\xi}_j) \right]
$$

$$
= \sum_{n=0}^{\infty} \sum_{k=1}^{N_{n,d}} \alpha_{n,k} \left[\sum_{i=1}^{q} \zeta_i Y_{n,k}(\boldsymbol{\xi}_i) \right]^2, \quad \boldsymbol{\zeta} \in \mathbb{R}^q. \quad (3.114)
$$

If the coefficients $\alpha_{n,k}$ are all non-negative, then (3.114) implies that the quadratic form for Φ is non-negative for all choices of $\boldsymbol{\zeta} \in \mathbb{R}^q$, all $q \geq 1$, and all $\{\boldsymbol{\xi}_1, \ldots, \boldsymbol{\xi}_q\} \subset \mathbb{S}^{d-1}$, thus proving Φ is positive semi-definite.

Now assume that all the coefficients $\alpha_{n,k} > 0$. To show that Φ is positive definite, assume

$$
\sum_{i,j=1}^{q} \Phi(\boldsymbol{\xi}_i, \boldsymbol{\xi}_j) \zeta_i \zeta_j = 0
$$

for some $q > 0$, some $\boldsymbol{\zeta} \in \mathbb{R}^q$, and some $\{\boldsymbol{\xi}_1, \ldots, \boldsymbol{\xi}_q\} \subset \mathbb{S}^{d-1}$. We need to show that $\boldsymbol{\zeta} = \mathbf{0}$. The formula (3.114) implies that

$$
\sum_{i=1}^{q} \zeta_i Y_{n,k}(\boldsymbol{\xi}_i) = 0
$$

for all of the basis functions $\{Y_{n,k} : 1 \leq k \leq N_n,\ n \geq 0\}$. By taking linear combinations, it follows that

$$
\sum_{i=1}^{q} \zeta_i p(\boldsymbol{\xi}_i) = 0
$$

for all spherical polynomials p. By using Lemma 3.27, we can choose for each i, $1 \leq i \leq q$, a polynomial p that satisfies $p(\boldsymbol{\xi}_j) = \delta_{ij}$, $j = 1, \ldots, q$. This then implies $\zeta_i = 0$ for each i, and thus $\boldsymbol{\zeta} = \mathbf{0}$.

It is more complicated to show the reverse implication, that Φ being positive semi-definite implies all $\alpha_{n,k}$ are non-negative, although (3.114) is suggestive of it. For this, we refer the reader to the discussion in Wendland [118, p. 312]. $\qquad\square$

To motivate the choice of the form in (3.109), introduce the integral operator $\mathcal{K}$,

$$\mathcal{K}f(\boldsymbol{\xi}) = \int_{\mathbb{S}^{d-1}} \Phi(\boldsymbol{\xi}, \boldsymbol{\eta})\, f(\boldsymbol{\eta})\, dS^{d-1}(\boldsymbol{\eta}), \quad \boldsymbol{\xi} \in \mathbb{S}^{d-1},$$

considered as an operator from $L^2(\mathbb{S}^{d-1})$ into $L^2(\mathbb{S}^{d-1})$, and assume its kernel function Φ is a symmetric function of its arguments. The operator $\mathcal{K}$ is called *positive* if

$$\begin{aligned}
(\mathcal{K}f, f) &\geq 0 \quad \forall f \in L^2(\mathbb{S}^{d-1}), \\
(\mathcal{K}f, f) &= 0 \quad \Longrightarrow \quad f = 0.
\end{aligned} \tag{3.115}$$

It follows from standard theory for integral operators that the eigenvalues $\{\lambda_j\}$ of $\mathcal{K}$ are real and positive, and the corresponding eigenfunctions $\{\psi_j\}$, $\mathcal{K}\psi_j = \lambda_j \psi_j$, can be chosen to be an orthonormal basis of $L^2(\mathbb{S}^{d-1})$. The following theorem is proved in Hochstadt [65, p. 90].

Theorem 3.29. (MERCER) *Assume the kernel function* $\Phi : \mathbb{S}^{d-1} \times \mathbb{S}^{d-1} \to \mathbb{R}$ *is continuous and symmetric, and assume* $\mathcal{K}$ *is a positive integral operator. Then*

$$\Phi(\boldsymbol{\xi}, \boldsymbol{\eta}) = \sum_{j=1}^{\infty} \lambda_j\, \psi_j(\boldsymbol{\xi})\, \psi_j(\boldsymbol{\eta}), \quad \boldsymbol{\xi}, \boldsymbol{\eta} \in \mathbb{S}^{d-1}, \tag{3.116}$$

and this series converges absolutely and uniformly over $\mathbb{S}^{d-1} \times \mathbb{S}^{d-1}$.

If the integral operator $\mathcal{K}$ is positive, then numerical integration of

$$(\mathcal{K}f, f) = \int_{\mathbb{S}^{d-1}} \int_{\mathbb{S}^{d-1}} \Phi(\boldsymbol{\xi}, \boldsymbol{\eta})\, f(\boldsymbol{\xi})\, f(\boldsymbol{\eta})\, dS^{d-1}(\boldsymbol{\eta})\, dS^{d-1}(\boldsymbol{\xi})$$

by an equal weight rule leads to statements of the form (3.106) for the particular quadrature nodes being used. Hence, (3.116) can be considered a motivation for the assumed form (3.109). Conversely, if Φ is a positive definite function of the form (3.109), then it is straightforward to show that the integral operator $\mathcal{K}$ with kernel function Φ is a positive operator on $L^2(\mathbb{S}^{d-1})$; moreover, the coefficients $\{\alpha_{n,k}\}$ are the eigenvalues of $\mathcal{K}$ and the spherical harmonics $\{Y_{n,k}\}$ are the corresponding eigenfunctions.

Chapter 4
Approximation Theory

For functions of a single variable, there is a rich literature on best approximations by ordinary polynomials and by trigonometric polynomials. In this chapter we discuss the extensions of these ideas to approximation on the sphere $\mathbb{S}^2$ in $\mathbb{R}^3$ and the unit disk $\mathbb{D}$ in $\mathbb{R}^2$. These results extend to higher dimensions, but most cases of interest in applications are for approximations on the unit sphere $\mathbb{S}^2$, the unit disk $\mathbb{D}$ in $\mathbb{R}^2$, and the unit ball $\mathbb{B}^3$ in $\mathbb{R}^3$.

At the center of approximation theory on $\mathbb{S}^2$ is the concept of "spherical harmonic", generalizing to $\mathbb{S}^2$ the trigonometric functions

$$\{1, \cos n\varphi, \sin n\varphi : n \geq 1\}$$

used in the univariate theory on $\mathbb{S}^1$. For developments of the one-variable theory, see, for example, Lorentz [77], Powell [91], or Rivlin [97]. Using spherical harmonics, the Fourier series representation of a periodic univariate function generalizes to the "Laplace series" or "Fourier–Laplace series" representation of a function defined on $\mathbb{S}^2$. As the name implies, the use of the Laplace series is an old idea, going back to P. Laplace and A. Legendre in the late 1700s. Their research arose from examining the function

$$\frac{1}{|\boldsymbol{x} - \boldsymbol{y}|}$$

for the potential of a gravitational force field at a point $\boldsymbol{x}$ arising from a unit mass at a point $\boldsymbol{y}$. This is still a useful topic and perspective, but other approaches to the subject are now more fruitful, especially when considering the approximation of functions defined on the sphere. A general introduction to the modern theory of spherical harmonics, spherical polynomials and approximation by them on $\mathbb{S}^d$, $d \geq 1$, is given in Chap. 2. In this chapter we expand on this material, while also restricting it to $\mathbb{S}^2$.

K. Atkinson and W. Han, *Spherical Harmonics and Approximations on the Unit Sphere: An Introduction*, Lecture Notes in Mathematics 2044, DOI 10.1007/978-3-642-25983-8_4, © Springer-Verlag Berlin Heidelberg 2012

Section 4.1 is a study of spherical polynomials, and Sect. 4.2 presents results from the research literature on the best approximation of functions by spherical polynomials. Section 4.3 presents analogous results on polynomials defined on the unit disk in $\mathbb{R}^2$. We restrict the presentation to real-valued functions, although there are usually straightforward extensions to complex-valued functions.

4.1   Spherical Polynomials

If $p(\boldsymbol{x})$ is a polynomial, then we call its restriction to $\mathbb{S}^2$ a "spherical polynomial". We define

$$\Pi_n(\mathbb{S}^2) := \{p|_{\mathbb{S}^2} : \deg(p) \le n\},$$

the space of spherical polynomials of degree $\le n$. Note that different polynomials p can result in the same spherical polynomial. For example, consider the two polynomials $p_1(\boldsymbol{x}) \equiv 1$ and $p_2(\boldsymbol{x}) \equiv |\boldsymbol{x}|^2 = x_1^2 + x_2^2 + x_3^2$, which lead to the same spherical polynomial. If we consider the corresponding approach using a polynomial $p(x_1, x_2)$ over $\mathbb{R}^2$, then restricting p to the unit circle results in the function $p(\cos\varphi, \sin\varphi)$, which is called a "trigonometric polynomial". Spherical polynomials can be considered as generalizations of trigonometric polynomials.

To better understand spherical polynomials, we specialize to $\mathbb{S}^2$ some results from Chap. 2. Begin with polynomials $p(\boldsymbol{x})$ over $\mathbb{R}^3$, letting

$$\Pi_n(\mathbb{R}^3) := \{p(\boldsymbol{x}) : \deg p \le n\}.$$

The space $\mathbb{H}_n(\mathbb{R}^3)$ consists of polynomials of degree n that are homogeneous. In the notation of Chap. 2, this space is denoted as $\mathbb{H}_n^3$. A simple basis for $\mathbb{H}_n(\mathbb{R}^3)$ is

$$\{\boldsymbol{x}^{\boldsymbol{\alpha}} \equiv x_1^{\alpha_1} x_2^{\alpha_2} x_3^{\alpha_3} : |\boldsymbol{\alpha}| \equiv \alpha_1 + \alpha_2 + \alpha_3 = n, \ \alpha_1, \alpha_2, \alpha_3 \ge 0\} \qquad (4.1)$$

and thus

$$\dim \mathbb{H}_n(\mathbb{R}^3) = \frac{1}{2}(n+1)(n+2);$$

see (2.3). Easily,

$$\Pi_n(\mathbb{R}^3) = \mathbb{H}_0(\mathbb{R}^3) + \mathbb{H}_1(\mathbb{R}^3) + \cdots + \mathbb{H}_n(\mathbb{R}^3). \qquad (4.2)$$

Also recall the space $\mathbb{Y}_n(\mathbb{R}^3)$ of harmonic homogeneous polynomials of degree n. The space $\mathbb{Y}_n \equiv \mathbb{Y}_n(\mathbb{S}^2)$ consists of the restrictions to $\mathbb{S}^2$ of the polynomials in $\mathbb{Y}_n(\mathbb{R}^3)$, and they are called spherical harmonics. Recalling (2.10), $\dim \mathbb{Y}_n(\mathbb{R}^3) = 2n + 1$.

The following result gives the relation of $\mathbb{H}_n(\mathbb{R}^3)$ to the spherical harmonics $\mathbb{Y}_n(\mathbb{R}^3)$. It is taken from [47, Theorem 2.2.2] and was proven earlier in Theorem 2.18 for the case of a general dimension d.

Lemma 4.1. *For $n \geq 2$, a polynomial $p \in \mathbb{H}_n(\mathbb{R}^3)$ can be decomposed as*

$$p(\boldsymbol{x}) = h(\boldsymbol{x}) + |\boldsymbol{x}|^2 q(\boldsymbol{x}) \tag{4.3}$$

with $h \in \mathbb{Y}_n(\mathbb{R}^3)$ and $q \in \mathbb{H}_{n-2}(\mathbb{R}^3)$. The choices of h and q are unique within the spaces $\mathbb{Y}_n(\mathbb{R}^3)$ and $\mathbb{H}_{n-2}(\mathbb{R}^3)$, respectively. Moreover, the polynomials $h(\boldsymbol{x})$ and $|\boldsymbol{x}|^2 q(\boldsymbol{x})$ are orthogonal using the inner product (2.6).

$$\mathbb{H}_n(\mathbb{R}^3) = \mathbb{Y}_n(\mathbb{R}^3) \oplus |\cdot|^2 \mathbb{H}_{n-2}(\mathbb{R}^3) \tag{4.4}$$

is an orthogonal decomposition of $\mathbb{H}_n(\mathbb{R}^3)$.

By letting $|\boldsymbol{x}| = 1$, the decomposition (4.3) implies that each spherical polynomial $p \in \Pi_n(\mathbb{S}^2)$, $n \geq 2$, can be written as the sum of a spherical harmonic of degree n and a spherical polynomial of degree $n - 2$. Also, $\Pi_n(\mathbb{S}^2) = \mathbb{Y}_n$ for $n = 0, 1$. Using (4.2) and induction with (4.3), it follows that the spherical harmonics can be used to create all spherical polynomials. If this is combined with Corollary 2.15, then we have

$$\Pi_n(\mathbb{S}^2) = \mathbb{Y}_0 \oplus \mathbb{Y}_1 \oplus \cdots \oplus \mathbb{Y}_n \tag{4.5}$$

is a decomposition of $\Pi_n(\mathbb{S}^2)$ into orthogonal subspaces, with the orthogonality based on the standard inner product for $L^2(\mathbb{S}^2)$,

$$(f, g) = \int_{\mathbb{S}^2} f(\boldsymbol{\eta}) g(\boldsymbol{\eta}) \, dS^2(\boldsymbol{\eta}). \tag{4.6}$$

This decomposition (4.5) is Corollary 2.19 for the case $d = 3$. For the remainder of this chapter, we will consider Π_n as denoting $\Pi_n(\mathbb{S}^2)$, unless stated explicitly otherwise.

With the orthogonal decomposition (4.5), a basis for Π_n can be introduced by giving a basis for each of the subspaces $\mathbb{Y}_k$, $k \geq 0$. The standard basis for spherical harmonics of degree n is (cf. Example 2.48)

$$\begin{aligned} Y_{n,1}(\boldsymbol{\xi}) &= c_n P_n(\cos \theta), \\ Y_{n,2m}(\boldsymbol{\xi}) &= c_{n,m} P_n^m(\cos \theta) \cos(m\phi), \\ Y_{n,2m+1}(\boldsymbol{\xi}) &= c_{n,m} P_n^m(\cos \theta) \sin(m\phi), \quad m = 1, \ldots, n \end{aligned} \tag{4.7}$$

with $\boldsymbol{\xi} = (\cos\phi\sin\theta, \sin\phi\sin\theta, \cos\theta)^T$. In this formula, $P_n(t)$ is a Legendre polynomial of degree n,

$$P_n(t) = \frac{1}{2^n n!} \frac{d^n}{dt^n}\left[(t^2 - 1)^n\right], \qquad (4.8)$$

see (2.70) with $d = 3$, and $P_n(t) = P_{n,3}(t)$. Also, $P_n^m(t)$ is an *associated Legendre function*,

$$P_n^m(t) = (-1)^m(1 - t^2)^{\frac{1}{2}m}\frac{d^m}{dt^m}P_n(t), \quad 1 \le m \le n, \qquad (4.9)$$

In the notation of Sect. 2.10,

$$P_n^m(t) \equiv (-1)^m\frac{(n+m)!}{n!}P_{n,3,m}(t).$$

Additional information on $P_n(t)$ is given later following (4.74). The constants in (4.7) are given by

$$c_n = \sqrt{\frac{2n+1}{4\pi}},$$

$$c_{n,m} = \sqrt{\frac{2n+1}{2\pi}\frac{(n-m)!}{(n+m)!}}.$$

Using the standard inner product of (4.6), the spherical harmonics given in (4.7) are an orthonormal basis of $\mathbb{Y}_n$. We obtain an orthonormal basis of Π_n using (4.7) and (4.5). For a development of the basis (4.7), see also MacRobert [78, Chaps. 4–7].

Although (4.7) appears to be the most popular choice of a basis for $\mathbb{Y}_n$, other orthonormal bases for $\mathbb{Y}_n$ and Π_n are possible, both for the inner product of (4.6) and for other weighted inner product spaces over $\mathbb{S}^2$. This is explored at great length in [121] and [44].

Recall the discussion in Sect. 2.8. For $f \in L^2(\mathbb{S}^2)$, the Fourier–Laplace series

$$f(\boldsymbol{\eta}) = \sum_{\ell=0}^{\infty} \mathcal{P}_\ell f(\boldsymbol{\eta}) \qquad (4.10)$$

is convergent in $L^2(\mathbb{S}^2)$ with the standard inner product (4.6) and associated norm. The term $\mathcal{P}_\ell f$ is the orthogonal projection of f into the subspace $\mathbb{Y}_\ell$ of spherical harmonics of order ℓ,

$$\mathcal{P}_\ell f(\boldsymbol{\eta}) = \sum_{j=1}^{2\ell+1} (f, Y_{\ell,j})\, Y_{\ell,j}(\boldsymbol{\eta}) \qquad (4.11)$$

using an orthonormal basis $\{Y_{\ell,j} : 1 \leq j \leq 2\ell + 1\}$ for $\mathbb{Y}_\ell$, for example (4.7). The convergence of (4.10) is discussed in Sect. 2.8.3. In the following section, we include a discussion of how the differentiability of f affects the rate of convergence.

As a part of our discussion of the convergence of the series in (4.10), we consider the orthogonal projection operator $\mathcal{Q}$, defined by

$$\mathcal{Q}_n f(\boldsymbol{\eta}) = \sum_{\ell=0}^{n} \mathcal{P}_\ell f(\boldsymbol{\eta}), \tag{4.12}$$

which maps $L^2(\mathbb{S}^2)$ onto $\Pi_n(\mathbb{S}^2)$. An important quantity in the consideration of the uniform convergence of (4.10) in $C(\mathbb{S}^2)$ is the norm

$$\|\mathcal{Q}_n\|_{C \to C}.$$

This is discussed later in Sect. 4.2.4.

4.2   Best Approximation on the Unit Sphere

For $1 \leq p < \infty$, recall that $L^p(\mathbb{S}^2)$ denotes the space of all measurable functions f on $\mathbb{S}^2$ for which

$$\|f\|_p = \left[\int_{\mathbb{S}^2} |f(\boldsymbol{\eta})|^p \, dS^2(\boldsymbol{\eta}) \right]^{\frac{1}{p}} < \infty.$$

For $p = \infty$, recall that $C(\mathbb{S}^2)$ denotes the space of all continuous functions f on $\mathbb{S}^2$, and we use the uniform norm

$$\|f\|_\infty = \max_{\boldsymbol{\eta} \in \mathbb{S}^2} |f(\boldsymbol{\eta})|.$$

Introduce the error of "best approximation" as follows. For $f \in L^p(\mathbb{S}^2)$ if $1 \leq p < \infty$ or $f \in C(\mathbb{S}^2)$ for $p = \infty$, define

$$E_{n,p}(f) = \min_{g \in \Pi_n} \|f - g\|_p. \tag{4.13}$$

A straightforward argument based on compactness and the Heine–Borel Theorem shows that the minimum is attained; the proof is omitted here.

There is a long history in the research literature for measuring the rate at which $E_{n,p}(f)$ decreases as a function of the "smoothness" of the function f. We note, in particular, the papers of Gronwall [54], Newman and Shapiro [88], Ragozin [93], Rustamov [99], Ditzian [39, 40], and Dai and Xu [36].

The latter paper contains both an excellent review of the literature and an extensive bibliography. A central concern of much, but not all, of the research literature of the past few decades has been the relationship of $E_{n,p}(f)$ to the smoothness of f. The researchers Rustamov, Ditzian, and Dai and Xu have given various ways of defining moduli of smoothness for f, and then, of showing an equivalence between the rates at which $E_{n,p}(f)$ and their moduli of smoothness converge to zero. These results are quite deep and are technically complicated to prove. We do not prove them in this text, instead just giving some of the background needed for understanding them. In addition, these papers prove results for the unit sphere $\mathbb{S}^d$, $d \geq 2$; we specialize these results to $\mathbb{S}^2$.

Ragozin [93] gives a bound on $E_{n,\infty}(f)$ for $f \in C(\mathbb{S}^2)$; the paper does not consider lower bounds for $E_{n,\infty}(f)$. Ganesh, Graham, and Sivaloganathan [50, Sect. 3] and Bagby, Bos, and Levenberg [18, Theorem 2] extend Ragozin's results to the related problem of simultaneously approximating f and its low order derivatives.

To bound $E_{n,p}(f)$, we present results of both Dai and Xu [36] and Ragozin [93].

4.2.1   The Approach to Best Approximation of Dai and Xu

The results in [36] are quite general, while also giving a definition of modulus of smoothness which relates more easily to our ordinary sense of differentiability on $\mathbb{S}^2$. We begin with a method for defining finite differences of a function f defined on $\mathbb{S}^2$.

Let e_1, e_2, e_3 denote the standard orthogonal basis in $\mathbb{R}^3$. Recall from Sect. 2.1 that $\mathbb{SO}^3$ denotes the set of all 3×3 orthogonal matrices with determinant 1. For $t \in \mathbb{R}$ and $1 \leq i \neq j \leq 3$, let $Q_{i,j,t} \in \mathbb{SO}^3$ denote the orthogonal matrix which causes a rotation of size t in the $x_i x_j$-plane, oriented such that the rotation from the vector e_i to the vector e_j is assumed to be positive. For example,

$$Q_{1,2,t}(\xi_1, \xi_2, \xi_3)^T = (s\cos(\phi+t), s\sin(\phi+t), \xi_3)^T,$$

where $(\xi_1, \xi_2) = s(\cos\phi, \sin\phi)$. It is known that every rotation $Q \in \mathbb{SO}^3$ can be written as

$$Q = Q_{1,2,\theta_1} Q_{2,3,\theta_2} Q_{1,2,\theta_3} \qquad (4.14)$$

for some $\theta_1, \theta_3 \in [0, 2\pi)$, $\theta_2 \in [0, \pi)$. The angles θ_1, θ_2 and θ_3 are the "Euler angles" of the rotation Q; see [117, p. 947].

For any $Q \in \mathbb{SO}^3$, denote

$$(T_Q f)(\boldsymbol{\xi}) := f(Q\boldsymbol{\xi}), \quad \boldsymbol{\xi} \in \mathbb{S}^2, \tag{4.15}$$

and denote the rth-order forward difference based on Q by

$$\Delta_Q^r f := (I - T_Q)^r f, \quad r \in \mathbb{N}. \tag{4.16}$$

As usual, $\Delta_Q^0 f = f$. There is no difficulty in interpreting this definition as an ordinary finite difference, e.g.

$$\Delta_Q f(\boldsymbol{\xi}) = f(\boldsymbol{\xi}) - f(Q\boldsymbol{\xi}).$$

To simplify the notation, let

$$\Delta_{i,j,t}^r f \equiv \Delta_{Q_{i,j,t}}^r f, \quad 1 \le i \ne j \le 3. \tag{4.17}$$

The result (4.14) shows the generality of the rotations $Q_{i,j,t}$, and thus we restrict the definition of the modulus of smoothness by referring only to these basic rotations. Define

$$\omega_r(f; \tau)_p := \sup_{|t| \le \tau} \max_{1 \le i < j \le 3} \|\Delta_{i,j,t}^r f\|_p, \quad r \in \mathbb{N}. \tag{4.18}$$

The case of $r = 1$ and $p = \infty$ is bounded by the usual modulus of continuity for a continuous function,

$$\omega_1(f; \tau)_\infty \le \omega(f; \tau) \equiv \sup_{\substack{\boldsymbol{\xi}, \boldsymbol{\eta} \in \mathbb{S}^2 \\ \theta(\boldsymbol{\xi}, \boldsymbol{\eta}) \le \tau}} |f(\boldsymbol{\xi}) - f(\boldsymbol{\eta})|. \tag{4.19}$$

As with the difference operators $\Delta_{Q_{i,j,t}}^r$ defined on the great circle intersection of $\mathbb{S}^2$ with the $x_i x_j$-plane in $\mathbb{R}^3$, we can also define derivative operators $D_{i,j}^r$. For example, consider differentiation in the great circle intersection in the $(1,2)$-plane. Define

$$D_{1,2}^r f(\boldsymbol{\xi}) = \left(-\frac{\partial}{\partial \phi}\right)^r f(s \cos \phi, s \sin \phi, x_3), \quad \boldsymbol{\xi} \in \mathbb{S}^2, \tag{4.20}$$

with $(\xi_1, \xi_2) = (s \cos \phi, s \sin \phi)$, for $r \in \mathbb{N}$. For $r = 0$, $D_{i,j}^r f = f$. Other ways of expressing these derivatives are discussed in [36, Sect. 2], [37, Sect. 3]; also, distributional derivatives $D_{1,2}^r f$ for general $f \in L^p(\mathbb{S}^2)$ are defined.

The properties of the rotations $Q_{i,j,t}$ and the modulus of continuity function ω_r are given in [36, Sect. 2]. Among these is the following important inequality that links $\Delta_{i,j,t}^r$ to the derivatives $D_{i,j}^r$ on $\mathbb{S}^2$: for $1 \le p \le \infty$,

$$\|\Delta_{i,j,t}^r f\|_p \le c\, |t|^r \, \|D_{i,j}^r f\|_p, \qquad r = 0,1,2,\ldots, \tag{4.21}$$

provided the function f is r-times differentiable over $\mathbb{S}^2$.

Theorem 4.2. (DAI AND XU) *For $f \in L^p(\mathbb{S}^2)$ if $1 \le p < \infty$ and $f \in C(\mathbb{S}^2)$ if $p = \infty$, it follows that*

$$E_{n,p}(f) \le c_r\, \omega_r\left(f; \frac{1}{n+1}\right)_p, \qquad n \ge 0 \tag{4.22}$$

for some constant $c_r > 0$.

Proof. This theorem is given in [36, Theorem 3.4]. The proof is complicated and it uses a number of results from the research literature which would take a great deal of space to develop. For that reason, we only give an outline of the proof; and some of the results are discussed for only $p = \infty$. The outline follows closely the proof given in [36, p. 1250].

Using the addition theorem (2.24), introduce the reproducing kernel integral operator

$$\mathcal{P}_\ell f(\boldsymbol\eta) = \frac{2\ell+1}{4\pi} \int_{\mathbb{S}^2} f(\boldsymbol\xi)\, P_\ell(\boldsymbol\eta \cdot \boldsymbol\xi)\, dS^2(\boldsymbol\xi), \qquad \boldsymbol\eta \in \mathbb{S}^2, \quad \ell \ge 0, \tag{4.23}$$

which is also the projection operator of (4.11). In this formula, $P_\ell(t)$ is the Legendre polynomial of degree ℓ on $[-1,1]$, defined in (4.8). Next, introduce a C^∞-function $\chi : [0, \infty) \to [0,1]$ with

$$\chi(t) = \begin{cases} 1, & 0 \le t \le 1, \\ 0, & t \ge 2, \end{cases}$$

and $0 \le \chi(t) < 1$ for $1 < t < 2$. Introduce the integral operator

$$\begin{aligned} V_n f(\boldsymbol\eta) &= \sum_{\ell=0}^{2n-1} \chi\left(\frac{\ell}{n}\right) \mathcal{P}_\ell f(\boldsymbol\eta) \\ &= \int_{\mathbb{S}^2} K_n(\boldsymbol\eta \cdot \boldsymbol\xi)\, f(\boldsymbol\xi)\, dS^2(\boldsymbol\xi), \qquad \boldsymbol\eta \in \mathbb{S}^2 \end{aligned} \tag{4.24}$$

with

$$K_n(t) = \sum_{\ell=0}^{2n-1} \chi\left(\frac{\ell}{n}\right) \frac{2\ell+1}{4\pi} P_\ell(\boldsymbol\eta \cdot \boldsymbol\xi).$$

Properties of the kernel K_n are given in [23, Lemma 3.3], and in particular, for any integer $k > 0$,

$$|K_n(\cos\theta)| \le \frac{c_k n^2}{(1+n\theta)^k}, \quad 0 \le \theta \le \pi, \tag{4.25}$$

with some constant $c_k > 0$.

Properties of the operator V_n are given in Rustamov [99, Lemma 3.1]; and in particular,

A1 $V_n f \in \Pi_{2n-1}(\mathbb{S}^2)$;
A2 $V_n f = f$ if $f \in \Pi_n(\mathbb{S}^2)$;
A3 For all $n \ge 1$,

$$\|V_n f\|_p \le \|V_n\|_p \|f\|_p, \quad f \in L^p(\mathbb{S}^2), \tag{4.26}$$

where $\|V_n\|_p$ denotes the operator norm of $V_n : L^p(\mathbb{S}^2) \to L^p(\mathbb{S}^2)$.

The first two properties are straightforward, whereas the property (4.26) is deeper and is complicated to prove; see [99, p. 316]. The operator V_n is used to define explicitly a polynomial of degree $\le n$ for which the error of approximating f is bounded by a constant multiple of $E_{n,p}(f)$.

Consider the polynomial $V_{[n/2]}f$, which is of degree $\le n$. Then by definition of the minimax error,

$$E_{n,p}(f) \le \left\| f - V_{[n/2]}f \right\|_p.$$

Using **A2** from above with the constant polynomial,

$$\int_{\mathbb{S}^2} K_n(\boldsymbol{\eta}{\cdot}\boldsymbol{\xi})\, dS^2(\boldsymbol{\xi}) \equiv 1.$$

We now restrict to the case $p = \infty$. Then

$$f(\boldsymbol{\eta}) - \left(V_{[n/2]}f\right)(\boldsymbol{\eta}) = \int_{\mathbb{S}^2} [f(\boldsymbol{\eta}) - f(\boldsymbol{\xi})]\, K_{[n/2]}(\boldsymbol{\eta}{\cdot}\boldsymbol{\xi})\, dS^2(\boldsymbol{\xi}),$$

$$\left|f(\boldsymbol{\eta}) - \left(V_{[n/2]}f\right)(\boldsymbol{\eta})\right| \le \int_{\mathbb{S}^2} |f(\boldsymbol{\eta}) - f(\boldsymbol{\xi})|\, \left|K_{[n/2]}(\boldsymbol{\eta}{\cdot}\boldsymbol{\xi})\right|\, dS^2(\boldsymbol{\xi}). \tag{4.27}$$

Denote by $\Gamma(\boldsymbol{\eta},\boldsymbol{\xi})$ the shorter of the great circle paths that connects $\boldsymbol{\eta}$ and $\boldsymbol{\xi}$, and recall the definition of the geodesic distance

$$\theta(\boldsymbol{\eta},\boldsymbol{\xi}) = \cos^{-1}(\boldsymbol{\eta}{\cdot}\boldsymbol{\xi}).$$

Using an argument similar to that in the proof of [36, Lemma 3.2], it can be shown that

$$|f(\boldsymbol{\eta}) - f(\boldsymbol{\xi})| \le c\omega_1(f, \theta(\boldsymbol{\eta}, \boldsymbol{\xi})), \quad \boldsymbol{\xi}, \boldsymbol{\eta} \in \mathbb{S}^2$$

for a suitably chosen constant $c > 0$. Combining this with (4.27),

$$\|f - V_{[n/2]}f\|_\infty \le c \max_{\boldsymbol{\eta} \in \mathbb{S}^2} \int_{\mathbb{S}^2} \omega_1(f, \theta(\boldsymbol{\eta}, \boldsymbol{\xi})) |K_{[n/2]}(\boldsymbol{\eta} \cdot \boldsymbol{\xi})| \, dS^2(\boldsymbol{\xi}).$$

Next, subdivide $\Gamma(\boldsymbol{\eta}, \boldsymbol{\xi})$ into subintervals of arc-length $\le 1/n$; this can be done with no more than $[n\,\theta(\boldsymbol{\eta}, \boldsymbol{\xi})] \le 1 + n\,\theta(\boldsymbol{\eta}, \boldsymbol{\xi})$ such subintervals. Then

$$\|f - V_{[n/2]}f\|_\infty \le c\omega(f, 1/n) \max_{\boldsymbol{\eta} \in \mathbb{S}^2} \int_{\mathbb{S}^2} (1 + n\,\theta(\boldsymbol{\eta}, \boldsymbol{\xi})) |K_{[n/2]}(\boldsymbol{\eta} \cdot \boldsymbol{\xi})| \, dS^2(\boldsymbol{\xi}).$$
$$(4.28)$$

The integral on the right side can be bounded independently of n. To do so, first note that the integral is constant as a function of $\boldsymbol{\eta}$ due to the rotational symmetry of the integrand and of $\mathbb{S}^2$. Thus, it is adequate to examine the integral for the special case of $\boldsymbol{\eta} = (0, 0, 1)^T$. Then $\boldsymbol{\eta} \cdot \boldsymbol{\xi} = \xi_3 = \cos\theta$, $\theta(\boldsymbol{\eta}, \boldsymbol{\xi}) = \theta$, $0 \le \theta \le \pi$. Also, use the bound (4.25) with $k = 4$. Then

$$\int_{\mathbb{S}^2} (1 + n\,\theta(\boldsymbol{\eta}, \boldsymbol{\xi})) |K_{[n/2]}(\boldsymbol{\eta} \cdot \boldsymbol{\xi})| \, dS^2(\boldsymbol{\xi})$$

$$= c_4 \int_0^{2\pi} \int_0^\pi (1 + n\,\theta) |K_{[n/2]}(\cos\theta)| \, d\theta \, d\phi$$

$$\le 2\pi c_4 \int_0^\pi (1 + n\,\theta) \frac{(n/2 + 1)^2}{(1 + n\,\theta/2)^4} \sin\theta \, d\theta$$

$$\le 2\pi (n/2 + 1)^2 c_4 \int_0^\pi \frac{(1 + n\,\theta)\,\theta}{(1 + n\,\theta/2)^4} \, d\theta,$$

where the latter inequality is obtained from

$$\sup_{0 < \theta \le \pi} \frac{\sin\theta}{\theta} = 1.$$

Change the variable of integration by letting $u = 1 + n\,\theta/2$. Then

$$\int_{\mathbb{S}^2} (1 + n\,\theta(\boldsymbol{\eta}, \boldsymbol{\xi})) |K_{[n/2]}(\boldsymbol{\eta} \cdot \boldsymbol{\xi})| \, dS^2(\boldsymbol{\xi})$$

$$\le 2\pi (1 + 2/n)^2 c_4 \int_1^{1 + \frac{1}{2}n\pi} \frac{(2u - 1)(u - 1)}{u^4} \, du$$

$$< 18\,\pi\,c_4 \int_1^\infty \frac{(2u-1)\,(u-1)}{u^4}\,du$$

Combining this with (4.28), we have

$$\left\| f - V_{[n/2]} f \right\|_\infty \le c\omega_1(f, 1/n)$$

for a constant $c > 0$. This shows (4.22) for $r = 1$ and $p = \infty$. For $1 \le p < \infty$, a similar but longer argument is given in [36, Lemma 3.2].

To show the result for $r > 1$, an inductive argument is given in [35, Sect. 4]. We omit it here. $\square$

By combining (4.22) with (4.18) and (4.21), we have the following corollary.

Corollary 4.3. *Let $r \ge 1$ be an integer. Assume f is r-times continuously differentiable over $\mathbb{S}^2$ with all such derivatives in $L^p(\mathbb{S}^2)$ for $0 \le p < \infty$ and in $C(\mathbb{S}^2)$ for $p = \infty$. Then*

$$E_{n,p}(f) \le \frac{c}{(n+1)^r}, \quad n \ge 0. \tag{4.29}$$

Results inverse to these are given in [36, Sect. 3.2]. We give one such result, taken from [36, Corollary 3.5].

Theorem 4.4. *For $0 < \alpha < r$ and $f \in L^p(\mathbb{S}^2)$ if $1 \le p < \infty$ and $f \in C(\mathbb{S}^2)$ if $p = \infty$, the statements $E_{n,p} \sim n^{-\alpha}$ and $\omega_r(f; t)_p \sim t^{-\alpha}$ are equivalent, where both are statements of proportionality.*

These results have been extended in [37], giving bounds using a Sobolev space setting. Define the Sobolev space $\mathcal{W}_p^r(\mathbb{S}^2)$ to be the space of functions $f \in L^p(\mathbb{S}^2)$ whose distributional derivatives $D_{i,j}^r f \in L^p(\mathbb{S}^2)$, $1 \le i < j \le 3$, and which satisfy

$$\|f\|_{\mathcal{W}_p^r} \equiv \|f\|_p + \sum_{1 \le i < j \le 3} \|D_{i,j}^r f\|_p < \infty. \tag{4.30}$$

For $p = \infty$, replace $L^\infty(\mathbb{S}^2)$ with $C(\mathbb{S}^2)$. To generalize the idea of a space of r-times differentiable Hölder continuous functions, introduce the set $\mathcal{W}_p^{r,\alpha}(\mathbb{S}^2)$, $\alpha \in (0, 1]$ and $r \ge 0$, to be the space of all $f \in \mathcal{W}_p^r(\mathbb{S}^2)$ for which

$$\|f\|_{\mathcal{W}_p^{r,\alpha}} \equiv \|f\|_p + \max_{1 \le i < j \le 3} \sup_{0 < |t| \le 1} \frac{\|\Delta_{i,j,t}(D_{i,j}^r f)\|_p}{|t|^\alpha} < \infty.$$

In [37, Sect. 3], the relationship of these spaces $\mathcal{W}_p^r(\mathbb{S}^2)$ and $\mathcal{W}_p^{r,\alpha}(\mathbb{S}^2)$ to other definitions are discussed; they are equivalent or closely related to other definitions of Sobolev spaces on $\mathbb{S}^2$.

Using this framework, the following result is proven. It provides a complete generalization to $\mathbb{S}^2$ of the Jackson type results for approximation on the unit circle $\mathbb{S}^1$.

Theorem 4.5. *Let* $r \in \mathbb{N}_0$, $\alpha \in (0,1]$, *and* $1 \leq p \leq \infty$. *Then for any* $f \in \mathcal{W}_p^{r,\alpha}(\mathbb{S}^2)$

$$E_{n,p}(f) \leq \frac{c}{(n+1)^{r+\alpha}} \|f\|_{\mathcal{W}_p^{r,\alpha}(\mathbb{S}^2)}, \quad n \geq 0.$$

Complete results for weighted best approximation over $\mathbb{S}^d$, $d \geq 2$, are also given in Dai and Xu [36, Sect. 4].

4.2.2   The Approach to Best Approximation of Ragozin

The paper of Newman and Shapiro [88] presents a constructive approximation that leads to the rate of convergence $\omega(f; 1/n)$ for $E_{n,\infty}(f)$. This construction was used by Ragozin [93, Theorem 3.3] to extend their result to functions that are several times continuously differentiable over $\mathbb{S}^2$, obtaining rates of convergence analogous to Theorem 4.5. The results in [93, Theorem 3.3] are for $\mathbb{S}^d$, $d \geq 2$; and we specialize to only $\mathbb{S}^2$. The analysis of Ragozin uses a framework from differential geometry to work with differentiation of functions f defined on $\mathbb{S}^2$. We briefly discuss below the ideas needed to understand this work. In doing so, we follow closely the presentation of Ganesh, Graham, and Sivaloganathan [50, Sect. 3].

We begin by considering the standard way of discussing the differentiability of a function defined on $\mathbb{S}^2$ (or on any smooth manifold in $\mathbb{R}^3$). At a point $\boldsymbol{x}_0$ on $\mathbb{S}^2$, form the tangent plane, say U. The plane U will be the basis of a local coordinate system for representing the nearby surface of $\mathbb{S}^2$. Select an open neighborhood of $\boldsymbol{x}_0$ on $\mathbb{S}^2$, say V, and project it orthogonally onto U, obtaining a planar region. Denote this mapping by $\varphi : V \to \varphi(V) \subset U$.

Functions defined on V can be reformulated as functions on U via the mapping φ. If $f \in C(V)$, then consider the function $\widehat{f}$ defined on $\varphi(V)$ by

$$\widehat{f}(\varphi(\boldsymbol{x})) = f(\boldsymbol{x}), \qquad \boldsymbol{x} \in V, \tag{4.31}$$

or

$$\widehat{f}(\boldsymbol{u}) = f(\varphi^{-1}(\boldsymbol{u})), \qquad \boldsymbol{u} \in \varphi(V).$$

The region $\varphi(V)$ is a subset of the plane U; we have our ordinary sense of partial differentiation when differentiating $\widehat{f}(\boldsymbol{u})$ based on a coordinate system in U. If the function $\widehat{f} \in C^k(\varphi(V))$, then we say $f \in C^k(V)$, $k \geq 0$.

We can choose a finite set of such tangent planes U so as to "cover" all of $\mathbb{S}^2$, having corresponding (overlapping) regions V. With the sphere $\mathbb{S}^2$, six such tangent planes are sufficient, and we denote them by $U_1, \ldots, U_6$. The corresponding (overlapping) regions on $\mathbb{S}^2$ are denoted by $V_1, \ldots, V_6$, with corresponding mappings $\varphi_1, \ldots, \varphi_6$. We say $f \in C^k(\mathbb{S}^2)$ if $\widehat{f}_i = f \circ \varphi_i^{-1} \in C^k(\varphi_i(V_i))$, $i = 1, \ldots, 6$. We define a norm on $C^k(\mathbb{S}^2)$ by

$$\|f\|_{C^k(\mathbb{S}^2)} = \max_{1 \leq i \leq 6} \max_{|\boldsymbol{\alpha}| \leq k} \|D^{\boldsymbol{\alpha}}(\widehat{f}_i)\|_{C^k(\varphi_i(V_i))}. \tag{4.32}$$

For the derivatives,

$$D^{\boldsymbol{\alpha}} g(\boldsymbol{u}) = \frac{\partial^{|\boldsymbol{\alpha}|} g(\boldsymbol{u})}{\partial u_1^{\alpha_1} \partial u_2^{\alpha_2}}, \qquad \boldsymbol{\alpha} = (\alpha_1, \alpha_2)$$

refers to differentiation within the associated tangent plane U_i, using a $u_1 u_2$-coordinate system in U. With this norm, $C^k(\mathbb{S}^2)$ is a Banach space. If $\gamma \in (0, 1]$, we can define similarly $C^{k,\gamma}(\mathbb{S}^2)$ as the set of functions $f \in C^k(\mathbb{S}^2)$ for which all derivatives $D^{\boldsymbol{\alpha}}(f \circ \varphi_i^{-1})$, $|\boldsymbol{\alpha}| = k$, are Hölder continuous on $\varphi_i(V_i)$ with exponent γ, $1 \leq i \leq 6$. A norm $\|f\|_{C^{k,\gamma}(\mathbb{S}^2)}$ can be defined analogously to (4.32).

This definition of $C^k(\mathbb{S}^2)$ is standard and it involves ordinary partial derivatives over planar regions, using local charts as described above. However, other ways of referring to the smoothness of functions $f \in C(\mathbb{S}^2)$ have been found necessary when analyzing the behaviour of $E_{n,p}(f)$, and this is reflected in the definition (4.18) used in the analysis of Dai and Xu [36]. The work of Ragozin [93] uses a differential geometry perspective, using Lie algebras and associated derivatives, as Lie derivatives provide a simple and global way to describe the smoothness of functions $f \in C(\mathbb{S}^2)$. It is also a framework for more general differentiable manifolds in $\mathbb{R}^d$, $d \geq 2$.

Recall that $\mathbb{SO}^3$ denotes the set of 3×3 real orthogonal matrices A with determinant equal to 1,

$$A^{\mathrm{T}} A = I, \qquad \det A = 1;$$

and let $\mathfrak{g}$ denote the set of 3×3 real skew-symmetric matrices D,

$$D^{\mathrm{T}} = -D.$$

The matrices in $\mathfrak{g}$ can be used to generate elements of $\mathbb{SO}^3$. In particular, for real t consider the one-parameter family of matrices

$$A(t) = e^{-tD} = \sum_{k=0}^{\infty} \frac{(-1)^k t^k D^k}{k!}.$$

Using elementary matrix algebra, it follows that

$$A(t)^{\mathrm{T}} = e^{-tD^{\mathrm{T}}} = e^{tD} = A(t)^{-1},$$

$$A(t)^{\mathrm{T}} A(t) = I,$$

showing $A(t)$ is orthogonal for all t. Trivially, $\det A(0) = 1$; and the orthogonality of $A(t)$ implies $\det A(t) = \pm 1$. The determinant function $\det B$ is a continuous function of its argument B. When combined with the earlier statements on $\det A(t)$, it follows that $\det A(t) = 1$ for all t.

To aid in calculating e^{-tD}, we use the following property of matrix exponentiation. Assume that a square matrix B has the canonical form

$$B = P^{-1} \Lambda P \tag{4.33}$$

with Λ a diagonal matrix and some nonsingular matrix P. Then

$$e^B = \sum_{k=0}^{\infty} \frac{B^k}{k!} = \sum_{k=0}^{\infty} \frac{\left(P^{-1} \Lambda P\right)^k}{k!} = \sum_{k=0}^{\infty} P^{-1} \frac{\Lambda^k}{k!} P$$

$$= P^{-1} e^{\Lambda} P. \tag{4.34}$$

The matrix e^{Λ} can be calculated using standard exponentiation.

As an important example, consider

$$D_1 = \begin{bmatrix} 0 & 0 & 0 \\ 0 & 0 & -1 \\ 0 & 1 & 0 \end{bmatrix}. \tag{4.35}$$

Then $D_1 = P^{-1} \Lambda P$ with $\Lambda = \mathrm{diag}\,[0, i, -i]$ and

$$P = \begin{bmatrix} 1 & 0 & 0 \\ 0 & 1 & i \\ 0 & 1 & -i \end{bmatrix}.$$

Using (4.34),

$$e^{-tD_1} = \begin{bmatrix} 1 & 0 & 0 \\ 0 & \cos t & -\sin t \\ 0 & \sin t & \cos t \end{bmatrix}.$$

The mapping $\boldsymbol{x} \mapsto e^{-tD_1} \boldsymbol{x}$ rotates $\boldsymbol{x}$ through an angle of t in the $x_2 x_3$-plane in $\mathbb{R}^3$. The matrices

$$D_2 = \begin{bmatrix} 0 & 0 & -1 \\ 0 & 0 & 0 \\ 1 & 0 & 0 \end{bmatrix}, \qquad D_3 = \begin{bmatrix} 0 & -1 & 0 \\ 1 & 0 & 0 \\ 0 & 0 & 0 \end{bmatrix} \qquad (4.36)$$

yield rotations in the $x_1 x_3$-plane and $x_1 x_2$-plane, respectively.

Some additional notation is needed. Given $f \in C(\mathbb{S}^2)$ and $E \in \mathbb{SO}^3$, the mapping $\boldsymbol{\xi} \mapsto f(E^{-1}\boldsymbol{\xi})$ is called the "action" of E on f; it is denoted by $E \circ f$ and belongs to $C(\mathbb{S}^2)$. Let $\mathcal{T}$ be a continuous linear mapping from $C(\mathbb{S}^2)$ to $C(\mathbb{S}^2)$. We say $\mathcal{T}$ is an *equivariant map* with respect to $\mathbb{SO}^3$ if

$$E \circ (\mathcal{T}f) = \mathcal{T}(E \circ f), \qquad \forall f \in C(\mathbb{S}^2), \ \forall E \in \mathbb{SO}^3.$$

The following lemma is important because the construction of a polynomial approximation in [88] is an equivariant mapping; see (4.46) below.

Lemma 4.6. *Let $p \in C[-1,1]$ and define*

$$\mathcal{T}f(\boldsymbol{\xi}) = \int_{\mathbb{S}^2} p(\boldsymbol{\xi} \cdot \boldsymbol{\eta}) \, f(\boldsymbol{\eta}) \, dS^2(\boldsymbol{\eta}), \quad \boldsymbol{\xi} \in \mathbb{S}^2.$$

Then $\mathcal{T} : C(\mathbb{S}^2) \to C(\mathbb{S}^2)$ is linear and continuous, and it is equivariant with respect to $\mathbb{SO}^3$.

Proof. It is straightforward that $f \in C(\mathbb{S}^2)$ implies $\mathcal{T}f \in C(\mathbb{S}^2)$, and also that $\mathcal{T} : C(\mathbb{S}^2) \to C(\mathbb{S}^2)$ is linear and continuous. To show $\mathcal{T}$ is equivariant, let $f \in C(\mathbb{S}^2)$ and $E \in \mathbb{SO}^3$. Then

$$(E \circ (\mathcal{T}f))(\boldsymbol{\xi}) = \int_{\mathbb{S}^2} p((E^{-1}\boldsymbol{\xi}) \cdot \boldsymbol{\eta}) \, f(\boldsymbol{\eta}) \, dS^2(\boldsymbol{\eta}) \qquad (4.37)$$

$$= \int_{\mathbb{S}^2} p((E^{-1}\boldsymbol{\xi}) \cdot (E^{-1}\boldsymbol{\eta})) \, f(E^{-1}\boldsymbol{\eta}) \, dS^2(\boldsymbol{\eta}) \qquad (4.38)$$

$$= \int_{\mathbb{S}^2} p(\boldsymbol{\xi} \cdot \boldsymbol{\eta}) \, f(E^{-1}\boldsymbol{\eta}) \, dS^2(\boldsymbol{\eta}) \qquad (4.39)$$

$$= (\mathcal{T}(E \circ f))(\boldsymbol{\xi}). \qquad (4.40)$$

Line (4.38) follows from the invariance under rotation of integrals over $\mathbb{S}^2$; and (4.39) follows from

$$(E^{-1}\boldsymbol{\xi}) \cdot (E^{-1}\boldsymbol{\eta}) = \boldsymbol{\xi} \cdot (E^{-T}E^{-1}\boldsymbol{\eta}) = \boldsymbol{\xi} \cdot \boldsymbol{\eta}$$

due to the orthogonality of E. $\qquad \square$

We now introduce another approach to differentiation of elements in $C(\mathbb{S}^2)$. Let $f \in C(\mathbb{S}^2)$. Given an element $D \in \mathfrak{g}$, we define an "algebraic derivative" of f, associated with D, by

$$Df(\boldsymbol{\xi}) = \lim_{t \to 0} \frac{f(e^{-tD}\boldsymbol{\xi}) - f(\boldsymbol{\xi})}{t} = \lim_{t \to 0} \frac{\left(e^{tD} \circ f\right)(\boldsymbol{\xi}) - f(\boldsymbol{\xi})}{t} \tag{4.41}$$

if it exists and is continuous over $\mathbb{S}^2$. This is defined globally over $\mathbb{S}^2$. This notation can be confusing; each $D \in \mathfrak{g}$ is to be associated with a differentiation of f over $\mathbb{S}^2$. The cases in (4.35)–(4.36) are of particular importance. The following lemma is critical to the analysis of the rate of convergence of $E_{n,\infty} \to 0$ in [93] and to extensions given in [50].

Lemma 4.7. *Let* $\mathcal{T} : C(\mathbb{S}^2) \to C(\mathbb{S}^2)$ *be a linear, continuous, and* $\mathbb{SO}^3$-*equivariant mapping. Let* $f \in C(\mathbb{S}^2)$ *and* $D \in \mathfrak{g}$. *If* Df *exists, then so does* $D(\mathcal{T}f)$, *and moreover,*

$$D(\mathcal{T}f) = \mathcal{T}(Df). \tag{4.42}$$

Proof.

$$D(\mathcal{T}f)(\boldsymbol{\xi}) = \lim_{t \to 0} \frac{\left(e^{tD}\right) \circ (\mathcal{T}f))(\boldsymbol{\xi}) - (\mathcal{T}f)(\boldsymbol{\xi})}{t}.$$

Using the linearity, continuity, and equivariant properties of $\mathcal{T}$,

$$\begin{aligned}
D\left(\mathcal{T}f\right)(\boldsymbol{\xi}) &= \lim_{t \to 0} \frac{\left(\mathcal{T}\left(e^{tD} \circ f\right)\right)(\boldsymbol{\xi}) - \mathcal{T}f(\boldsymbol{\xi})}{t} \\
&= \mathcal{T} \lim_{t \to 0} \frac{\left(e^{tD} \circ f\right)(\boldsymbol{\xi}) - f(\boldsymbol{\xi})}{t} \\
&= \mathcal{T}Df(\boldsymbol{\xi}). \qquad \qquad \square
\end{aligned}$$

A new space of differentiable functions is introduced in [50, p. 1396].

$$C_{\text{alg}}^1 = \left\{ f \in C(\mathbb{S}^2) : Df \in C(\mathbb{S}^2) \text{ for all } D \in \mathfrak{g} \right\}.$$

Inductively, for $k \geq 1$,

$$C_{\text{alg}}^k = \left\{ f \in C_{\text{alg}}^{k-1} : Df \in C^{k-1}(\mathbb{S}^2) \text{ for all } D \in \mathfrak{g} \right\} \tag{4.43}$$

with $C_{\text{alg}}^0 = C(\mathbb{S}^2)$. A norm is introduced based on the algebraic derivatives associated with D_1, D_2, and D_3 from (4.35)–(4.36). For a multi-index $\alpha = (\alpha_1, \alpha_2, \alpha_3)$, consider algebraic derivatives

$$D^\alpha = D_1^{\alpha_1} D_2^{\alpha_2} D_3^{\alpha_3}. \tag{4.44}$$

For $f \in C_{\text{alg}}^k$, define

$$\|f\|_{k,\text{alg}} = \max \left\{ \|D^\alpha f\|_\infty : |\alpha| \leq k \right\},$$

where the derivatives $D^\alpha f$ are restricted to those of the form in (4.44). The following is shown in [50, Theorem 3.4]; we omit the proof.

Theorem 4.8. *For $k \geq 0$, $C_{\mathrm{alg}}^k = C^k(\mathbb{S}^2)$. In addition, there are constants $c_k, d_k > 0$ such that*

$$c_k \|f\|_{k,\mathrm{alg}} \leq \|f\|_{C^k(\mathbb{S}^2)} \leq d_k \|f\|_{k,\mathrm{alg}} \quad \forall f \in C^k(\mathbb{S}^2). \tag{4.45}$$

This says that smoothness of a function f in C_{alg}^k is equivalent to smoothness in the sense of the classical space $C^k(\mathbb{S}^2)$. The space of rotations $\mathbb{SO}^3$ is called the *full orthogonal group for* $\mathbb{R}^3$ and it is an example of a *Lie group*. The space $\mathfrak{g}$ is a *Lie algebra*, and the derivative in (4.41) is a *Lie derivative*.

We return to the problem of bounding the minimax error $E_{n,\infty}(f)$. Newman and Shapiro [88, p. 216] created a special operator for generating good polynomial approximations. They did so for only even degrees n, say $n = 2m$. Since $E_{n,\infty}(f)$ is monotone nonincreasing with $E_{n,\infty} \to 0$ as $n \to \infty$, there is no problem with such a restriction on n. Let $P_{m+1}(t)$ be the Legendre polynomial of degree $m+1$ on $[-1, 1]$, and let λ_{m+1} denote its largest root. Define

$$K_n(t) = \left(\frac{P_{m+1}(t)}{t - \lambda_{m+1}} \right)^2.$$

This is a polynomial of degree $n = 2m$. For $f \in C(\mathbb{S}^2)$, define

$$\mathcal{T}_n f(\boldsymbol{\xi}) = \frac{\displaystyle\int_{\mathbb{S}^2} K_n(\boldsymbol{\xi} \cdot \boldsymbol{\eta})\, f(\boldsymbol{\eta})\, dS^2(\boldsymbol{\eta})}{\displaystyle\int_{\mathbb{S}^2} K_n(\boldsymbol{\xi} \cdot \boldsymbol{\eta})\, dS^2(\boldsymbol{\eta})}, \qquad \boldsymbol{\xi} \in \mathbb{S}^2. \tag{4.46}$$

The denominator is independent of $\boldsymbol{\xi}$,

$$\int_{\mathbb{S}^2} K_n(\boldsymbol{\xi} \cdot \boldsymbol{\eta})\, dS^2(\boldsymbol{\eta}) = 2\pi \int_{-1}^{1} K_n(t)\, dt.$$

The function $\mathcal{T}_n f(\boldsymbol{\xi})$ is a spherical polynomial of degree $\leq n$; and $\mathcal{T}_n : C(\mathbb{S}^2) \to \Pi_n$ is a linear, continuous, and equivariant mapping. The following result is proven in [88, pp. 213–216].

Theorem 4.9. (Newman and Shapiro) *Assume $f \in C(\mathbb{S}^2)$. Then*

$$\|f - \mathcal{T}_n f\|_\infty \leq c\, \omega\left(f; \frac{1}{n+1} \right), \quad n \geq 0 \tag{4.47}$$

for a suitable constant $c > 0$ that is independent of n and f.

Note that $E_{n,\infty}(f)$ is the minimum of all approximation errors, and consequently,

$$E_{n,\infty}(f) \leq \|f - \mathcal{T}_n f\|_\infty .$$

Equation (4.47) bounds $E_{n,\infty}(f)$. This applies also to the bound given below in Theorem 4.11.

Ragozin showed how to use this construction to obtain improved bounds for $E_{n,\infty}(f)$ when f has additional smoothness. In particular the following theorem is proven in [93, Theorem 2.1]. We begin with his definition of modulus of continuity for algebraic derivatives. Refer to the algebraic derivatives associated with D_1, D_2, and D_3 from (4.35)–(4.36). For a function $f \in C^{k+1}(\mathbb{S}^2)$, define

$$\omega^R(f^{(1)}; h) := \sum_{j=1}^{3} \omega(D_j f; h),$$

$$\omega^R(f^{(k+1)}; h) := \sum_{j=1}^{3} \omega^R((D_j f)^{(k)}; h), \quad k \geq 1.$$

By means of Theorem 4.8, we can interpret these as standard moduli of continuity for functions in $C^k(\mathbb{S}^2)$, $k \geq 1$.

Theorem 4.10. (RAGOZIN) *Let* $\mathcal{T} : C(\mathbb{S}^2) \to C(\mathbb{S}^2)$ *be a continuous, equivariant linear mapping. Suppose that* $\mathcal{T}$ *satisfies*

$$\|f - \mathcal{T}f\|_\infty \leq A\omega(f; h) \quad \forall f \in C(\mathbb{S}^2)$$

with some constant $A > 0$. *Then* $_k\mathcal{T} := I - (I - \mathcal{T})^{k+1}$ *satisfies*

$$\|f -_k \mathcal{T}f\|_\infty \leq A^{k+1} h^k \omega^R(f^{(k)}; h) \quad \forall f \in C^k(\mathbb{S}^2).$$

Moreover, $\mathrm{Range}(_k\mathcal{T}) \subset \mathrm{Range}(\mathcal{T})$.

Using this, Ragozin [93, Theorem 3.3] proved the following.

Theorem 4.11. *Let* $\mathcal{T}_n$ *be the operator defined in* (4.46), *and let* $_k\mathcal{T}_n := I - (I - \mathcal{T}_n)^{k+1}$. *Then*

$$\|f -_k \mathcal{T}_n f\|_\infty \leq A_k n^{-k} \omega^R\left(f^{(k)}; \frac{1}{n+1}\right), \quad n \geq 0, \forall f \in C^k(\mathbb{S}^2) \quad (4.48)$$

for some constant $A_k > 0$.

As a corollary, assume $f \in C^{k,\gamma}(\mathbb{S}^2)$ for some $\gamma \in (0,1]$, meaning all of the kth-order derivatives of f satisfy

$$|g(\boldsymbol{\xi}) - g(\boldsymbol{\eta})| \leq c_{k,\gamma}(f) |\boldsymbol{\xi} - \boldsymbol{\eta}|^\gamma , \qquad \boldsymbol{\xi}, \boldsymbol{\eta} \in \mathbb{S}^2,$$

where g is a generic designation of all such derivatives and $c_{k,\gamma}(f) \geq 0$ is a constant. Then

$$E_{n,\infty}(f) \leq \frac{B_k c_{k,\gamma}(f)}{(n+1)^{k+\gamma}}, \quad n \geq 0. \tag{4.49}$$

The constant $B_k > 0$ is independent of f and n. This bound is of the same order as that given earlier in Theorem 4.5.

4.2.3 Best Simultaneous Approximation Including Derivatives

Theorem 4.11 on the rate of uniform convergence of best approximations to a function $f \in C(\mathbb{S}^2)$ has been extended to the case of simultaneous approximation of f and its low-order derivatives. Ganesh, Graham, and Sivaloganathan [50, Theorem 3.5] used the above framework to study the simultaneous approximation of a function f and its derivatives up to a given order ℓ.

Theorem 4.12. (GANESH, GRAHAM, AND SIVALOGANATHAN) *Let $\ell \geq 0$. For $n \geq 0$, let $_\ell\mathcal{T}_n := I - (I - \mathcal{T}_n)^{\ell+1}$ with $\mathcal{T}_n$ as defined in (4.46). Then there exists a constant $c_\ell > 0$ such that*

$$\|f - {}_\ell\mathcal{T}_n f\|_k \leq c_\ell n^{k-\ell} \|f\|_{C^\ell(\mathbb{S}^2)}, \quad 0 \leq k \leq \ell, \ \forall f \in C^\ell(\mathbb{S}^2). \tag{4.50}$$

The proof is omitted except for the note that a crucial role is played by $\mathcal{T}_n$ being equivariant, namely that

$$\|D^\alpha(f - {}_\ell\mathcal{T}_n f)\|_\infty = \|D^\alpha f - {}_\ell\mathcal{T}_n D^\alpha f\|_\infty$$

This permits the application of Ragozin's Theorem 4.11 to $D^\alpha f$.

An alternative derivation of the bound (4.50) is given in Bagby, Bos, and Levenberg [18, Theorem 2]. To carry it out, we first discuss an extension of f to an open neighborhood of $\mathbb{S}^2$. Given $f \in C(\mathbb{S}^2)$, define $\widehat{f}$ as follows:

$$\widehat{f}(\boldsymbol{x}) = f\left(\frac{\boldsymbol{x}}{|\boldsymbol{x}|}\right), \quad 1 - \varepsilon < |\boldsymbol{x}| < 1 + \varepsilon \tag{4.51}$$

for some $0 < \varepsilon < 1$. Denote this open neighborhood of $\mathbb{S}^2$ by Ω_ε. The mapping $\widehat{f}$ extends f as a constant in the direction orthogonal to $\mathbb{S}^2$.

Recall the multi-index notation $\boldsymbol{\alpha} = (\alpha_1, \alpha_2, \alpha_3)$ used in (4.1). With it we recall the derivative notation

$$D^\alpha \widehat{f}(\boldsymbol{x}) = \frac{\partial^{|\alpha|} \widehat{f}(x_1, x_2, x_3)}{\partial x_1^{\alpha_1} \partial x_2^{\alpha_2} \partial x_3^{\alpha_3}}. \tag{4.52}$$

It is straightforward to show that

$$f \in C^r(\mathbb{S}^2) \implies \hat{f} \in C^r(\Omega_\varepsilon), \quad r \in \mathbb{N}_0. \tag{4.53}$$

The following is proven in [18, Theorem 2].

Theorem 4.13. *Assume $f \in C^r(\mathbb{S}^2)$ for some $r \in \mathbb{N}$. Then for each $n \in \mathbb{N}_0$, there is a polynomial $p_n \in \Pi_n(\mathbb{R}^3)$ for which*

$$\left\| D^{\boldsymbol{\alpha}} \hat{f} - D^{\boldsymbol{\alpha}} p_n \right\|_{C(\Omega_\varepsilon)} \leq \frac{c}{n^{\ell - |\boldsymbol{\alpha}|}} \sum_{|\gamma| \leq \ell} \| D^\gamma \hat{f} \|_{C(\Omega_\varepsilon)}, \quad |\boldsymbol{\alpha}| \leq \min\{\ell, n\} \tag{4.54}$$

for some constant $c > 0$.

The bound (4.54) for spatial derivatives of $\hat{f}$ can be related back to associated surface derivatives of f via (4.51) and to the norm for $C^k(\mathbb{S}^2)$ given in (4.32). This is omitted here.

4.2.4 Lebesgue Constants

The Laplace expansion (4.10) of $f \in L^2(\mathbb{S}^2)$ is given by

$$f(\boldsymbol{\eta}) = \sum_{k=0}^{\infty} \sum_{\ell=1}^{2k+1} (f, Y_{k,\ell}) Y_{k,\ell}(\boldsymbol{\eta}), \qquad \boldsymbol{\eta} \in \mathbb{S}^2. \tag{4.55}$$

The truncation of this to terms of order $\leq n$ is given by

$$\mathcal{Q}_n f(\boldsymbol{\eta}) = \sum_{\ell=0}^{n} \mathcal{P}_\ell f(\boldsymbol{\eta})$$

$$= \sum_{k=0}^{n} \sum_{\ell=1}^{2k+1} (f, Y_{k,\ell}) Y_{k,\ell}(\boldsymbol{\eta}), \qquad \boldsymbol{\eta} \in \mathbb{S}^2; \tag{4.56}$$

see (4.12). Recalling (4.23), the projection $\mathcal{P}_\ell f$ of (4.11) is given by

$$\mathcal{P}_\ell f(\boldsymbol{\eta}) = \frac{2\ell+1}{4\pi} (f, P_\ell(\boldsymbol{\eta} \cdot \circ))$$

$$= \frac{2\ell+1}{4\pi} \int_{\mathbb{S}^2} f(\boldsymbol{\xi}) P_\ell(\boldsymbol{\eta} \cdot \boldsymbol{\xi}) \, dS^2(\boldsymbol{\xi})$$

with $P_\ell(t)$ the Legendre polynomial of degree ℓ on $[-1, 1]$, defined in (4.8). From [50, p. 1399],

$$\mathcal{Q}_n f(\boldsymbol{\eta}) = \frac{n+1}{4\pi}\left(f, P_n^{(1,0)}(\boldsymbol{\eta}\cdot\circ)\right)$$

$$= \frac{n+1}{4\pi}\int_{\mathbb{S}^2} f(\boldsymbol{\xi})P_n^{(1,0)}(\boldsymbol{\eta}\cdot\boldsymbol{\xi})\,dS^2(\boldsymbol{\xi}),\qquad \boldsymbol{\eta}\in\mathbb{S}^2. \qquad (4.57)$$

The function $P_n^{(1,0)}(t)$ is the Jacobi polynomial of degree n associated with the weight function $w(t) = 1 - t$ on the interval $[-1,1]$; see (4.73) below. The function $\mathcal{Q}_n f$ is the orthogonal projection of f onto the subspace Π_n of $L^2(\mathbb{S}^2)$ and it satisfies

$$\|f - \mathcal{Q}_n f\|_{L^2(\mathbb{S}^2)} = \min_{g\in\Pi_n}\|f - g\|_{L^2(\mathbb{S}^2)}. \qquad (4.58)$$

The orthogonal projection operator $\mathcal{Q}_n$ has norm 1 when regarded as a mapping from $L^2(\mathbb{S}^2)$ to $L^2(\mathbb{S}^2)$.

Let $f \in L^2(\mathbb{S}^2)$. Recalling the definition of best approximation in (4.13), let $p_n \in \Pi_n$ be a best approximation to f in $C(\mathbb{S}^2)$ using the uniform norm. Then

$$f - \mathcal{Q}_n f = f - p_n - \mathcal{Q}_n(f - p_n)$$

$$= (I - \mathcal{Q}_n)(f - p_n). \qquad (4.59)$$

The operator $I - \mathcal{Q}_n$ is also an orthogonal projection and thus has norm 1 as an operator from $L^2(\mathbb{S}^2)$ to $L^2(\mathbb{S}^2)$. Applying the $L^2(\mathbb{S}^2)$ norm and using (4.58),

$$\|f - \mathcal{Q}_n f\|_{L^2(\mathbb{S}^2)} \le \|f - p_n\|_{L^2(\mathbb{S}^2)}$$

$$\le 2\sqrt{\pi}\,\|f - p_n\|_\infty$$

$$\le 2\sqrt{\pi}\,E_{n,\infty}(f). \qquad (4.60)$$

Apply Theorem 4.5 or Theorem 4.11 to obtain a bound on the rate of convergence of $\mathcal{Q}_n f$ to f in $L^2(\mathbb{S}^2)$.

In order to discuss the uniform convergence of $\mathcal{Q}_n f$ on $\mathbb{S}^2$, use the uniform norm $\|\cdot\|_\infty$ and, as before, bound $f - \mathcal{Q}_n f$ via (4.59). Then

$$\|f - \mathcal{Q}_n f\|_\infty \le (1 + \|\mathcal{Q}_n\|_{C\to C})\,\|f - p_n\|_\infty,$$

$$\|f - \mathcal{Q}_n f\|_\infty \le (1 + \|\mathcal{Q}_n\|_{C\to C})\,E_{n,\infty}(f). \qquad (4.61)$$

Thus we obtain uniform bounds for the error in the truncated Laplace series for f by finding a bound for $\|\mathcal{Q}_n\|_{C\to C}$. This is often called the "Lebesgue constant" for the approximation $\mathcal{Q}_n f$.

Recalling (4.57), we notice the projection $\mathcal{Q}_n$ is an integral operator. Bounding $\|\mathcal{Q}_n\|_{C\to C}$ is equivalent to finding the operator norm of this integral operator from $C(\mathbb{S}^2)$ to $C(\mathbb{S}^2)$, a well-known process. In this case,

$$\|\mathcal{Q}_n\|_{C \to C} = \frac{n+1}{4\pi} \max_{\boldsymbol{\eta} \in \mathbb{S}^2} \int_{\mathbb{S}^2} \left| P_n^{(1,0)}(\boldsymbol{\eta} \cdot \boldsymbol{\xi}) \right| dS^2(\boldsymbol{\xi}).$$

Using symmetry in $\boldsymbol{\xi} \in \mathbb{S}^2$, this integral over $\mathbb{S}^2$ is constant and thus it simplifies to

$$\|\mathcal{Q}_n\|_{C \to C} = \frac{n+1}{4\pi} \int_{\mathbb{S}^2} \left| P_n^{(1,0)}(\eta_3) \right| dS^2(\boldsymbol{\eta})$$

$$= \frac{n+1}{2} \int_{-1}^{1} \left| P_n^{(1,0)}(z) \right| dz. \tag{4.62}$$

In Gronwall [54], this is shown to lead to

$$\lim_{n \to \infty} \frac{1}{\sqrt{n}} \|\mathcal{Q}_n\|_{C \to C} = 2 \sqrt{\frac{2}{\pi}}. \tag{4.63}$$

Hence,

$$\|\mathcal{Q}_n\|_{C \sim C} \approx 2 \sqrt{\frac{2}{\pi}} \sqrt{n} \qquad \text{for } n \text{ sufficiently large.} \tag{4.64}$$

The generalization of this to $\mathbb{S}^d$ for any $d \geq 2$ is given in [94].

Corollary 4.14. *Assume that $f \in C^{k,\gamma}(\mathbb{S}^2)$ for some $k \geq 0$ and some $\gamma \in (0,1]$, and further assume that $k + \gamma > \frac{1}{2}$. Then*

$$\|f - \mathcal{Q}_n f\|_\infty \leq \frac{c}{n^{k+\gamma-1/2}}$$

for a suitable constant $c > 0$. Thus the Laplace series (4.55) is uniformly convergent in $C(\mathbb{S}^2)$.

Proof. It is a straightforward combination of (4.49), (4.61), and (4.63). □

Evaluating $\mathcal{Q}_n f$ requires evaluating the integral coefficients $(f, Y_{k,\ell})$. A numerical method for doing so is given in Sect. 5.7.1.

4.2.5 Best Approximation for a Parameterized Family

Consider approximating a function $u(\boldsymbol{\xi}, t)$ with $0 \leq t \leq T$ for some $T > 0$. Approximating such functions occurs when solving problems over $\mathbb{S}^2$ with a dependence on time t. Error bounds often depend on some norm involving both the spatial variable $\boldsymbol{\xi}$ and the time variable t. The earlier error bounds for best approximation can be extended to such approximation over $\mathbb{S}^2$, obtaining polynomials over $\mathbb{S}^2$ that vary with t. We present one approach to doing so by using the construction from Theorem 4.11. As notation, let $u_t = u(\cdot, t)$, $t \in [0, T]$.

Using (4.46), consider

$$\mathcal{T}_n u_t(\boldsymbol{\xi}) = \frac{\displaystyle\int_{\mathbb{S}^2} K_n(\boldsymbol{\xi}\cdot\boldsymbol{\eta})\, u(\boldsymbol{\eta}, t)\, dS^2(\boldsymbol{\eta})}{\displaystyle\int_{\mathbb{S}^2} K_n(\boldsymbol{\xi}\cdot\boldsymbol{\eta})\, dS^2(\boldsymbol{\eta})}, \qquad \boldsymbol{\xi} \in \mathbb{S}^2. \tag{4.65}$$

Then $\mathcal{T}_n u_t \in \Pi_n$ and thus has coefficients that are functions of t. Note that if u is continuously differentiable with respect to t, then we can differentiate $\mathcal{T}_n u_t$ with respect to t, obtaining yet another polynomial in Π_n,

$$\frac{\partial}{\partial t}\left(\mathcal{T}_n u_t(\boldsymbol{\xi})\right) = \frac{\displaystyle\int_{\mathbb{S}^2} K_n(\boldsymbol{\xi}\cdot\boldsymbol{\eta})\, \frac{\partial u(\boldsymbol{\eta}, t)}{\partial t}\, dS^2(\boldsymbol{\eta})}{\displaystyle\int_{\mathbb{S}^2} K_n(\boldsymbol{\xi}\cdot\boldsymbol{\eta})\, dS^2(\boldsymbol{\eta})}, \qquad \boldsymbol{\xi} \in \mathbb{S}^2.$$

This same process can be applied to powers of $\mathcal{T}_n$, and thus to the operator $_k\mathcal{T}_n = I - (I - \mathcal{T}_n)^{k+1}$ introduced in Theorem 4.11. The quantity $_k\mathcal{T}_n u_t$ is a polynomial in Π_n for all $t \in [0, T]$, as is its derivative $\partial\left(_k\mathcal{T}_n u_t\right)/\partial t$.

We can apply (4.48) from Theorem 4.11 to obtain error bounds for $\mathcal{T}_n u_t$ and its derivatives with respect to t:

$$\begin{aligned}
\|u_t -\,_k\mathcal{T}_n u_t\|_\infty &\leq A_k n^{-k} \omega^R\left(u_t^{(k)}; \frac{1}{n+1}\right), \\
\left\|\frac{\partial u_t}{\partial t} - \frac{\partial\left(_k\mathcal{T}_n u_t\right)}{\partial t}\right\|_\infty &\leq A_k n^{-k} \omega^R\left(\left(\frac{\partial u_t}{\partial t}\right)^{(k)}; \frac{1}{n+1}\right).
\end{aligned} \tag{4.66}$$

Using (4.45), we can relate these bounds to the differentiability over $\mathbb{S}^2$ of u_t and $\partial u_t/\partial t$. Assume, u_t, $\partial u_t/\partial t \in C^{k,\gamma}\left(\mathbb{S}^2\right)$, $k \geq 0$ and some $\gamma \in (0, 1]$, for $0 \leq t \leq T$. Further assume that the kth-order derivatives of u_t over $\mathbb{S}^2$ satisfy

$$|g_t(\boldsymbol{\xi}) - g_t(\boldsymbol{\eta})| \leq \widehat{c}_{k,\gamma} \, |\boldsymbol{\xi} - \boldsymbol{\eta}|^\gamma, \qquad \boldsymbol{\xi}, \boldsymbol{\eta} \in \mathbb{S}^2, \quad 0 \leq t \leq T,$$

$$\left|\frac{\partial g_t(\boldsymbol{\xi})}{\partial t} - \frac{\partial g_t(\boldsymbol{\eta})}{\partial t}\right| \leq \widehat{c}_{k,\gamma} \, |\boldsymbol{\xi} - \boldsymbol{\eta}|^\gamma, \qquad \boldsymbol{\xi}, \boldsymbol{\eta} \in \mathbb{S}^2, \quad 0 \leq t \leq T,$$

where g_t is a generic such derivative and $\widehat{c}_{k,\gamma} > 0$ is a constant dependent on u, k, and γ. Combining (4.66) and (4.45),

$$\max\left[\|u_t -\,_k\mathcal{T}_n u_t\|_\infty, \left\|\frac{\partial u_t}{\partial t} - \frac{\partial\left(_k\mathcal{T}_n u_t\right)}{\partial t}\right\|_\infty\right] \leq \frac{B_k \widehat{c}_{k,\gamma}}{(n+1)^{k+\gamma}}, \qquad n \geq 0,\ 0 \leq t \leq T \tag{4.67}$$

for a constant $B_k > 0$ that is independent of u and n.

4.3 Approximation on the Unit Disk

We consider the approximation by multivariate polynomials of functions defined on the unit disk $\mathbb{B}^2 = \{x \in \mathbb{R}^2 : |x| < 1\}$. Let $\Pi_n(\mathbb{R}^2)$ denote that space of all polynomials p in the variable $x = (x_1, x_2)$ of degree less than or equal to n,

$$p(x) = \sum_{i+j \leq n} a_{i,j} x_1^i x_2^j.$$

It is straightforward to show that

$$\dim \Pi_n(\mathbb{R}^2) = \frac{1}{2}(n+1)(n+2).$$

Later in this section we discuss bases for $\Pi_n(\mathbb{R}^2)$ that are orthogonal over $\mathbb{B}^2$.

Given $f \in C(\overline{\mathbb{B}}^2)$, define

$$E_n(f) = \min_{g \in \Pi_n(\mathbb{R}^2)} \|f - g\|_\infty \tag{4.68}$$

Using the Weierstrass Theorem, $E_n(f) \to 0$ as $n \to \infty$. A natural question is to ask how this convergence is affected by the differentiability of f. One answer to this was given in Ragozin [93]. We begin with some notation. For partial derivatives of a function f, we proceed in analogy with (4.52):

$$D^\alpha f(x) = \frac{\partial^{|\alpha|} f(x_1, x_2)}{\partial x_1^{\alpha_1} \partial x_2^{\alpha_2}}, \qquad \alpha = (\alpha_1, \alpha_2).$$

Define

$$\|f\|_r = \sum_{|\alpha| \leq r} \|D^\alpha f\|_\infty,$$

$$\omega(f; h) = \sup_{|x-y| \leq h} |f(x) - f(y)|,$$

$$\omega(f^{(r)}; h) = \sum_{|\alpha|=r} \omega(D^\alpha f; h).$$

The following result is from Ragozin [93, Theorem 3.4].

Theorem 4.15. *Given* $f \in C^r(\overline{\mathbb{B}}^2)$,

$$E_n(f) \leq \frac{C(r)}{n^r} \left(\frac{\|f\|_r}{n} + \omega\left(f^{(r)}; \frac{1}{n}\right) \right), \qquad n \geq 1, \tag{4.69}$$

with $C(r) \geq 0$ *dependent only on* r.

If all derivatives $D^\alpha f$ with $|\alpha| = r$ satisfy a Hölder condition with exponent $\gamma \in (0, 1]$, then it follows immediately from (4.69) that

$$E_n(f) = \mathcal{O}(n^{-(r+\gamma)}). \qquad (4.70)$$

Additional results on best approximation in $L^p(\mathbb{B}^2)$ are given in Dai and Xu [36, Part II], and an extension of (4.70) is given in [37, Corollary 5.9].

For simultaneous approximation of f and some its low order derivatives, we have the following result from Bagby, Bos, and Levenberg [18, Theorem 1].

Theorem 4.16. *Assume $f \in C^{r+m}(\overline{\mathbb{B}}^2)$ for some $r, m \geq 0$. Then*

$$\inf_{p \in \Pi_n(\mathbb{R}^2)} \max_{|\alpha| \leq r} \|D^\alpha f - D^\alpha p\|_\infty \leq \frac{c(f, m)}{n^m} \omega_{m+r}(f, 1/n), \qquad n \geq 1$$

with

$$\omega_{m+r}(f, \delta) = \max_{|\alpha| = m+r} \sup_{|\boldsymbol{x}-\boldsymbol{y}| \leq \delta} |D^\alpha f(\boldsymbol{x}) - D^\alpha f(\boldsymbol{y})|.$$

4.3.1 Orthogonal Polynomials

We proceed in analogy with the decomposition (4.5) for $\Pi_n(\mathbb{S}^2)$. Let $\mathcal{V}_0$ denote the set of constant functions on $\mathbb{B}^2$, and define

$$\mathcal{V}_n = \left\{ p \in \Pi_n(\mathbb{R}^2) : p \perp \Pi_{n-1}(\mathbb{R}^2) \right\}, \quad n \geq 1,$$

$$p \perp \Pi_{n-1}(\mathbb{R}^2) \iff (p, q) = 0 \quad \forall q \in \Pi_{n-1}(\mathbb{R}^2),$$

$$(p, q) = \int_{\mathbb{B}^2} p(\boldsymbol{y})\, q(\boldsymbol{y})\, d\boldsymbol{y}. \qquad (4.71)$$

Then we can write

$$\Pi_n(\mathbb{R}^2) = \mathcal{V}_0 \oplus \mathcal{V}_1 \oplus \cdots \oplus \mathcal{V}_n. \qquad (4.72)$$

To generate an orthonormal basis for $\Pi_n(\mathbb{B}^2)$ using the inner product of (4.71), we must generate an orthonormal basis for the space $\mathcal{V}_k$ and show it is orthogonal to $\Pi_{k-1}(\mathbb{B}^2)$, $k \geq 1$.

The dimension of $\mathcal{V}_k$ is

$$\frac{1}{2}(k+1)(k+2) - \frac{1}{2}k(k+1) = k+1.$$

For $k \geq 1$, there is no unique orthonormal basis for $\mathcal{V}_k$. There are a number of well-known orthonormal bases, and we give two of them. An excellent development of this topic is given in Xu [121], along with connections to orthonormal bases over other standard regions, including the sphere, and to weighted L_2 spaces over $\mathbb{B}^2$.

We begin with some notation and results for some important orthogonal polynomials of one variable. As additional references, we refer the reader to [5, Chap. 5], [9, Sect. 4.4], [52, 115]. In addition, see the handbooks [1, Chap. 22] and [89, Chap. 18].

The *Jacobi polynomials* $\left\{ P_n^{(\alpha,\beta)} : n \geq 0 \right\}$ are the polynomials that are *orthogonal* on $[-1, 1]$ with respect to the inner product

$$(f, g) = \int_{-1}^{1} f(t) \, g(t) \, (1 - t)^{\alpha} \, (1 + t)^{\beta} \, dt.$$

It is assumed that $\alpha, \beta > -1$. There are various ways of writing the Jacobi polynomials. For example,

$$P_n^{(\alpha,\beta)}(t) = \frac{(-1)^n}{2^n \, n!} \, (1 - t)^{-\alpha} \, (1 + t)^{-\beta} \frac{d^n}{dt^n} \left[(1 - t)^{\alpha+n} \, (1 + t)^{\beta+n} \right]. \quad (4.73)$$

The Jacobi polynomials are normalized by requiring

$$P_n^{(\alpha,\beta)}(1) = \binom{n + \alpha}{n}, \quad n = 0, 1, \ldots.$$

We have

$$\int_{-1}^{1} \left[P_n^{(\alpha,\beta)}(t) \right]^2 (1 - t)^{\alpha} \, (1 + t)^{\beta} \, dt$$

$$= \frac{2^{\alpha+\beta+1}}{2n + \alpha + \beta + 1} \frac{\Gamma(n + \alpha + 1) \, \Gamma(n + \beta + 1)}{n! \, \Gamma(n + \alpha + \beta + 1)}.$$

There are several important special cases, with all being a constant multiple of an appropriate Jacobi polynomial of the same degree.

- The *Legendre polynomials*, $P_n(t)$. Take $\alpha = \beta = 0$. These are discussed in extensive detail in Sect. 2.7, and we review here only a very few of their properties. The inner product associated with the definition of orthogonality is simply

$$(f, g) = \int_{-1}^{1} f(t) \, g(t) \, dt.$$

Then

$$P_n(t) = \frac{(-1)^n}{2^n \, n!} \frac{d^n}{dt^n} \left[\left(1 - t^2\right)^n \right], \qquad n = 1, 2, \ldots \qquad (4.74)$$

with $P_0(t) \equiv 1$; recall (4.8). Note that $P_n(t) \equiv P_{n,3}(t)$ in the notation of Chap. 2. They are normalized with $P_n(1) = 1$, $n \geq 0$. Also,

$$\int_{-1}^{1} [P_n(t)]^2 \, dt = \frac{2}{2n+1}, \qquad n = 0, 1, \ldots.$$

The triple recursion relation is

$$P_{n+1}(t) = \frac{2n+1}{n+1} t P_n(t) - \frac{n}{n+1} P_{n-1}(t), \qquad n = 1, 2, \ldots.$$

- The *Chebyshev polynomials of the first kind*, $T_n(t)$. Take $\alpha = \beta = -\frac{1}{2}$. The inner product is

$$(f, g) = \int_{-1}^{1} \frac{f(t) \, g(t)}{\sqrt{1 - t^2}} \, dt.$$

Then

$$T_n(t) = \cos(n\theta), \qquad t = \cos\theta, \qquad n = 0, 1, \ldots.$$

Note that $T_n(t) \equiv P_{n,2}(t)$ in the notation of Chap. 2. They are normalized with $T_n(1) = 1$. Also,

$$\int_{-1}^{1} \frac{[T_n(t)]^2}{\sqrt{1 - t^2}} \, dt = \begin{cases} \frac{1}{2}\pi, & n \geq 1, \\ \pi, & n = 0, \end{cases}$$

$$T_n(t) = \frac{P_n^{(-1/2,-1/2)}(t)}{P_n^{(-1/2,-1/2)}(1)}.$$

The triple recursion relation is

$$T_{n+1}(t) = 2t \, T_n(t) - T_{n-1}(t), \qquad n = 1, 2, \ldots.$$

- The *Chebyshev polynomials of the second kind*, $U_n(t)$. Take $\alpha = \beta = \frac{1}{2}$. The inner product is

$$(f, g) = \int_{-1}^{1} f(t) \, g(t) \sqrt{1 - t^2} \, dt.$$

Then·

$$U_n(t) = \frac{\sin(n+1)\theta}{\sin\theta}, \quad t = \cos\theta, \ n = 0, 1, \ldots.$$

They are normalized with $U_n(1) = n + 1$ for $n \geq 0$. Also,

$$\int_{-1}^{1} [U_n(t)]^2 \sqrt{1 - t^2}\, dt = \frac{1}{2}\pi, \qquad n = 0, 1, \ldots,$$

$$U_n(t) = (n+1)\frac{P_n^{(1/2,1/2)}(t)}{P_n^{(1/2,1/2)}(1)}.$$

The triple recursion relation is

$$U_{n+1}(t) = 2t\, U_n(t) - U_{n-1}(t), \qquad n = 1, 2, \ldots. \tag{4.75}$$

- The *Gegenbauer polynomials*, $C_{n,\lambda}(t)$ with $\lambda > -\frac{1}{2}$, $\lambda \neq 0$. These polynomials are briefly discussed in Sect. 2.9. Take $\alpha = \beta = \lambda - \frac{1}{2}$. The inner product is

$$(f, g) = \int_{-1}^{1} f(t)\, g(t) \left(1 - t^2\right)^{\lambda - \frac{1}{2}}\, dt$$

Then define

$$C_{n,\lambda}(t) = \frac{(2\lambda)_n}{\left(\lambda + \frac{1}{2}\right)_n} P_n^{(\lambda - 1/2, \lambda - 1/2)}(t)$$

$$= \frac{(2\lambda)_n}{(-2)^n \left(\lambda + \frac{1}{2}\right)_n n!} \left(1 - t^2\right)^{\frac{1}{2} - \lambda} \frac{d^n}{dt^n}\left[\left(1 - t^2\right)^{\lambda - \frac{1}{2} + n}\right],$$

$n = 0, 1, \ldots$, where Pochhammer's symbol $(\gamma)_n$ is used. They are normalized with the condition

$$C_{n,\lambda}(1) = \binom{n + 2\lambda - 1}{n}$$

for $n \geq 1$; and $C_{0,\lambda}(1) = 1$. Also,

$$\int_{-1}^{1} \left(1 - t^2\right)^{\lambda - \frac{1}{2}} [C_{n,\lambda}(t)]^2\, dt = \frac{\pi 2^{1-2\lambda}\Gamma(n + 2\lambda)}{n!\, (n + \lambda)\, [\Gamma(\lambda)]^2}.$$

The triple recursion relation is

$$C_{n+1,\lambda}(t) = \frac{2(n+\lambda)}{n+1} C_{n,\lambda}(t) - \frac{n + 2\lambda - 1}{n+1} C_{n-1,\lambda}(t), \quad n = 1, 2, \ldots.$$

Example 4.17. Returning to the definition of an orthonormal basis for the subspace for $\mathcal{V}_n$, we generally denote it by $\{\varphi_0^n, \varphi_1^n, \ldots, \varphi_n^n\}$.

For $n = 0, 1, \ldots$ and $k = 0, 1, \ldots, n$, define

$$\varphi_k^n(\boldsymbol{x}) = h_{k,n} C_{n-k,k+1}(x_1) \left(1 - x_1^2\right)^{k/2} C_{k,1/2}\left(\frac{x_2}{\sqrt{1 - x_1^2}}\right), \qquad (4.76)$$

with $h_{k,n}$ so chosen that $\|\varphi_k^n\|_{L^2} = 1$,

$$h_{k,n} = 2^k k! \left[\frac{(n-k)!\,(2k+1)\,(n+1)}{\pi\,(n+k+1)!}\right]^{\frac{1}{2}}.$$

To prove that

$$(\varphi_k^n, \varphi_\ell^m) = 0, \quad (n, k) \neq (m, \ell),$$

and to obtain the formula for $h_{k,n}$, apply the following identity suggested by [121, p. 138],

$$\int_{\mathbb{B}^2} f(\boldsymbol{x})\, d\boldsymbol{x} = \int_{-1}^{1} \int_{-\sqrt{1-x_1^2}}^{\sqrt{1-x_1^2}} f(x_1, x_2)\, dx_1\, dx_2$$

$$= \int_{-1}^{1} \int_{-1}^{1} f\left(x_1, s\sqrt{1 - x_1^2}\right) \sqrt{1 - x_1^2}\, dx_1\, ds$$

and use the identities given earlier for the Gegenbauer polynomials. Figure 4.1 contains a graph of $\varphi_4^5(\boldsymbol{x})$ over $\mathbb{B}^2$. □

Example 4.18. For $n = 0, 1, \ldots$ and $k = 0, 1, \ldots, n$, define

$$\varphi_k^n(\boldsymbol{x}) = \frac{1}{\sqrt{\pi}} U_n\left(x_1 \cos\frac{k\pi}{n+1} + x_2 \sin\frac{k\pi}{n+1}\right). \qquad (4.77)$$

This family of polynomials was introduced in [76]. These are sometimes called "ridge polynomials" as they are constant along the lines

$$x_1 \cos\frac{k\pi}{n+1} + x_2 \sin\frac{k\pi}{n+1} = constant,$$

looking oscillatory along a perpendicular to this line. As an illustrative example, Fig. 4.2 contains a graph of $\varphi_4^5(\boldsymbol{x})$ over $\mathbb{B}^2$. This family is particularly useful when there is a need to calculate partial derivatives of the basis functions. Recurrence formulas for low order derivatives of U_n can be obtained by differentiating the recurrence relation (4.75). □

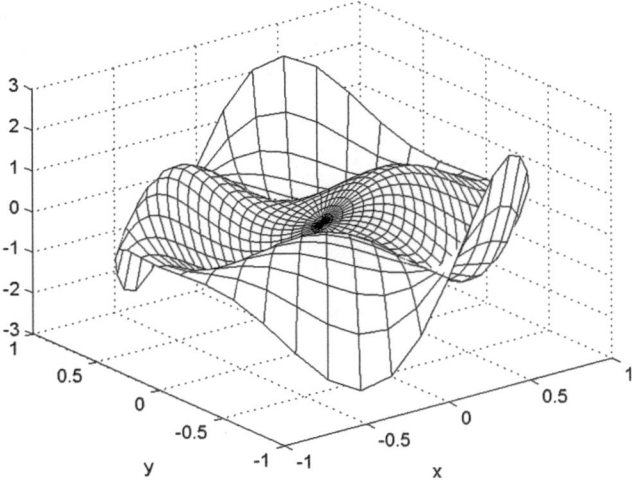

Fig. 4.1 The polynomial $\varphi_4^5(\boldsymbol{x})$ defined in (4.76)

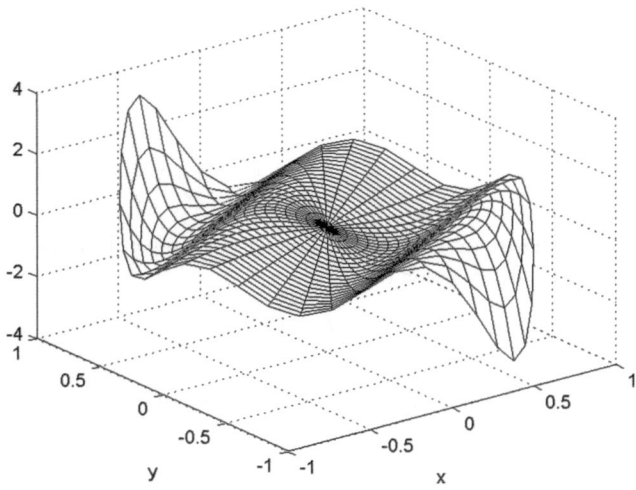

Fig. 4.2 The ridge polynomial $\varphi_4^5(\boldsymbol{x})$

Another family of orthonormal polynomials over $\mathbb{B}^2$ is given in [121, Sect. 1.2.2]. And yet another family of orthonormal polynomials are the Zernike polynomials; see [117, p. 3234].

4.3.2 Properties of Orthogonal Polynomials over $\mathbb{B}^2$

Orthogonal polynomials of one variable have a number of special properties; for example, they satisfy a triple recurrence relation, as is illustrated with the orthogonal polynomials given earlier. Some of these properties generalize to the multivariable case, and a thorough discussion of this topic is given in Xu [121]. We present one such generalization from [121]: the triple recursion relation.

We begin by introducing some additional notation. Let

$$\Phi_n(\boldsymbol{x}) = [\varphi_0^n, \varphi_1^n, \dots, \varphi_n^n]^{\mathrm{T}}, \quad n = 0, 1, \dots.$$

It is important to note that even though there are many possible bases for each subspace $\mathcal{V}_n$, the subspace decomposition (4.72) for $\Pi_n(\mathbb{R}^2)$ is invariant. By looking for a relationship between the subspaces $\mathcal{V}_n$, it is possible to generalize the proof that is used often in deriving the triple recurrence relation for univariate orthogonal polynomials.

For $n \geq 1$, multiply $\Phi_n(\boldsymbol{x})$ by x_i for $i = 1, 2$. Then each component of $x_i\Phi_n(\boldsymbol{x})$ will be a polynomial of degree $n + 1$. As such, each component can be expanded using the orthonormal basis $\{\varphi_\ell^m : 0 \leq \ell \leq m, \ 0 \leq m \leq n+1\}$ for $\Pi_{n+1}(\mathbb{R}^2)$. Using the orthonormality of our basis functions and the decomposition

$$\Pi_{n+1}(\mathbb{R}^2) = \mathcal{V}_0 \oplus \mathcal{V}_1 \oplus \cdots \oplus \mathcal{V}_{n+1},$$

we have

$$x_i\varphi_k^n(\boldsymbol{x}) = \sum_{m=0}^{n+1} \sum_{\ell=0}^{m} (x_i\varphi_k^n, \varphi_\ell^m)\, \varphi_\ell^m(\boldsymbol{x}). \tag{4.78}$$

Consider each coefficient

$$(x_i\varphi_k^n, \varphi_\ell^m) = \int_{\mathbb{B}^2} x_i\varphi_k^n(\boldsymbol{x})\, \varphi_\ell^m(\boldsymbol{x})\, d\boldsymbol{x}.$$

The function $x_i\varphi_\ell^m(\boldsymbol{x})$ is a polynomial of degree $m + 1$. For $m \leq n - 2$, use the orthogonality of $\varphi_k^n(\boldsymbol{x})$ to $\mathcal{V}_0, \dots, \mathcal{V}_{n-1}$ to obtain

$$(x_i\varphi_k^n, \varphi_\ell^m) = 0, \qquad 0 \leq \ell \leq m, \ 0 \leq m \leq n - 2.$$

Returning to (4.78), we have

$$x_i\varphi_k^n(\boldsymbol{x}) = \sum_{m=n-1}^{n+1} \sum_{\ell=0}^{m} (x_i\varphi_k^n, \varphi_\ell^m)\, \varphi_\ell^m(\boldsymbol{x}).$$

In vector notation,

$$x_i \Phi_n(\boldsymbol{x}) = A_{n,i} \Phi_{n+1}(\boldsymbol{x}) + B_{n,i} \Phi_n(\boldsymbol{x}) + C_{n,i} \Phi_{n-1}(\boldsymbol{x}), \qquad i = 1, 2. \quad (4.79)$$

The matrix coefficients are given by

$$A_{n,i} = \int_{\mathbb{B}^2} x_i \Phi_n(\boldsymbol{x}) \, \Phi_{n+1}^{\mathrm{T}}(\boldsymbol{x}) \, d\boldsymbol{x},$$

$$B_{n,i} = \int_{\mathbb{B}^2} x_i \Phi_n(\boldsymbol{x}) \, \Phi_n^{\mathrm{T}}(\boldsymbol{x}) \, d\boldsymbol{x},$$

$$C_{n,i} = \int_{\mathbb{B}^2} x_i \Phi_n(\boldsymbol{x}) \, \Phi_{n-1}^{\mathrm{T}}(\boldsymbol{x}) \, d\boldsymbol{x} = A_{n-1,i}^{\mathrm{T}}.$$

See Dunkl and Xu [44] and Xu [121] for a complete introduction to multivariable orthogonal polynomials on the unit sphere, the unit ball, and the unit simplex in $\mathbb{R}^d$, $d \geq 2$.

4.3.3 Orthogonal Expansions

The set of all polynomials over $\mathbb{R}^2$ are dense in $L^2(\mathbb{B}^2)$; this follows from the Weierstrass theorem and the denseness of $C(\overline{\mathbb{B}}^2)$ in $L^2(\mathbb{B}^2)$. As a consequence, we consider the orthonormal expansion

$$f(\boldsymbol{x}) = \sum_{m=0}^{\infty} \sum_{k=0}^{m} (f, \varphi_k^m) \, \varphi_k^m(\boldsymbol{x}), \quad \boldsymbol{x} \in \mathbb{B}^2. \quad (4.80)$$

This is convergent in the norm of $L^2(\mathbb{B}^2)$. The function

$$\mathcal{Q}_n f(\boldsymbol{x}) = \sum_{m=0}^{n} \sum_{k=0}^{m} (f, \varphi_k^m) \, \varphi_k^m(\boldsymbol{x}), \quad \boldsymbol{x} \in \mathbb{B}^2 \quad (4.81)$$

is the orthogonal projection of f onto the subspace $\Pi_n(\mathbb{R}^2)$ of $L^2(\mathbb{B}^2)$. This means that

$$\|f - \mathcal{Q}_n f\|_{L^2(\mathbb{B}^2)} = \min_{g \in \Pi_n(\mathbb{R}^2)} \|f - g\|_{L^2(\mathbb{B}^2)}. \quad (4.82)$$

The projection $\mathcal{Q}_n$ has norm 1 as an operator from $L^2(\mathbb{B}^2)$ to $L^2(\mathbb{B}^2)$.

Assume $f \in C(\mathbb{B}^2)$. Recalling (4.68), let $p_n \in \Pi_n(\mathbb{R}^2)$ denote a best approximation for which $E_{n,\infty}(f) = \|f - p_n\|_\infty$. Then

$$f - \mathcal{Q}_n f = (I - \mathcal{Q}_n)(f - p_n) \quad (4.83)$$

because $\mathcal{Q}_n p_n = p_n$. The operator $I - \mathcal{Q}_n$ is also an orthogonal projection and has norm 1. Applying the $L^2(\mathbb{B}^2)$-norm and using (4.82),

$$\|f - \mathcal{Q}_n f\|_{L^2(\mathbb{B}^2)} \leq \|f - p_n\|_{L^2(\mathbb{B}^2)}$$
$$\leq \sqrt{\pi}\,\|f - p_n\|_\infty$$
$$= \sqrt{\pi}\,E_{n,\infty}(f).$$

Theorem 4.15 can then be applied to bound the order of convergence of $\mathcal{Q}_n f$ to f in $L^2(\mathbb{B}^2)$.

To obtain a bound for the rate of uniform convergence of $\mathcal{Q}_n f$ to f, we proceed similarly. In particular, we have from (4.83) that

$$\|f - \mathcal{Q}_n f\|_\infty \leq (1 + \|\mathcal{Q}_n\|_{C \to C})\, E_{n,\infty}(f). \tag{4.84}$$

We need to bound $\|\mathcal{Q}_n\|_{C \to C}$, the operator norm for $\mathcal{Q}_n$ when it is regarded as a linear operator from $C(\mathbb{B}^2)$ to $C(\mathbb{B}^2)$. From [120, (1.2)],

$$\mathcal{Q}_n f(\boldsymbol{x}) = \int_{\mathbb{B}^2} G_n(\boldsymbol{x}, \boldsymbol{y})\, f(\boldsymbol{y})\, d\boldsymbol{y}, \qquad \boldsymbol{x} \in \mathbb{B}^2,$$

$$G_n(\boldsymbol{x}, \boldsymbol{y}) = \frac{\sqrt{\pi}\,(n+2)!}{4\,\Gamma(n+\frac{3}{2})} \int_0^\pi P_n^{(\frac{3}{2},\frac{1}{2})}\left(\boldsymbol{x} \cdot \boldsymbol{y} + \sqrt{1 - |\boldsymbol{x}|^2}\sqrt{1 - |\boldsymbol{y}|^2}\cos\psi\right) d\psi.$$

In addition,

$$\|\mathcal{Q}_n\|_{C \to C} = \max_{\boldsymbol{x} \in \mathbb{B}^2} \int_{\mathbb{B}^2} |G_n(\boldsymbol{x}, \boldsymbol{y})|\, d\boldsymbol{y}$$
$$= \mathcal{O}(n) \tag{4.85}$$

with the latter result taken from [120, Theorem 1.1].

Corollary 4.19. *Assume $f \in C^r(\mathbb{B}^2)$ with $r \geq 1$; and further assume that the rth-derivatives satisfy a Hölder condition with exponent $\alpha \in (0, 1]$. Then $\mathcal{Q}_n f$ converges uniformly to f on $\mathbb{S}^2$, and*

$$\|f - \mathcal{Q}_n f\|_\infty \leq \frac{c}{n^{r-1+\alpha}} \tag{4.86}$$

for some constant $c \geq 0$.

Proof. Combine (4.70), (4.84), and (4.85). □

Evaluating $\mathcal{Q}_n f$ requires computing the integral coefficients (f, φ_k^m). A numerical method for doing so is given in Sect. 5.7.2.

Chapter 5
Numerical Quadrature

In this chapter we discuss numerical approximation of the integral

$$I(f) = \int_{\mathbb{S}^2} f(\boldsymbol{\eta}) \, dS^2(\boldsymbol{\eta}). \tag{5.1}$$

The integrand f can be well-behaved or singular, although our initial development assumes f is continuous and, usually, several times continuously differentiable. Such integrals occur in a wide variety of physical applications; and the calculation of the coefficients in a Laplace series expansion of a given function (see (4.55)) requires evaluating such integrals.

Another important source of such integrals (5.1) comes from transforming an integral over a more general surface to an integral over $\mathbb{S}^2$. Assume we have a mapping

$$\mathcal{M} : \mathbb{S}^2 \xrightarrow[onto]{1-1} \partial\Omega, \tag{5.2}$$

where Ω is a simply connected region in $\mathbb{R}^3$ and $\partial\Omega$ denotes its boundary. This mapping is usually assumed to be continuously differentiable, thus eliminating regions Ω for which the boundary has corners or edges. Using this mapping, the integral of a function g over $\partial\Omega$ can be transformed to an integral over $\mathbb{S}^2$. In particular,

$$\int_{\partial\Omega} g(\boldsymbol{x}) \, d\sigma(\boldsymbol{x}) = \int_{\mathbb{S}^2} g(\mathcal{M}(\boldsymbol{\eta})) \, J_{\mathcal{M}}(\boldsymbol{\eta}) \, dS^2(\boldsymbol{\eta}), \tag{5.3}$$

where $J_{\mathcal{M}}(\boldsymbol{\eta})$ denotes the absolute value of the Jacobian of the mapping $\mathcal{M}$. As an aid to the construction of this Jacobian, see the appendix in [11].

In addition to the notation $\boldsymbol{x} = (x_1, x_2, x_3)^T$, we will also write $\boldsymbol{x} = (x, y, z)^T$ on occasions; and in two dimensions, we use $\boldsymbol{x} = (x_1, x_2)^T$ or $\boldsymbol{x} = (x, y)^T$.

K. Atkinson and W. Han, *Spherical Harmonics and Approximations on the Unit Sphere: An Introduction*, Lecture Notes in Mathematics 2044, DOI 10.1007/978-3-642-25983-8_5, © Springer-Verlag Berlin Heidelberg 2012

Example 5.1. Let Ω denote the ellipsoid with boundary

$$\left(\frac{x}{a}\right)^2 + \left(\frac{y}{b}\right)^2 + \left(\frac{z}{c}\right)^2 = 1 \qquad (5.4)$$

for some constants $a, b, c > 0$. The mapping $\mathcal{M}$ is given by

$$\mathcal{M} : \boldsymbol{\eta} \mapsto (a\eta_1, b\eta_2, c\eta_3), \quad \boldsymbol{\eta} \in \mathbb{S}^2. \qquad (5.5)$$

Its Jacobian is given by

$$J_{\mathcal{M}}(\boldsymbol{\eta}) = \sqrt{(bc\eta_1)^2 + (ac\eta_2)^2 + (ab\eta_3)^2}. \qquad \square$$

We refer to formulas for the numerical approximation of $I(f)$ as quadrature formulas or numerical integration formulas. There are a variety of approaches to developing such quadrature formulas and we examine several of them in this chapter. We begin in Sect. 5.1 with methods based on $I(f)$ being represented as a double integral using spherical coordinates, followed by application of suitably chosen single variable quadrature schemes. In Sect. 5.2 we look at composite methods that are based on giving a mesh on $\mathbb{S}^2$ and then using a simple formula for quadrature over each element of the mesh. In Sect. 5.3 we discuss high order methods that generalize the concept of Gaussian quadrature for functions of a single variable. In Sect. 5.4 we give a simple approach to the numerical integration of empirical data. In Sect. 5.5 we discuss a numerical integration method for integrands containing a point singularity, and in Sect. 5.6 a brief discussion is given of numerical approximation of integrals defined on the unit disk in the plane. The chapter concludes in Sect. 5.7 with a discussion of the use of numerical integration to approximate the truncated Laplace expansion on $\mathbb{S}^2$ and the truncated orthogonal polynomial expansion on the unit disk $\mathbb{B}^2$ in $\mathbb{R}^2$, which are concepts introduced in Chap. 4. In this chapter we consider mainly the approximation of integrals on $\mathbb{S}^2$, and for that reason, we simplify our notation for spherical polynomials by using Π_n rather than $\Pi_n(\mathbb{S}^2)$.

For a summary of multivariate quadrature, see the book [111] and the papers [32], [33].

5.1 The Use of Univariate Formulas

The easiest approach to approximating $I(f)$ begins by using spherical coordinates,

$$\boldsymbol{\eta} \mapsto (\cos\phi\sin\theta, \sin\phi\sin\theta, \cos\theta), \quad 0 \le \phi \le 2\pi, \quad 0 \le \theta \le \pi,$$

leading to

$$I(f) = \int_0^{2\pi} \int_0^{\pi} f(\cos\phi\sin\theta, \sin\phi\sin\theta, \cos\theta) \sin\theta \, d\theta \, d\phi. \qquad (5.6)$$

We can now use single variable numerical integration on each of the iterated integrals.

Since the integrand is periodic in ϕ with period 2π, it is sensible to use the trapezoidal rule with uniform spacing,

$$\tilde{I}(g) \equiv \int_0^{2\pi} g(\phi)\, d\phi \approx \tilde{I}_m(g) \equiv h \sum_{j=0}^{m}{}'' g(jh), \quad h = \frac{2\pi}{m}. \tag{5.7}$$

The double-prime notation implies that the first and last terms should be halved before the summation is computed. Since g is periodic over $[0, 2\pi]$, this simplifies to

$$\tilde{I}_m(g) \equiv h \sum_{j=0}^{m-1} g(jh) = h \sum_{j=1}^{m} g(jh).$$

Later we need the following standard result for the trapezoidal method.

Lemma 5.2. *For $m \geq 2$, $k \geq 0$,*

$$\int_0^{2\pi} \cos(k\phi)\, d\phi = \begin{cases} 2\pi, & k = 0, \\ 0, & k > 0, \end{cases} \tag{5.8}$$

$$\frac{2\pi}{m} \sum_{j=0}^{m-1} \cos\left(k\frac{2j\pi}{m}\right) = \begin{cases} 2\pi, & k = 0, m, 2m, \ldots, \\ 0, & \text{otherwise,} \end{cases} \tag{5.9}$$

$$\int_0^{2\pi} \sin(k\phi)\, d\phi = \frac{2\pi}{m} \sum_{j=1}^{m-1} \sin\left(k\frac{2j\pi}{m}\right) = 0, \qquad k \geq 0.$$

Proof. For the trapezoidal rule sums, combine the identities

$$\cos(\omega) = \frac{1}{2}\left(e^{i\omega} + e^{-i\omega}\right),$$

$$\sin(\omega) = \frac{1}{2i}\left(e^{i\omega} - e^{-i\omega}\right)$$

with the summation formula for a finite geometric series.   □

The trapezoidal rule converges quite rapidly in the case that g is a smooth function. To discuss convergence for periodic functions, we introduce suitable function spaces. Begin by introducing the Fourier series of a function $f \in L^2(0, 2\pi)$:

$$f(\phi) = \sum_{k=-\infty}^{\infty} a_k \psi_k(\phi), \qquad \psi_k(\phi) = \frac{1}{\sqrt{2\pi}} e^{ik\phi},$$

$$a_k = \int_0^{2\pi} f(\omega) \, \overline{\psi_k(\omega)} \, d\omega. \tag{5.10}$$

For any real number $q \geq 0$, define $H^q(2\pi)$ to be the set of all functions $g \in L^2(0, 2\pi)$ for which

$$\|f\|_q \equiv \sqrt{|a_0|^2 + \sum_{\substack{k=-\infty \\ k \neq 0}}^{\infty} |k|^{2q} \, |a_k|^2} < \infty,$$

where (5.10) is the Fourier series for f. The space $H^q(2\pi)$ is a Hilbert space, and the inner product associated with it is given by

$$(f, g)_q = a_0 \overline{b}_0 + \sum_{\substack{k=-\infty \\ k \neq 0}}^{\infty} |k|^{2q} \, a_k \overline{b}_k,$$

where f and g have the Fourier series with coefficients $\{a_k\}$ and $\{b_k\}$, respectively. When q is an integer, the norm for $H^q(2\pi)$ is equivalent to the norm for $H^q(0, 2\pi)$. The following is a well-known result; for a proof, see [13, p. 316].

Theorem 5.3. *Assume $q > \frac{1}{2}$, and let $g \in H^q(2\pi)$. Then*

$$\left| \widetilde{I}(g) - \widetilde{I}_m(g) \right| \leq \frac{\sqrt{4\pi \zeta(2q)}}{m^q} \|g\|_q, \quad m \geq 1, \tag{5.11}$$

where ζ denotes the zeta function,

$$\zeta(s) = \sum_{j=1}^{\infty} \frac{1}{j^s}.$$

The remaining integral in (5.6) for $0 \leq \theta \leq \pi$ is more problematic. To obtain an efficient numerical integration method for (5.6), the integration in θ must be chosen with some care. Begin by using the change of variable $z = \cos \theta$ in (5.6),

$$I(f) = \int_0^{2\pi} \int_{-1}^{1} f\left(\cos \phi \sqrt{1 - z^2}, \sin \phi \sqrt{1 - z^2}, z \right) dz \, d\phi.$$

Then apply Gauss–Legendre quadrature to the integration over $-1 \leq z \leq 1$.

More precisely, given $n > 1$, apply the trapezoidal rule with $m = 2n$ subdivisions to the integration in ϕ, and apply Gauss–Legendre quadrature with n nodes to the integration in z over $[-1, 1]$. Let

$$h = \frac{\pi}{n}, \qquad \phi_j = jh, \quad j = 0, 1, \ldots, 2n.$$

Let $\{z_1, \ldots, z_n\}$ and $\{w_1, \ldots, w_n\}$ denote the Gauss–Legendre nodes and weights, respectively, over $[-1, 1]$. Then define

$$I_n(f) = h \sum_{j=0}^{2n-1} \sum_{k=1}^{n} w_k f\left(\cos\phi_j \sqrt{1 - z_k^2}, \sin\phi_j \sqrt{1 - z_k^2}, z_k\right)$$

$$= h \sum_{j=0}^{2n-1} \sum_{k=1}^{n} w_k f(\cos\phi_j \sin\theta_k, \sin\phi_j \sin\theta_k, \cos\theta_k), \tag{5.12}$$

where $z_j = \cos\theta_j$, $j = 1, \ldots, n$. We call this a "product Gaussian quadrature formula".

Theorem 5.4. *Assume $f \in \Pi_{2n-1}$, a spherical polynomial of degree less than or equal to $2n-1$. Then $I(f) = I_n(f)$. In addition, for $f(x, y, z) = z^{2n}$, $I(f) \neq I_n(f)$.*

Proof. Let

$$f(x, y, z) = x^r y^s z^t, \quad r + s + t \leq 2n - 1.$$

Using spherical coordinates,

$$(x, y, z) = (\cos\phi \sin\theta, \sin\phi \sin\theta, \cos\theta),$$

we obtain

$$I = \int_{\mathbb{S}^2} x^r y^s z^t \, dU = J^{r,s} K^{r,s,t},$$

$$J^{r,s} = \int_0^{2\pi} \cos^r \phi \sin^s \phi \, d\phi,$$

$$K^{r,s,t} = \int_0^{\pi} \sin^{r+s+1} \theta \cos^t \theta \, d\theta.$$

For the corresponding numerical integral, we have

$$I_m = \sum_{j=1}^{2n} \sum_{k=1}^{n} w_k x_{j,k}^r y_{j,k}^s z_k^t = J_n^{r,s} K_n^{r,s,t},$$

$$J_n^{r,s} = h \sum_{j=1}^{2n} \cos^r \phi_j \sin^s \phi_j,$$

$$K_n^{r,s,t} = \sum_{k=1}^{n} w_k \sin^{r+s+1} \theta_k \cos^t \theta_k.$$

The numbers $\{x_{j,k}, y_{j,k}, z_k\}$ refer to the cartesian coordinates of the points on the sphere obtained by using the spherical coordinates $\{\phi_j\}$ and $\{\theta_k\}$,

$$(x_{j,k}, y_{j,k}, z_k) = (\cos\phi_j \sin\theta_k, \sin\phi_j \sin\theta_k, \cos\theta_k).$$

We are interested in analyzing the error

$$E_n = I - I_n = J^{r,s} K^{r,s,t} - J_n^{r,s} K_n^{r,s,t}.$$

At this point we break into cases.

Assume now that r is odd and s is even. Using the properties

$$\cos^r(\pi + \omega) = \cos^r(\pi - \omega),$$

$$\sin^s(\pi + \omega) = \sin^s(\omega),$$

$$\sin^s\left(\frac{\pi}{2} + \omega\right) = \sin^s\left(\frac{\pi}{2} - \omega\right),$$

$$\cos^r\left(\frac{\pi}{2} + \omega\right) = -\cos^r\left(\frac{\pi}{2} - \omega\right),$$

we have

$$J^{r,s} = \int_0^{2\pi} \cos^r\phi \, \sin^s\phi \, d\phi = 0.$$

The same properties will show that

$$J_n^{r,s} = h\sum_{j=1}^{2n} \cos^r\phi_j \, \sin^s\phi_j = 0.$$

Similar arguments show that when r is even and s is odd, or when both are odd, $J^{r,s} = J_n^{r,s} = 0$. Therefore, $I - I_n = 0$ in these cases.

For handling $J^{r,s}$, we begin by writing

$$\cos^r\phi \, \sin^s\phi = (\cos^r\phi)\left(1 - \cos^2\phi\right)^{s/2}.$$

This is a polynomial in powers of $\cos^2\phi$, with the highest degree term being $\cos^{r+s}\phi$.

To look at even powers of $\cos\phi$, say 2ℓ for $\ell \geq 1$, use the identity

$$(\cos\phi)^{2\ell} = 4^{-\ell}\left[\binom{2\ell}{\ell} + 2\sum_{j=1}^{\ell}\binom{2\ell}{\ell+j}\cos(2j\phi)\right].$$

This can be proven by expanding

$$(\cos\phi)^{2\ell} = \left(\frac{e^{i\phi}+e^{-i\phi}}{2}\right)^{2\ell}.$$

Using (5.8) from Lemma 5.2 with $m = 2\ell$, it follows that

$$\int_0^{2\pi}(\cos\phi)^{2\ell}\,d\phi = h\sum_{j=0}^{2n-1}(\cos\phi_j)^{2\ell}$$

for $0 \le 2\ell \le 2n-2$. This then proves $J^{r,s} = J_n^{r,s}$ for r and s even, $r+s+t \le 2n-1$. For the error $E_{n,\infty}$, we can now write

$$E_n = J^{r,s}\left(K^{r,s,t} - K_n^{r,s,t}\right)$$

and only the error in the Gauss–Legendre quadrature must be considered.

Consider now $K^{r,s,t}$, again with r and s even. Let $z = \cos\theta$,

$$K^{r,s,t} = \int_0^\pi \sin^{r+s+1}\theta\,\cos^t\theta\,d\theta = \int_{-1}^1 z^t\left(1-z^2\right)^{\frac{1}{2}(r+s)}\,dz.$$

This integrand is a polynomial in z of degree $r+s+t \le 2n-1$. Since Gauss–Legendre quadrature with n nodes is used in defining $K_n^{r,s,t}$, we have $K^{r,s,t} = K_n^{r,s,t}$. Thus $I - I_n = 0$.

Combining these results, we have $I = I_n$ for any monomial $f = x^r y^s z^t$ with $r+s+t \le 2n-1$, $r,s,t \ge 0$, completing the proof that the degree of precision of I_n in (5.12) is at least $2n-1$. If we now consider $f(x,y,z) = z^{2n}$, then $I - I_n$ is simply the error in the Gaussian quadrature formula for integrating z^{2n}, and this is well-known to be nonzero (see [9, (5.3.16)]).   □

We now look at an error bound for the product Gaussian quadrature formula (5.12). Recall the results from Theorem 4.5 on best uniform approximation by spherical polynomials. The minimax error for the uniform approximation of a function $f \in C(\mathbb{S}^2)$ by a polynomial from Π_m is defined by

$$E_{m,\infty}(f) := \min_{p\in\Pi_m}\|f - p\|_\infty.$$

Let p_m^* denote a polynomial from Π_m for which this minimax error is attained. An elementary argument shows the existence of such a polynomial p_m^*. For the formula (5.12), we have

$$I(p_{2n-1}^*) = I_n(p_{2n-1}^*).$$

Table 5.1 Product
Gaussian quadrature
method

n	Nodes	Error
2	8	$-1.17\mathrm{E}-02$
3	18	$-4.00\mathrm{E}-04$
4	32	$-4.91\mathrm{E}-07$
5	50	$-3.84\mathrm{E}-09$
6	72	$-2.21\mathrm{E}-12$

Also, note that for any $g \in C(\mathbb{S}^2)$,

$$|I(g)| \leq 4\pi\|g\|_\infty, \qquad |I_n(g)| \leq 4\pi\|g\|_\infty.$$

Then,

$$I(f) - I_n(f) = I(f - p_{2n-1}^*) - I_n(f - p_{2n-1}^*),$$
$$|I(f) - I_n(f)| \leq \left|I(f - p_{2n-1}^*)\right| + \left|I_n(f - p_{2n-1}^*)\right|$$
$$\leq 8\pi\|f - p_{2n-1}^*\|_\infty.$$

Therefore,

$$|I(f) - I_n(f)| \leq 8\pi\, E_{2n-1,\infty}(f). \tag{5.13}$$

We now apply the bounds for the minimax uniform error given in (4.49), with the rate of convergence to zero depending on the smoothness of the function f.

Example 5.5. Consider evaluating the integral

$$I = \int_{\mathbb{S}^2} e^x dS \doteq 14.7680137457653. \tag{5.14}$$

The numerical results are given in Table 5.1, where the column *Error* gives $I - I_n$. The convergence is very rapid.   □

5.2 Composite Methods

We define quadrature methods over $\mathbb{S}^2$ that are the generalization of composite piecewise single-variable numerical integration rules over an interval $[a, b]$. Recall the schema for the single variable case.

- Subdivide the integration region $[a, b]$ into smaller subintervals.
- Apply a simple rule to perform the integration on each such subinterval.

The trapezoidal rule (5.7) and Simpson's rule are examples.

For approximating the spherical integral $I(f)$, begin by subdividing $\mathbb{S}^2$,

$$\mathbb{S}^2 = \bigcup_{i=1}^{N} \Delta_i$$

for elements $\Delta_1, \ldots, \Delta_N$ with non-overlapping interiors. Then write

$$I(f) = \int_{\mathbb{S}^2} f(\boldsymbol{\eta}) \, dS^2(\boldsymbol{\eta}) = \sum_{i=1}^{N} \int_{\Delta_i} f(\boldsymbol{\eta}) \, dS^2(\boldsymbol{\eta}). \tag{5.15}$$

Next apply a simple quadrature formula to each sub-integral. There are two aspects to this, both important in obtaining an accurate and convenient formula.

1. How to choose the mesh $\{\Delta_i\}_{i=1}^{N}$?
2. How to choose the quadrature formula for the sub-integral?

There is a large literature on creating meshes for domains in the plane, space, and surfaces. We restrict ourselves to schema that seem desirable when doing numerical quadrature over the sphere. In our presentation in this section, we restrict the elements to being spherical triangles. The choice of how to create a triangular mesh is important, as we illustrate later with a numerical example.

We begin with an initial triangulation of $\mathbb{S}^2$, say

$$\mathcal{T}_0 = \{\Delta_{0,1}, \ldots, \Delta_{0,N_0}\}.$$

We subdivide or refine it, obtaining ever finer meshes, resulting in a sequence of triangulations,

$$\mathcal{T}_\ell = \{\Delta_{\ell,1}, \ldots, \Delta_{\ell,N_\ell}\}, \qquad \mathbb{S}^2 = \bigcup_{i=1}^{N_\ell} \Delta_{\ell,i}$$

for $\ell = 0, 1, \ldots$. For notation, we let $N_\ell = |\mathcal{T}_\ell|$ denote the number of elements in $\mathcal{T}_\ell$.

To refine a spherical triangle Δ, we connect the midpoints of each side of Δ with a great circle path, thus producing four new spherical triangles. This is illustrated in Fig. 5.1. With this method of refinement, the number of elements in $\mathcal{T}_\ell$ is four times the number in $\mathcal{T}_{\ell-1}$, or $N_\ell = 4N_{\ell-1}$. A desirable feature of this refinement method is that the angles in a newly-formed refinement triangle will not vary a great deal from those in the original triangle.

For the initial triangulation, we begin with one of the regular polyhedrons with triangular faces: the 4-sided *tetrahedron*, the 8-sided *octahedron*, and the 20-sided *icosahedron*. Inscribe one of these into $\mathbb{S}^2$, and then project outward

Fig. 5.1 A spherical
triangle Δ and its
refinement by connecting
the midpoints of its sides

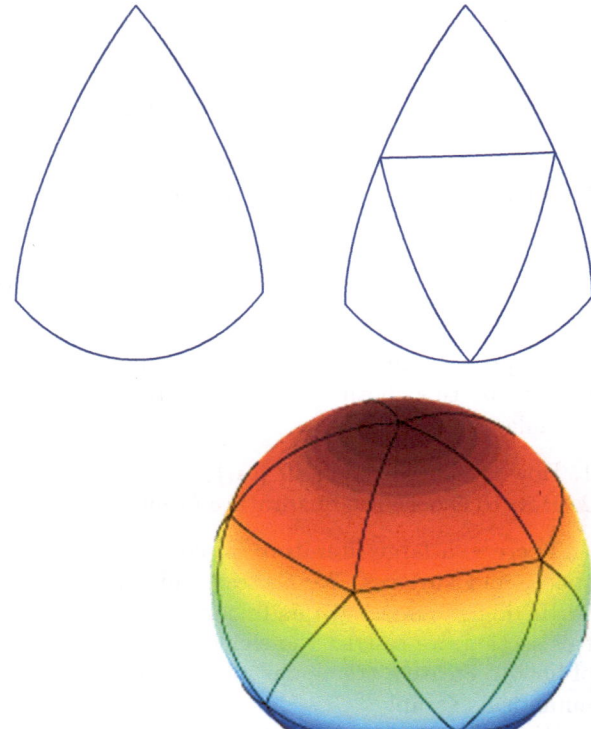

Fig. 5.2 Icosahedral
triangulation of sphere

from the polyhedron onto $\mathbb{S}^2$. Figure 5.2 shows the use of the icosahedron to
generate $\mathcal{T}_0$.

We write

$$I(f) = \sum_{j=1}^{N_\ell} \int_{\Delta_{\ell,j}} f(\boldsymbol{\eta}) \, dS^2(\boldsymbol{\eta})$$

and we approximate each of the sub-integrals by a simple low-order method.

5.2.1   The Centroid Method

Consider the generic sub-integral

$$\int_\Delta f(\boldsymbol{\eta}) \, dS^2(\boldsymbol{\eta}). \tag{5.16}$$

The simplest approximation is to replace $f(\boldsymbol{\eta})$ by the constant function equal to $f(\boldsymbol{\eta}^{(c)})$, with $\boldsymbol{\eta}^{(c)}$ the centroid of Δ, and follow this by integrating exactly the resulting sub-integral. Let Δ have vertices $\{\boldsymbol{v}_1, \boldsymbol{v}_2, \boldsymbol{v}_3\}$. The centroid is defined by

$$\boldsymbol{\eta}^{(c)} = \frac{\boldsymbol{v}_1 + \boldsymbol{v}_2 + \boldsymbol{v}_3}{|\boldsymbol{v}_1 + \boldsymbol{v}_2 + \boldsymbol{v}_3|}.$$

Thus we have

$$\int_\Delta f(\boldsymbol{\eta}) \, dS^2(\boldsymbol{\eta}) \approx f\left(\boldsymbol{\eta}^{(c)}\right) \int_\Delta dS^2(\boldsymbol{\eta}) = f\left(\boldsymbol{\eta}^{(c)}\right) \text{area}(\Delta).$$

For such a spherical triangle Δ, "Girard's Theorem" states that

$$\text{area}(\Delta) = \theta_1 + \theta_2 + \theta_3 - \pi \qquad (5.17)$$

with θ_i the angle of Δ at the vertex $\boldsymbol{v}_i$ [117, p. 1196]. The "centroid rule" for approximating $I(f)$ is given by

$$I(f) \approx I_{N_\ell}(f) \equiv \sum_{j=1}^{N_\ell} f\left(\boldsymbol{\eta}_j^{(c)}\right) \text{area}(\Delta_{\ell,j}) \qquad (5.18)$$

with $\boldsymbol{\eta}_j^{(c)}$ the centroid of $\Delta_{\ell,j}$.

Using an error analysis that we discuss later and assuming the refinement process illustrated above in Fig. 5.1, with $N_\ell = 4N_{\ell-1}$, it can be shown that

$$I(f) - I_{N_\ell}(f) = \mathcal{O}\left(\frac{1}{N_\ell}\right), \qquad (5.19)$$

provided f is twice-continuously differentiable over $\mathbb{S}^2$. The error should decrease by a factor of approximately 4 when comparing the use of $\mathcal{T}_\ell$ with that of $\mathcal{T}_{\ell-1}$.

Example 5.6. Consider the integral (5.14) used in Example 5.5. In Table 5.2, we give the errors when using triangulations based on the tetrahedral, octahedral, and icosahedral triangulations. We also give the ratios of successive errors, to measure the rate at which the error is decreasing. With each triangulation scheme, the error has an error ratio of approximately 4 for larger values of N_ℓ, and this is in agreement with (5.19). As stated earlier, the type of triangulation can make a significant difference in the accuracy of the approximation scheme. The icosahedral triangulation leads to a much smaller error when compared to the other triangulations with a comparable number of nodes. A partial explanation of this is given in Sect. 5.3.2. $\qquad\square$

Table 5.2 Centroid method errors for (5.14) for varying triangulations

Tetrahedral			Octahedral			Icosahedral		
N	Error	Ratio	N	Error	Ratio	N	Error	Ratio
4	3.35E − 1		8	4.84E − 2		20	−9.55E − 5	
16	5.81E − 2	5.8	32	4.70E − 4	103	80	6.71E − 7	−142
64	2.11E − 2	2.7	128	4.27E − 4	1.1	320	5.01E − 8	13.4
256	4.58E − 3	4.6	512	1.00E − 4	4.3	1, 280	1.01E − 8	4.9
1,024	1.12E − 3	4.1	2, 048	2.47E − 5	4.1			
4,096	2.77E − 4	4.0						

5.2.2   General Composite Methods

For a more general approach to approximating the generic sub-integral in (5.16), we begin by using a change of variables to transform the integration region Δ to the unit simplex σ in the plane,

$$\sigma = \{(s,t) : 0 \le s, t, s + t \le 1\}. \tag{5.20}$$

See Fig. 5.3. Given a spherical triangle Δ_k with vertices $\{v_{k,1}, v_{k,2}, v_{k,3}\}$, introduce

$$\boldsymbol{p}_k(s,t) = v_{k,1} + t\,(v_{k,2} - v_{k,1}) + s\,(v_{k,3} - v_{k,1}) \tag{5.21}$$

$$= u v_{k,1} + t v_{k,2} + s v_{k,3},$$

$$\boldsymbol{m}_k(s,t) = \frac{\boldsymbol{p}_k(s,t)}{|\boldsymbol{p}_k(s,t)|}, \tag{5.22}$$

with $u = 1 - s - t$. The image of $\boldsymbol{p}_k(s,t)$ as (s,t) varies over σ is the planar triangle joining the three vertices $\{v_{k,1}, v_{k,2}, v_{k,3}\}$. The image of $\boldsymbol{m}_k(s,t)$ is the spherical triangle Δ_k. This defines a mapping

$$\boldsymbol{m}_k : \ \sigma \xrightarrow[onto]{1-1} \Delta_k.$$

The integral of f over Δ_k becomes

$$\int_{\Delta_k} f(\boldsymbol{\eta})\, dS^2(\boldsymbol{\eta}) = \int_\sigma f(\boldsymbol{m}_k(s,t))\, |D_s \boldsymbol{m}_k \times D_t \boldsymbol{m}_k|\, d\sigma. \tag{5.23}$$

For notation,

$$D_s \boldsymbol{m}_k(s,t) = \frac{\partial \boldsymbol{m}_k(s,t)}{\partial s}, \quad D_t \boldsymbol{m}_k(s,t) = \frac{\partial \boldsymbol{m}_k(s,t)}{\partial t},$$

Fig. 5.3 The unit simplex σ. The symbol "*" indicates the centroid of σ

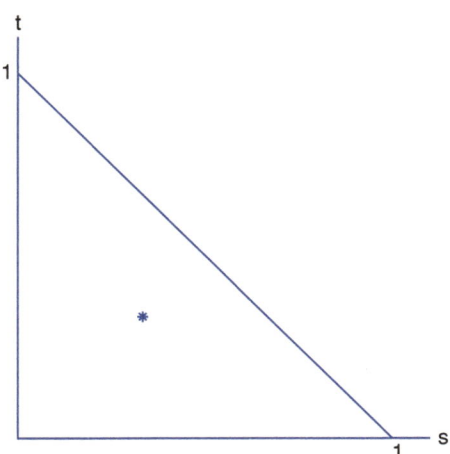

and $|D_s\boldsymbol{m}_k \times D_t\boldsymbol{m}_k|$ denotes the Euclidean norm of the cross-product of $D_s\boldsymbol{m}_k$ and $D_t\boldsymbol{m}_k$. Use numerical integration approximations over σ to approximate integrals over Δ_k. Then combine these to approximate the integral of f over $\mathbb{S}^2$.

There is a large literature on numerical approximations to integrals over σ,

$$\int_\sigma g(s,t)\,d\sigma.$$

We give a few such rules and their degree of precision.

R1. Degree of precision 2.

$$\int_\sigma g(s,t)\,d\sigma \approx \tfrac{1}{6}\left[g\big(0,\tfrac{1}{2}\big) + g\big(\tfrac{1}{2},0\big) + g\big(\tfrac{1}{2},\tfrac{1}{2}\big)\right]. \qquad (5.24)$$

This is based on integrating the quadratic polynomial in (s,t) that interpolates g at the three corner points $\{(0,0),(0,1),(1,0)\}$ and at the midpoints of the three sides of σ, namely $\left\{\big(0,\tfrac{1}{2}\big),\big(\tfrac{1}{2},0\big),\big(\tfrac{1}{2},\tfrac{1}{2}\big)\right\}$.

R2. Degree of precision 2.

$$\int_\sigma g(s,t)\,d\sigma \approx \tfrac{1}{6}\left[g\big(\tfrac{1}{6},\tfrac{1}{6}\big) + g\big(\tfrac{1}{6},\tfrac{2}{3}\big) + g\big(\tfrac{2}{3},\tfrac{1}{6}\big)\right]. \qquad (5.25)$$

Rule **R2** is based on integrating the linear polynomial that interpolates g at the points $\left\{\big(\tfrac{1}{6},\tfrac{1}{6}\big),\big(\tfrac{1}{6},\tfrac{2}{3}\big),\big(\tfrac{2}{3},\tfrac{1}{6}\big)\right\}$. From its definition, it will integrate exactly any linear polynomial in (s,t). It can then be checked that it also integrates exactly any quadratic polynomial in (s,t).

R3. Degree of precision 5.

$$\int_\sigma g(s,t)\, d\sigma \approx \tfrac{9}{80} g(\tfrac{1}{3}, \tfrac{1}{3}) + B\left[g(\alpha, \alpha) + g(\alpha, \beta) + g(\beta, \alpha)\right]$$

$$+ C\left[g(\gamma, \gamma) + g(\gamma, \delta) + g(\delta, \gamma)\right] \tag{5.26}$$

with

$$\alpha = \frac{6 - \sqrt{15}}{21}, \quad \beta = \frac{9 + 2\sqrt{15}}{21},$$

$$\gamma = \frac{6 + \sqrt{15}}{21}, \quad \delta = \frac{9 - 2\sqrt{15}}{21},$$

$$B = \frac{155 - \sqrt{15}}{2400}, \quad C = \frac{155 + \sqrt{15}}{2400}.$$

The formulas **R1** and **R2** are the two cases of formula T_n:2-1 from Stroud [111, p. 307], and **R3** is formula T2:5-1 from [111, p. 314]. When used over all of $\mathbb{S}^2$, as in (5.15), and with the method of refinement given above (as illustrated in Fig. 5.1), the rules **R1** and **R2** have an effective degree of precision of 3. An argument for this uses a certain symmetry in the mesh due to the refinement being used here; see [10, Theorem 5.3.4]. For other quadrature rules over σ, see Stroud [111, p. 306] and Lyness and Jespersen [75].

Example 5.7. Consider again the numerical integration of the integral (5.14), but now apply composite quadrature based on each of the rules **R1–R3**. In light of the superior performance with the icosahedral triangulation seen in Table 5.2, we use only it for illustrating the composite rules based on **R1–R3**. The numerical results with varying $N = N_\ell$ are given in Table 5.3. The ratios imply that the composite methods based on **R1** and **R2** have an empirical convergence rate of $\mathcal{O}(N^{-2})$, whereas the composite method based on **R3** has an empirical rate of $\mathcal{O}(N^{-3})$. We explain these rates of convergence in the error analysis given below.   □

5.2.3 Error Analysis

To analyze the error in a composite quadrature method, we begin with a short discussion of Taylor's theorem for functions of two variables.

Lemma 5.8. *Assume $g \in C^{q+1}(\sigma)$ for some integer $q > 0$, with σ the unit simplex of (5.20). Choose a point of expansion $(a, b) \in \sigma$. Then*

$$g(a + \xi, b + \eta) = \mathcal{T}_q(g) + \mathcal{R}_q(g), \qquad (a + \xi, b + \eta) \in \sigma, \tag{5.27}$$

Table 5.3 Composite rules **R1**–**R3** with an icosahedral triangulation

N	Rule R1			Rule R2			Rule R3		
	Nodes	Error	Ratio	Nodes	Error	Ratio	Nodes	Error	Ratio
20	30	$2.42\mathrm{E}-1$		60	$2.42\mathrm{E}-1$		140	$-2.33\mathrm{E}-2$	
80	120	$1.41\mathrm{E}-2$	17.1	240	$1.62\mathrm{E}-2$	15.0	560	$-4.25\mathrm{E}-4$	54.9
320	480	$8.83\mathrm{E}-4$	16.0	960	$1.01\mathrm{E}-3$	16.0	2,240	$-6.78\mathrm{E}-6$	62.7
1,280	1,920	$5.52\mathrm{E}-5$	16.0	2,880	$6.33\mathrm{E}-5$	16.0	8,960	$-1.06\mathrm{E}-7$	63.8

$$T_q(g) = \sum_{k=0}^{q} \frac{1}{k!} \left(\xi \frac{\partial}{\partial s} + \eta \frac{\partial}{\partial t} \right)^k g(s,t) \Bigg|_{(s,t)=(a,b)} , \tag{5.28}$$

$$\mathcal{R}_q(g) = \frac{1}{q!} \int_0^1 (1-\tau)^q \left(\xi \frac{\partial}{\partial s} + \eta \frac{\partial}{\partial t} \right)^{q+1} g(s,t)\, d\tau \tag{5.29}$$

$$= \frac{1}{(q+1)!} \left(\xi \frac{\partial}{\partial s} + \eta \frac{\partial}{\partial t} \right)^{q+1} g(s,t) \Bigg|_{(s,t)=(a+\theta\xi,b+\theta\eta)} . \tag{5.30}$$

Following the differentiation in (5.29), *let* $(s,t) = (a+\tau\xi, b+\tau\eta)$. *In* (5.30), θ *is some number satisfying* $0 < \theta < 1$.

Proof. As is standard in the derivation of this formula, introduce

$$F(\tau) = g(a+\tau\xi, b+\tau\eta), \quad 0 \le \tau \le 1.$$

Using the one-variable Taylor's theorem with remainder,

$$F(1) = \sum_{k=0}^{q} \frac{1}{k!} F^{(k)}(0) + \frac{1}{q!} \int_0^1 (1-\tau)^q\, F^{(q+1)}(\tau)\, d\tau$$

$$= \sum_{k=0}^{q} \frac{1}{k!} F^{(k)}(0) + \frac{1}{(q+1)!} F^{(q+1)}(\theta).$$

Note that

$$F(0) = g(a,b),$$
$$F(1) = g(a+\xi, b+\eta).$$

To relate this more directly to $g(a+\tau\xi, b+\tau\eta)$, note that

$$F^{(k)}(\tau) = \left(\xi \frac{\partial}{\partial s} + \eta \frac{\partial}{\partial t} \right)^k g(s,t) \Bigg|_{(s,t)=(a+\tau\xi,b+\tau\eta)}$$

for $k \ge 1$, and thus $T_q(g)$ is an expansion about (a,b) in terms of products of powers of ξ and η. Combining these results proves the lemma. $\qquad \square$

For (5.27) we commonly use $(a,b) = (\frac{1}{3}, \frac{1}{3})$, the centroid of σ, or $(a,b) = (0,0)$. Also, we obtain an expansion for $g(s,t)$ by applying (5.27)–(5.29) with $(\xi, \eta) = (s-a, t-b)$.

We now turn to an analysis of the error for quadrature rules over σ, such as **R1–R3** given earlier. Let L denote a linear functional on $C(\sigma)$. For example, the rule **R2** can be written as

$$L(g) = \tfrac{1}{6} \left[g\left(\tfrac{1}{6}, \tfrac{1}{6}\right) + g\left(\tfrac{1}{6}, \tfrac{2}{3}\right) + g\left(\tfrac{2}{3}, \tfrac{1}{6}\right) \right]. \tag{5.31}$$

Consider the error functional

$$\mathcal{E}(g) = \int_\sigma g(s,t)\, d\sigma - L(g).$$

Assume that

$$\mathcal{E}(g) = 0 \quad \text{for any polynomial } g \text{ with } \deg(g) \leq q.$$

For the above L in (5.31), $q = 2$. Recall the Taylor expansion (5.27). Since $\mathcal{E}$ is linear,

$$\mathcal{E}(g) = \mathcal{E}(\mathcal{T}_q(g)) + \mathcal{E}(\mathcal{R}_q(g)).$$

The term $\mathcal{E}(\mathcal{T}_q(g)) = 0$ since $\mathcal{T}_q(F)$ is a polynomial in (s,t) of degree $\leq q$. Thus

$$\mathcal{E}(g) = \int_\sigma \mathcal{R}_q(g)\, d\sigma - L(\mathcal{R}_q(g)). \tag{5.32}$$

The error in integrating g over σ equals that of the error in integrating the Taylor series error $\mathcal{R}_q(g)$.

Return to the remainder formula (5.29) with the derivatives of g evaluated at $(s,t) = (a + \tau\xi, b + \tau\eta)$. To bound the error

$$\mathcal{E}(g) = \int_\sigma \mathcal{R}_q(g)\, d\sigma - L(\mathcal{R}_q(g)),$$

we must apply both the integral over σ and the linear functional L to $\mathcal{R}_q(g)$. The important thing to note is that all terms will involve

$$\frac{\partial^{q+1} g(s,t)}{\partial s^j \partial t^k}, \quad 0 \leq j, k \leq q+1, \quad j+k = q+1.$$

We now apply this to

$$\int_{\Delta_k} f(\boldsymbol{\eta})\, dS^2(\boldsymbol{\eta}) = \int_\sigma f(\boldsymbol{m}_k(s,t))\, |D_s \boldsymbol{m}_k \times D_t \boldsymbol{m}_k|\, d\sigma \tag{5.33}$$

and thus we must look at the derivatives of

$$g(s,t) = f(\boldsymbol{m}_k(s,t))\, |D_s \boldsymbol{m}_k \times D_t \boldsymbol{m}_k|. \tag{5.34}$$

To do this, we must examine the derivatives of $\boldsymbol{m}_k(s,t)$. Recall

$$p_k(s,t) = v_{k,1} + t(v_{k,2} - v_{k,1}) + s(v_{k,3} - v_{k,1})$$
$$= u v_{k,1} + t v_{k,2} + s v_{k,3}$$

with $u = 1 - s - t$. As (s,t) varies over σ, the image $p_k(\sigma)$ gives the planar triangle with vertices $\{v_{k,1}, v_{k,2}, v_{k,3}\}$, and

$$m_k(s,t) = \frac{p_k(s,t)}{|p_k(s,t)|}$$

has Δ_k as its image. To simplify the notation, consider

$$p(s,t) = v_1 + t(v_2 - v_1) + s(v_3 - v_1),$$
$$m(s,t) = \frac{p(s,t)}{|p(s,t)|}. \tag{5.35}$$

Also, let $p_s = \partial p / \partial s$ and $p_t = \partial p / \partial t$, and do similarly for m_s and m_t. Immediately,

$$p_s = v_3 - v_1, \quad p_t = v_2 - v_1.$$

We also need the partial derivatives with respect to s and t of $|p(s,t)|$. First,

$$\frac{\partial}{\partial s}\left(|p|^2\right) = \frac{\partial}{\partial s}(p \cdot p) = 2 p_s \cdot p.$$

Also,

$$\frac{\partial}{\partial s}\left(|p|^2\right) = 2|p| \frac{\partial}{\partial s}(|p|).$$

Thus,

$$\frac{\partial}{\partial s}(|p|) = \frac{p_s \cdot p}{|p|}.$$

Returning to the derivative of $m(s,t)$ in (5.35),

$$m_s = \frac{p_s}{|p|} - \frac{p}{|p|^2} \frac{\partial}{\partial s}(|p|) = \frac{p_s}{|p|} - \frac{p}{|p|^2} \frac{p_s \cdot p}{|p|}$$
$$= \frac{p_s}{|p|} - m\left(m \cdot \frac{p_s}{|p|}\right).$$

There is an analogous formula for m_t.

From the definition of $\boldsymbol{p}_k(\sigma)$ as the planar triangle determined by the vertices $\{\boldsymbol{v}_{k,1}, \boldsymbol{v}_{k,2}, \boldsymbol{v}_{k,3}\}$, it is easy to visualize that $\boldsymbol{p}_k(s,t)$ is bounded away from 0 for all of our polyhedral triangulations,

$$c \le |\boldsymbol{p}_k(s,t)| \le 1, \quad (s,t) \in \sigma$$

for some $c > 0$. Consequently,

$$|\boldsymbol{m}_s(s,t)| \le \frac{|\boldsymbol{p}_s|}{|\boldsymbol{p}|} + |\boldsymbol{m}| \left(|\boldsymbol{m}| \frac{|\boldsymbol{p}_s|}{|\boldsymbol{p}|} \right)$$

$$\le \frac{2}{c} |\boldsymbol{v}_3 - \boldsymbol{v}_1|, \quad (s,t) \in \sigma,$$

using $|\boldsymbol{m}| = 1$. Analogously,

$$|\boldsymbol{m}_t(s,t)| \le \frac{2}{c} |\boldsymbol{v}_2 - \boldsymbol{v}_1|, \quad (s,t) \in \sigma.$$

This argument can be continued inductively to yield

$$\left| \frac{\partial^{q+1} \boldsymbol{m}(s,t)}{\partial s^j \partial t^k} \right| \le c_q h^{q+1}, \quad 0 \le j, k \le q+1, \quad j+k = q+1$$

with $h = \max \{|\boldsymbol{v}_2 - \boldsymbol{v}_1|, |\boldsymbol{v}_3 - \boldsymbol{v}_1|\}$. Note also that

$$|D_s \boldsymbol{m} \times D_t \boldsymbol{m}| \le \frac{4}{c^2} h^2. \tag{5.36}$$

Return to the integral

$$\int_{\Delta_k} f(\boldsymbol{\eta}) \, dS^2(\boldsymbol{\eta}) = \int_\sigma f(\boldsymbol{m}_k(s,t)) \, |D_s \boldsymbol{m}_k \times D_t \boldsymbol{m}_k| \, d\sigma$$

in which we apply the above with

$$g(s,t) = f(\boldsymbol{m}_k(s,t)) \, |D_s \boldsymbol{m}_k \times D_t \boldsymbol{m}_k|.$$

We are using a sequence of triangulations

$$\mathcal{T}_\ell = \{\Delta_{\ell,1}, \ldots, \Delta_{\ell,N_\ell}\}.$$

Let

$$h_\ell = \max_{\Delta \in \mathcal{T}_\ell} \text{diam}(\Delta).$$

We assume a regularity in the triangulation $\mathcal{T}_\ell$ by means of

$$h_\ell^2 N_\ell = \mathcal{O}(1), \tag{5.37}$$

where $N_\ell = |\mathcal{T}_\ell|$. This is easily shown to be true for our polyhedral-based triangulations.

As earlier, we assume $f \in C^{q+1}(\mathbb{S}^2)$. By applying the earlier bounds, it is straightforward to obtain

$$\left| \frac{\partial^{q+1} g(s,t)}{\partial s^j \partial t^k} \right| \leq c_q h^{q+3}, \quad 0 \leq j, k \leq q+1, \quad j+k = q+1. \tag{5.38}$$

We apply this to the integral (5.33), with g defined in (5.34). Doing this leads to an error of size $\mathcal{O}\left(h_\ell^{q+3}\right)$ for the single integral over Δ_k.

Theorem 5.9. *Assume the triangulation scheme $\mathcal{T}_0, \mathcal{T}_2, \ldots$ satisfies (5.37). For a function $f \in C(\mathbb{S}^2)$, let $I_{N_\ell}(f)$ denote the composite quadrature formula obtained by applying the basic quadrature formula L to each integral*

$$\int_{\Delta_k} f(\boldsymbol{\eta}) \, dS^2(\boldsymbol{\eta}) = \int_\sigma f(\boldsymbol{m}_k(s,t)) \, |D_s \boldsymbol{m}_k \times D_t \boldsymbol{m}_k| \, d\sigma, \tag{5.39}$$

where L has degree of precision $q \geq 0$ over σ. Assume $f \in C^{q+1}(\mathbb{S}^2)$. Then

$$I(f) - I_{N_\ell}(f) = \mathcal{O}\left(h_\ell^{q+1}\right). \tag{5.40}$$

Proof. Use (5.38) for each of the N_ℓ sub-integrals (5.39), obtaining

$$I(f) - I_{N_\ell}(f) = \mathcal{O}\left(N_\ell h_\ell^{q+3}\right).$$

Applying (5.37) completes the proof.   □

The result (5.40) agrees with the empirical results shown in Table 5.3. What is lacking in this discussion is a proof that the composite formulas based on **R1** and **R2** have an effective degree of precision $q = 3$ rather than the degree of precision of $q = 2$ that is true over just σ. As noted earlier, this requires an additional analysis, and we refer the reader to [10, Theorem 5.3.4]. Again, this is a consequence of the refinement process illustrated in Fig. 5.1. As an analogy, recall how quadratic interpolation is used to define the 3-point Simpson's quadrature rule for functions of one variable. A fortuitous cancellation occurs when the middle interpolation point is the midpoint of the remaining two interpolation points, leading to a degree of precision of 3, whereas the use of quadratic interpolation would lead one to expect a degree of precision of only 2.

For the centroid rule, the error functional over a single spherical triangle Δ is

$$\mathcal{E}(f) = \int_\Delta f(\boldsymbol{\eta}) \, dS^2(\boldsymbol{\eta}) - f\left(\boldsymbol{\eta}^{(c)}\right) \int_\Delta dS(\boldsymbol{\eta})$$

$$= \int_\sigma \left[f(\boldsymbol{m}(s,t)) - f\left(\boldsymbol{m}\left(\tfrac{1}{3}, \tfrac{1}{3}\right)\right) \right] |D_s\boldsymbol{m} \times D_t\boldsymbol{m}| \, d\sigma.$$

For f a constant function, $\mathcal{E}(f) = 0$, and thus the degree of precision $q \geq 0$; and it can be shown to be exactly $q = 0$. In analogy with the preceding paragraph for the composite formulas based on **R1** and **R2**, the centroid method can be shown to have an effective degree of precision of $q = 1$ when used with the type of refinement illustrated in Fig. 5.1. Again, use the type of analysis given in [10, Theorem 5.3.4]; it is omitted here. With $q = 1$, we obtain the convergence asserted earlier in (5.19).

5.3 High Order Gauss-Type Methods

For one-variable integration, a Gaussian quadrature formula is based on asking that the formula be exact for polynomials of as large a degree as possible. Recall that with n nodes, it is possible to have a degree of precision of $2n - 1$. This approach generalizes to multivariable integration. Consider a formula

$$I(f) \equiv \int_{\mathbb{S}^2} f(\boldsymbol{\eta}) \, dS^2(\boldsymbol{\eta}) \approx I_N(f) \equiv \sum_{k=1}^N w_k f(\boldsymbol{\eta}_k). \tag{5.41}$$

The nodes $\{\boldsymbol{\eta}_k\}$ and weights $\{w_k\}$ are to be so chosen that the formula is exact for spherical polynomials of as large a degree as possible.

To simplify the development of such formulas, much such work has begun with the following important theorem of S. Sobolev [109, Theorem 1]. It involves the notion of a group $\mathcal{G}$ of rotations on $\mathbb{S}^2$. If $\gamma \in \mathcal{G}$, then $\gamma : \mathbb{S}^2 \xrightarrow[\text{onto}]{1-1} \mathbb{S}^2$ is some rotation of the sphere. We also allow "inversions" of the sphere in which points $\boldsymbol{\eta} \in \mathbb{S}^2$ are reversed through a given plane containing the origin. If $f \in C(\mathbb{S}^2)$, then $f_\gamma \in C(\mathbb{S}^2)$ is defined by $f_\gamma(\boldsymbol{\eta}) = f(\gamma(\boldsymbol{\eta}))$, $\boldsymbol{\eta} \in \mathbb{S}^2$. We say f is "invariant under $\mathcal{G}$" if $f_\gamma = f$ for all $\gamma \in \mathcal{G}$.

Theorem 5.10. *Let $\mathcal{G}$ denote a finite rotation group on the sphere, possibly including inversion. Consider a numerical quadrature scheme (5.41). Assume the numerical scheme is invariant under the actions of the rotation group,*

$$I_N(f_\gamma) = I_N(f) \quad \forall \gamma \in \mathcal{G}. \tag{5.42}$$

This implies

$$\{\boldsymbol{\eta}_i : i = 1, \ldots, N\} = \{\gamma(\boldsymbol{\eta}_i) : i = 1, \ldots, N\}$$

for all $\gamma \in \mathcal{G}$. In addition, for each node $\boldsymbol{\eta}_i$, $i = 1, \ldots, N$, consider $\mathcal{S}_i = \{\gamma(\boldsymbol{\eta}_i) : \gamma \in \mathcal{G}\}$. Then (5.42) implies that the weights $\{w_j\}$ associated with the nodes in $\mathcal{S}_i$ are to be equal. With such an invariant quadrature scheme, in order that $I(f) = I_N(f)$ for each $f \in \Pi_d$, it is necessary and sufficient that $I(f) = I_N(f)$ for each $f \in \Pi_d$ that is invariant under the action of all elements of $\mathcal{G}$. We then say that $I_N(f)$ has "degree of precision d".

Proof. Because of the rotational symmetry of $\mathbb{S}^2$, we have

$$I(f) = \int_{\mathbb{S}^2} f(\boldsymbol{\eta}) \, dS^2(\boldsymbol{\eta}) = \int_{\mathbb{S}^2} f_\gamma(\boldsymbol{\eta}) \, dS^2(\boldsymbol{\eta}) = I(f_\gamma). \qquad (5.43)$$

Define

$$f^* = \frac{1}{|\mathcal{G}|} \sum_{\gamma \in \mathcal{G}} f_\gamma,$$

where $|\mathcal{G}|$ denotes the number of elements in $\mathcal{G}$. It is straightforward to show that f^* is invariant under $\mathcal{G}$ (i.e. $f^*_\gamma = f^*$ for all $\gamma \in \mathcal{G}$). Moreover, from (5.43), $I(f) = I(f^*)$. Similarly, from (5.42), $I_N(f) = I_N(f^*)$. Combining,

$$I(f) - I_N(f) = I(f^*) - I_N(f^*).$$

Thus if we want $I(f) = I_N(f)$ for all $f \in \Pi_d$, we need only show that the result is true for all $f \in \Pi_d$ that are invariant with respect to $\mathcal{G}$. $\qquad \square$

As earlier with the product Gauss formulae in (5.13), we can show

$$\left| I(f) - I_{N(d)}(f) \right| \le 8\pi \, E_d(f). \qquad (5.44)$$

And as before, the minimax error given in (4.49) can be applied to give a bound on the rate of convergence of $I_N(f)$ to $I(f)$, depending on the smoothness of the function f.

To obtain the nodes $\{\boldsymbol{\eta}_i\}$ and weights $\{w_i\}$ is quite complicated. The nodes and weights in the quadrature formula (5.41) must be chosen to be invariant under the actions in $\mathcal{G}$, and the polynomials that are invariant under $\mathcal{G}$ must be identified. The error $I(f) - I_N(f)$ must be zero for the invariant polynomials f of as large a degree as possible, and this leads to solving a system of nonlinear algebraic equations. The number of such algebraic equations is given in [109, Theorem 2].

To illustrate the use of Sobolev's theorem, we consider a formula

$$I_{32}(f) = A \sum_{k=1}^{12} f\left(\boldsymbol{\eta}_k^{(v)}\right) + B \sum_{k=1}^{20} f\left(\boldsymbol{\eta}_k^{(f)}\right) \qquad (5.45)$$

in which the nodes $\left\{\boldsymbol{\eta}_k^{(v)}\right\}$ are the vertices of an icosahedron inscribed in $\mathbb{S}^2$ and $\left\{\boldsymbol{\eta}_k^{(f)}\right\}$ are the centroids of its faces. The two lowest-degree spherical polynomials that are invariant under the icosahedral group $\mathcal{G}_{20}^*$ with inversions are 1 and

$$5x^4z^2 + 5y^4z^2 + z^6 + 10x^2y^2z^2 - 5x^2z^4 - 5y^2z^4 + 2x^5z + 10xy^4z - 20x^3y^2z.$$

The formula (5.45) is applied to these two polynomials and is equated to the true integrals. The resulting formula has the weights

$$A = \frac{5\pi}{42}, \qquad B = \frac{9\pi}{70}. \qquad (5.46)$$

The next higher degree invariant polynomial is of degree 10 and the quadrature (5.45)–(5.46) is not exact for this case. Consequently, the formula (5.45)–(5.46) has degree of precision 9. This is formula U_3:9-1 in Stroud [111, p. 299].

Lebedev [73] applied Sobolev's theorem using $\mathcal{G} = G_8^*$, the octahedral rotation group with inversion. Elements of this group leave invariant the octahedron in $\mathbb{R}^3$. A spherical polynomial is invariant under G_8^* if and only if it is a polynomial in τ_1 and τ_2,

$$\tau_1 = x^2y^2 + x^2z^2 + y^2z^2,$$

$$\tau_2 = x^2y^2z^2.$$

In his paper, Lebedev gives a family of formulas

$$I_N(f) = \sum_{i=1}^{N(d)} w_i f(P_i) \qquad (5.47)$$

for various degrees of precision $d \le 29$. The case $d = 23$ is given explicitly; it uses 194 nodes, and the weights w_i are positive. For details, see [73].

Finding such quadrature formulas $I_N(f)$ for the icosahedral group with inversion is discussed by Ahrens and Beylkin [3]. We give an example of their use later in this section, for the integral (5.50). Other related approaches to developing high order formulas are given in Keast and Diaz [67], McLaren [79], and Stroud [111].

Table 5.4 High degree of
precision quadrature

d	Nodes	Error
5	12	1.72E − 04
9	32	3.40E − 09
14	72	1.24E − 14

Example 5.11. Consider again the numerical integration of the integral
(5.14), but now apply the integration formulas from Stroud [111, Sect. 8.6]
that have a high degree of precision. The first is U_3:5-1 and uses as nodes the
vertices of an inscribed icosahedron (degree of precision $d = 5$), the second
is U_3:9-1 ($d = 9$), and the third one is U_3:14-1 ($d = 14$). The results are
given in Table 5.4. Compare these results with those of the product Gauss
quadrature given earlier in Table 5.1.   □

5.3.1 Efficiency of a High-Order Formula

For the N-point formula (5.41), there are $4N$ parameters to be determined,
namely the weights $\{w_i\}$ and the components of the nodes $\{\boldsymbol{\eta}_i\}$. The nodes
are subject to the constraints $|\boldsymbol{\eta}_i| = 1$, $i = 1, \ldots, N$. Thus there are essentially
$3N$ parameters to be determined. If we wish to have $I_N(f) = I(f)$ for all $f \in
\Pi_L$, then the number of constraints on the formula is equal to the dimension
of Π_L, namely $(L + 1)^2$. As a measure of the efficiency of the quadrature
method (5.41), it was suggested in McLaren [79] that the ratio

$$R \equiv \frac{(L+1)^2}{3N} \qquad (5.48)$$

be used.

In general we would expect this ratio to be bounded by 1, although there
are a few quadrature formulas for which E is slightly larger than 1. As an
illustration of the latter, $r = 100/96$ for (5.45)–(5.46). The methods described
in [3, 73] generally have $R \approx 1$, especially for larger values of N and L.
In contrast, the product Gauss formula I_n of (5.12) has $L = 2n - 1$ and
$N = 2n^2$, and thus it has the efficiency

$$R = \frac{(2n)^2}{3(2n^2)} = \frac{2}{3}. \qquad (5.49)$$

For a comparable degree of precision, the product Gauss formula (5.12)
uses approximately 50% more function evaluations than do the methods of
this section which have an efficiency ratio of approximately 1. However, the

Table 5.5 Comparison of product Gauss and Ahrens–Beylkin formulas

Product Gauss			Ahrens–Beylkin			
deg	Nodes	Error	deg	nodes	Error	Ratio
27	392	2.92E − 05	27	312	3.15E − 06	1.26
37	722	−2.96E − 08	36	492	1.02E − 08	1.47
41	882	−3.91E − 09	40	612	−4.51E − 10	1.44
45	1,058	3.00E − 10	45	732	2.55E − 10	1.45
49	1,250	−1.16E − 11	50	912	1.54E − 12	1.37
57	1,682	8.88E − 16	54	1,032	−8.88E − 16	1.63

product Gauss formula has the virtue that its nodes and weights are very easy to construct.

Example 5.12. Consider approximating the integral

$$I = \int_{\mathbb{S}^2} \sin^2\left(\pi\left(\eta_1 + \eta_2^2 + \eta_3^3\right)\right) dS^2(\boldsymbol{\eta}) \doteq 4.373708291416826. \qquad (5.50)$$

We compare the product Gauss formula (5.12) and the optimal formulas of C. Ahrens and G. Beylkin [3, Icosahedral-based optimal-order nodes and weights, private communication, 2010]. Table 5.5 contains numerical results for increasing degrees of precision *deg*. Rows 1 through 5 are for comparable degrees of precision. Row 6 is the lowest degree of precision in which full double-precision accuracy was attained. The final column gives the ratio of the number of nodes used with the product Gauss formula as compared to the number of nodes used with the Ahrens–Beylkin formula. The results are consistent with the results stated following (5.49).   □

5.3.2   The Centroid Method

Recall the centroid method (5.18) and its numerical illustration for the integral (5.14), with the numerical results given in Table 5.2. The numerical results varied significantly with the triangulation, with the accuracy improving from the tetrahedral to the octahedral to the icosahedral triangulations. We can now give an explanation of this. First, note that the centroid method is invariant under the rotation group associated with its triangulation. As a consequence, the centroid method has a degree of precision that can be computed using Sobolev's theorem by considering only those polynomials that are invariant under the rotation group. The centroid rule is exact for constant functions with any triangulation. With the tetrahedral, octahedral, and icosahedral groups (with inversion), the next higher degree invariant spherical polynomials have degrees 3, 4, and 6, respectively. The centroid method with any of these regular polyhedral triangulations is not exact for the latter associated invariant polynomial. As a consequence, the centroid method with the tetrahedral, octahedral, and icosahedral triangulations has a degree of precision of 2, 3, and 5, respectively.

Denote the degree of precision by d, and let p_d^* denote the best uniform approximation of the integrand f by spherical polynomials from Π_d. Then the error in the centroid method when integrating f satisfies

$$I(f) - I_n(f) = I(f - p_d^*) - I_n(f - p_d^*),$$
$$|I(f) - I_n(f)| \le 8\,\pi\,E_{d,\infty}(f).$$

Thus the error is actually based on the function $f - p_d^*$ rather than on f. Because $\|f - p_d^*\|_\infty$ decreases as d increases, it is likely that the centroid error will be less as d increases. This is exactly what is seen in Table 5.2.

5.3.3 An Alternative Approach

Another approach to creating quadrature methods with a large degree of precision is to begin by seeking a suitable polynomial interpolation formula and to then integrate it. This is a topic that also has application to extending empirical data and to other approximation problems on the sphere. Integrating the resulting interpolation formula leads to a numerical integration formula with a high degree of precision. For a desired degree of precision $d > 0$, the dimension of Π_d is

$$N = (d+1)^2.$$

Given a set of basis functions $\{\varphi_1, \ldots, \varphi_N\}$ for Π_d, consider maximizing the quantity

$$\det \begin{bmatrix} \varphi_1(\boldsymbol{\eta}_1) & \cdots & \varphi_k(\boldsymbol{\eta}_1) & \cdots & \varphi_N(\boldsymbol{\eta}_1) \\ \vdots & \ddots & \vdots & \ddots & \vdots \\ \varphi_1(\boldsymbol{\eta}_j) & \cdots & \varphi_k(\boldsymbol{\eta}_j) & \cdots & \varphi_N(\boldsymbol{\eta}_j) \\ \vdots & \ddots & \vdots & \ddots & \vdots \\ \varphi_1(\boldsymbol{\eta}_N) & \cdots & \varphi_k(\boldsymbol{\eta}_N) & \cdots & \varphi_N(\boldsymbol{\eta}_N) \end{bmatrix} \tag{5.51}$$

as $\{\boldsymbol{\eta}_1, \ldots, \boldsymbol{\eta}_N\}$ is allowed to vary over $\mathbb{S}^2$, and of course, such a maximum value will be positive. Sets of such points $\{\boldsymbol{\eta}_i\}$ for which this determinant is nonzero are called "fundamental systems" for polynomial interpolation over $\mathbb{S}^2$. With such a choice of points, we can define "Lagrange interpolation basis functions",

$$
\ell_j(\boldsymbol{\eta}) = \frac{\det \begin{bmatrix} \varphi_1(\boldsymbol{\eta}_1) & \cdots & \varphi_k(\boldsymbol{\eta}_1) & \cdots & \varphi_N(\boldsymbol{\eta}_1) \\ \vdots & \ddots & \vdots & \ddots & \vdots \\ \varphi_1(\boldsymbol{\eta}_{j-1}) & \cdots & \varphi_k(\boldsymbol{\eta}_{j-1}) & \cdots & \varphi_N(\boldsymbol{\eta}_{j-1}) \\ \varphi_1(\boldsymbol{\eta}) & \cdots & \varphi_k(\boldsymbol{\eta}) & \cdots & \varphi_N(\boldsymbol{\eta}) \\ \varphi_1(\boldsymbol{\eta}_{j+1}) & \cdots & \varphi_k(\boldsymbol{\eta}_{j+1}) & \cdots & \varphi_N(\boldsymbol{\eta}_{j+1}) \\ \vdots & \ddots & \vdots & \ddots & \vdots \\ \varphi_1(\boldsymbol{\eta}_N) & \cdots & \varphi_k(\boldsymbol{\eta}_N) & \cdots & \varphi_N(\boldsymbol{\eta}_N) \end{bmatrix}}{\det \begin{bmatrix} \varphi_1(\boldsymbol{\eta}_1) & \cdots & \varphi_k(\boldsymbol{\eta}_1) & \cdots & \varphi_N(\boldsymbol{\eta}_1) \\ \vdots & \ddots & \vdots & \ddots & \vdots \\ \varphi_1(\boldsymbol{\eta}_j) & \cdots & \varphi_k(\boldsymbol{\eta}_j) & \cdots & \varphi_N(\boldsymbol{\eta}_j) \\ \vdots & \ddots & \vdots & \ddots & \vdots \\ \varphi_1(\boldsymbol{\eta}_N) & \cdots & \varphi_k(\boldsymbol{\eta}_N) & \cdots & \varphi_N(\boldsymbol{\eta}_N) \end{bmatrix}}
\tag{5.52}
$$

for $j = 1, \ldots, N$. These satisfy $\ell_j(\boldsymbol{\eta}_i) = \delta_{i,j}$, $i, j = 1, \ldots, N$. With the optimality property for the nodes we have

$$
|\ell_j(\boldsymbol{\eta})| \le 1, \quad \boldsymbol{\eta} \in \mathbb{S}^2, \quad j = 1, \ldots, N.
\tag{5.53}
$$

The polynomial

$$
p(\boldsymbol{\eta}) = \sum_{j=1}^{N} f(\boldsymbol{\eta}_j)\, \ell_j(\boldsymbol{\eta})
\tag{5.54}
$$

belongs to Π_d and it interpolates $f(\boldsymbol{\eta})$ at the points $\{\boldsymbol{\eta}_1, \ldots, \boldsymbol{\eta}_N\}$. The property (5.53) provides a type of stability for the interpolation.

For the integration formula, use

$$
I(f) \approx I_d(f) = \sum_{j=1}^{N} f(\boldsymbol{\eta}_j)\, w_j,
\tag{5.55}
$$

$$
w_j = \int_{\mathbb{S}^2} \ell_j(\boldsymbol{\eta})\, dS^2(\boldsymbol{\eta}).
$$

The maximizing of (5.51) and its subsequent use in defining interpolation and quadrature formulas have been studied by Sloan and Womersley [107, 108]. They use the basis functions

$$
\varphi_i(\boldsymbol{\eta}) = G_n(\boldsymbol{\eta}, \boldsymbol{\eta}_i),
$$

Table 5.6 Comparison of Sloan–Womersley and Ahrens–Beylkin formulas

Degree of precision	Sloan–Womersley			Ahrens–Beylkin	
	Nodes	Error	Λ_d	Nodes	Error
27	784	$-8.14\mathrm{E}-6$	30.78	312	$3.15\mathrm{E}-06$
36	1,369	$-7.07\mathrm{E}-9$	49.03	492	$1.02\mathrm{E}-08$
40	1,681	$6.60\mathrm{E}-10$	48.91	612	$-4.51\mathrm{E}-10$
45	2,116	$3.02\mathrm{E}-12$	57.88	732	$2.55\mathrm{E}-10$
50	2,601	$-2.66\mathrm{E}-13$	69.47	912	$1.54\mathrm{E}-12$

where $G_n(\boldsymbol{\xi}, \boldsymbol{\eta})$ is the reproducing kernel for Π_n. These basis functions are discussed below following (5.87). Formulas with high degrees of precision are found for which all weights w_k are positive. The positivity of the weights is important in guaranteeing stability in the quadrature formula as a function of N, as is illustrated in (5.57)–(5.58) in the following section. Tables of nodes and weights based on maximizing (5.51) are given in [119].

Example 5.13. We compare this method with the results given in Example 5.12 for the integral (5.50); see Table 5.6. For the same degree of precision, the errors are comparable for the Sloan–Womersley and Ahrens–Beylkin formulas. However, the former is less efficient because of the much larger number of nodes needed. For the integration formula I_d of (5.55), the efficiency ratio of (5.48) is

$$R = \frac{(d+1)^2}{3(d+1)^2} = \frac{1}{3}.$$

For an equivalent degree of precision, the cost in function evaluations is three times that for the Ahrens–Beylkin formulas and it is two times the cost of the product Gauss formula (5.12). However, the Sloan–Womersley choice of points has good properties as regards interpolation on $\mathbb{S}^2$. Define the interpolatory projection operator

$$\mathcal{P}_d f(\boldsymbol{\eta}) = \sum_{k=1}^{N} f(\boldsymbol{\eta}_k) \, \ell_k(\boldsymbol{\eta})$$

which maps $C(\mathbb{S}^2)$ onto Π_d. Then the property (5.53) implies

$$\Lambda_d := \|\mathcal{P}_d\|_{C(\mathbb{S}^2) \to \Pi_d} \leq (d+1)^2.$$

These are called the Lebesgue constants for the interpolation method. Their actual values are listed in Table 5.6; they increase faster than d, much less rapidly than $(d+1)^2$. $\square$

5.4   Integration of Scattered Data

An important source of quadrature problems on the sphere is the integration of empirical data obtained from scattered nodes on the unit sphere. To handle such problems and associated topics in the interpolation and approximation of functions using scattered data, the area of "meshless discretization methods" has been developed. This is a very large topic and we do not attempt to cover it here; instead we give a simple and low-order approach to the numerical integration of scattered data over the sphere. For introductions to the general area of meshless discretization methods, see Buhmann [25], Fasshauer [45], and Wendland [118]. For a survey of the approximation and interpolation of meshless data over the unit sphere, including the use of radial basis function methods, wavelets, and spline functions, see the review of Fasshauer and Schumaker [46]; and for an introduction to quadrature of scattered data using radial basis functions, see Sommariva and Womersley [110].

Assume we are given node points $P = \{\boldsymbol{\eta}_1, \ldots, \boldsymbol{\eta}_N\}$ and associated approximate function values $\{f_i : i = 1, \ldots, N\}$, $f_i \approx f(\boldsymbol{\eta}_i)$. We want to approximate

$$I(f) = \int_{\mathbb{S}^2} f(\boldsymbol{\eta})\, dS^2(\boldsymbol{\eta}).$$

The data values f_i will often contain experimental error. We give a simple and straightforward method for estimating $I(f)$.

Let $\mathcal{T}_N = \{\Delta_1, \ldots, \Delta_{M(N)}\}$ denote a triangulation of $\mathbb{S}^2$ in which the vertices of the triangulation are exactly the nodes P. Then

$$I(f) = \sum_{k=1}^{M} \int_{\Delta_k} f(\boldsymbol{\eta})\, dS^2(\boldsymbol{\eta}).$$

To approximate the integral over Δ_k, use

$$\int_{\Delta_k} f(\boldsymbol{\eta})\, dS^2(\boldsymbol{\eta}) \approx \frac{1}{3} \left[f(\boldsymbol{\eta}_{k,1}) + f(\boldsymbol{\eta}_{k,2}) + f(\boldsymbol{\eta}_{k,3}) \right] \operatorname{area}(\Delta_k)$$

in which $\{\boldsymbol{\eta}_{k,1}, \boldsymbol{\eta}_{k,2}, \boldsymbol{\eta}_{k,3}\}$ denotes the vertices of Δ_k. Denote by $I_N(f)$ the resulting approximation of $I(f)$,

$$I_N(f) = \frac{1}{3} \sum_{k=1}^{M} \left[f(\boldsymbol{\eta}_{k,1}) + f(\boldsymbol{\eta}_{k,2}) + f(\boldsymbol{\eta}_{k,3}) \right] \operatorname{area}(\Delta_k). \qquad (5.56)$$

For $\operatorname{area}(\Delta_k)$, see (5.17).

An error analysis can be given similar to that in Theorem 5.9 for the methods (5.24)–(5.26). The method (5.56) has a degree of precision of 0, and as a consequence,

$$|I(f) - I_N(f)| \leq 4\pi c_f h$$

with

$$h = \max_{\Delta \in \mathcal{T}_N} \text{diam}(\Delta).$$

This assumes only that the function f is Lipschitz continuous over $\mathbb{S}^2$ with a Lipschitz constant c_f.

Assume there is error in the function values $\{f(\boldsymbol{\eta}_i)\}$, say

$$f(\boldsymbol{\eta}_i) = \widehat{f}_i + \varepsilon_i, \quad i = 1, \ldots, N, \tag{5.57}$$

and let $I_N(\widehat{f})$ denote the numerical integral based on using the approximate values $\left\{\widehat{f}_i\right\}$. Then it is straightforward to show

$$\left|I_N(f) - I_N(\widehat{f})\right| \leq 4\pi \max_{1 \leq i \leq N} |\varepsilon_i|. \tag{5.58}$$

How is the triangulation $\mathcal{T}_N$ to be chosen? A popular choice is the Delaunay triangulation [21, Chap. 9], in part because it satisfies certain optimality conditions when used for planar data. For a discussion of alternative triangulations, see [46, p. 135].

Example 5.14. We generate a set of nodes that is somewhat uniformly distributed over $\mathbb{S}^2$, while still having some randomness. This is to reflect the sometimes random choice of measurement points $\boldsymbol{\eta}$ in practice. For our example, we begin with an icosahedral-based triangulation, and then we generate one node point $\boldsymbol{\eta}_i$ randomly within each face. For the construction of the Delaunay triangulation, we use MATLAB codes from Burkardt [26]; also, Fortran codes are given in Renka [96]. A graph of a resulting Delaunay triangulation is given in Fig. 5.4 with $N = 80$. Using this triangulation, we apply the quadrature formula (5.56) with $N = 80$ to the integral (5.14) used in our earlier examples. The error is -2.01×10^{-2}, with a relative error of -1.36×10^{-3}.   □

A related problem is to determine a set of nodes $\{\boldsymbol{\eta}_1, \ldots, \boldsymbol{\eta}_N\}$ at which the experimental data is to be calculated. Often this is combined with a desire to estimate $I(f)$ using

$$I(f) \approx I_N(f) \equiv \frac{4\pi}{N} \sum_{k=1}^{N} f(\boldsymbol{\eta}_k).$$

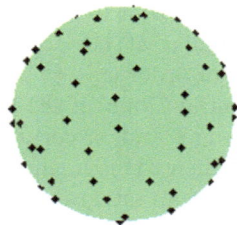 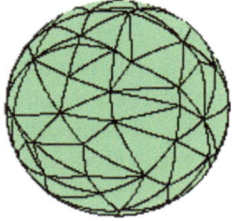

Fig. 5.4 Pseudo-random points on the sphere (*left*) and the Delaunay triangulation of those points (*right*)

If the nodes are determined such that this formula has a degree of precision d, then this formula is called a "spherical d-design". For introductions and extensive discussions of the spherical d-designs, see Bannai and Bannai [20] and Cui and Freeden [34].

In general, it is believed to be more efficient when doing empirical measurements if the points $\{\boldsymbol{\eta}_1, \ldots, \boldsymbol{\eta}_N\}$ at which data is to be measured are "equidistributed" over the sphere. There are many different ways of defining this concept, and there is a large literature on determining such equidistributed point sets. For example, one approach is to compute a set of points $\{\boldsymbol{\eta}_1, \ldots, \boldsymbol{\eta}_N\}$ that maximizes the quantity

$$\min_{1 \le i < j \le N} \left| \boldsymbol{\eta}_i - \boldsymbol{\eta}_j \right|,$$

thus maximizing the distance between the points $\boldsymbol{\eta}_k$. See Saff and Kuijlaars [100] and Hardin and Saff [59] for a discussion of this problem. For a recent construction of spherical d-designs that have a good equidistribution of nodes and are useful for polynomial interpolation over Π_n, see An, Chen, Sloan, and Womersley [4].

5.5 Integration of Singular Functions

Consider the numerical approximation of integrals with a point-singularity in the integrand. Two examples from potential theory are the single-layer and double-layer potentials,

$$\mathcal{S}\rho(\boldsymbol{x}) = \int_{\partial\Omega} \frac{\rho(\boldsymbol{y})}{|\boldsymbol{x} - \boldsymbol{y}|} \, d\sigma(\boldsymbol{y}), \quad \boldsymbol{x} \in \partial\Omega,$$

$$\mathcal{D}\rho(\boldsymbol{x}) = \int_{\partial\Omega} \rho(\boldsymbol{y}) \frac{\partial}{\partial\boldsymbol{\nu_y}} \left(\frac{1}{|\boldsymbol{x} - \boldsymbol{y}|} \right) d\sigma(\boldsymbol{y}), \quad \boldsymbol{x} \in \partial\Omega,$$

respectively. In these integrals, $\partial\Omega$ is the boundary of a simply-connected bounded region in $\mathbb{R}^3$. The integrands are singular at $\boldsymbol{y} = \boldsymbol{x}$. The single-layer potential $\mathcal{S}\rho(\boldsymbol{x})$ has a singular behaviour of order $|\boldsymbol{x} - \boldsymbol{y}|^{-1}$ for $\boldsymbol{y} \approx \boldsymbol{x}$; and the same is true for the double-layer potential $\mathcal{D}\rho(\boldsymbol{x})$ if $\partial\Omega$ is a smooth surface. (If $\partial\Omega$ is not smooth, e.g. when it has edges and/or corners, then $\mathcal{D}\rho$ has a more complicated singular behaviour.) When $\partial\Omega$ is a smooth surface, a change of variables will convert such integrals to an integral over $\mathbb{S}^2$ with a singularity point on the sphere corresponding to $\boldsymbol{x}$; for an example of such a mapping, see Example 5.1.

We begin by first considering the case in which the integrand f in

$$I(f) = \int_{\mathbb{S}^2} f(\boldsymbol{\eta}) \, dS^2(\boldsymbol{\eta})$$

is well-behaved, defining a numerical method which we extend later to singular integrands. Before applying a numerical method, we apply a transformation

$$L : \ \mathbb{S}^2 \xrightarrow[onto]{1-1} \mathbb{S}^2.$$

Using spherical coordinates, define

$$L : \boldsymbol{\eta} = (\cos\phi\sin\theta, \sin\phi\sin\theta, \cos\theta)$$

$$\mapsto \widetilde{\boldsymbol{\eta}} = \frac{(\cos\phi\,\sin^q\theta, \sin\phi\,\sin^q\theta, \cos\theta)}{\sqrt{\cos^2\theta + \sin^{2q}\theta}} \equiv L(\phi, \theta). \qquad (5.59)$$

In this transformation, $q \geq 1$ is a "grading parameter". The north and south poles of U remain fixed, while the region around them is distorted by the mapping.

The integral $I(f)$ becomes

$$I(f) = \int_{\mathbb{S}^2} f(\widetilde{\boldsymbol{\eta}}) \, dS^2(\widetilde{\boldsymbol{\eta}})$$

$$= \int_{\mathbb{S}^2} f(L(\widetilde{\boldsymbol{\eta}})) \, J_L(\widetilde{\boldsymbol{\eta}}) \, dS^2(\widetilde{\boldsymbol{\eta}}) \qquad (5.60)$$

with $J_L(\widetilde{\boldsymbol{\eta}})$ the Jacobian of the mapping L,

$$J_L(\widetilde{\boldsymbol{\eta}}) = |D_\phi L(\phi, \theta) \times D_\theta L(\phi, \theta)| = \frac{\sin^{2q-1}\theta \left(q\cos^2\theta + \sin^2\theta\right)}{\left(\sin^{2q}\theta + \cos^2\theta\right)^{\frac{3}{2}}}. \qquad (5.61)$$

In spherical coordinates,

$$I(f) = \int_0^\pi \frac{\sin^{2q-1}\theta \left(q\cos^2\theta + \sin^2\theta\right)}{\left(\sin^{2q}\theta + \cos^2\theta\right)^{\frac{3}{2}}} \int_0^{2\pi} f(\xi, \eta, \zeta) \, d\phi \, d\theta,$$

$$(\xi, \eta, \zeta) = \frac{(\cos\phi\sin^q\theta, \sin\phi\sin^q\theta, \cos\theta)}{\sqrt{\sin^{2q}\theta + \cos^2\theta}}. \tag{5.62}$$

For $n \geq 1$, let $h = \pi/n$, and

$$\theta_j = jh, \quad 0 \leq j \leq n; \qquad \phi_j = jh, \quad 0 \leq j \leq 2n.$$

For a generic function g, introduce the bivariate trapezoidal approximation

$$\int_0^\pi \int_0^{2\pi} g(\sin\theta, \cos\theta, \sin\phi, \cos\phi)\, d\phi\, d\theta$$

$$\approx h^2 \sum_{k=0}^n{}'' \sum_{j=0}^{2n}{}'' g(\sin\theta_k, \cos\theta_k, \sin\phi_j, \cos\phi_j)$$

in which the superscript notation $''$ means to multiply the first and last terms by $\frac{1}{2}$ before summing. Apply this to (5.62). Note that the integrand is zero for $\theta = 0, \pi$ and that the integrand has period 2π in ϕ. Therefore

$$\int_0^\pi \int_0^{2\pi} g(\sin\theta, \cos\theta, \sin\phi, \cos\phi)\, d\phi\, d\theta$$

$$\approx h^2 \sum_{k=1}^{n-1} \sum_{j=1}^{2n} g(\sin\theta_k, \cos\theta_k, \sin\phi_j, \cos\phi_j) \equiv I_n, \tag{5.63}$$

$$g = \frac{\sin^{2q-1}\theta\left(q\cos^2\theta + \sin^2\theta\right)}{\left(\sin^{2q}\theta + \cos^2\theta\right)^{\frac{3}{2}}} f(\xi, \eta, \zeta),$$

with (ξ, η, ζ) as in (5.62).

To analyze the convergence of this method, we apply the following result; see [9, p. 285] or most other numerical analysis textbooks for a discussion of this theorem.

Theorem 5.15. (EULER–MACLAURIN FORMULA) *Let $m \geq 0$, $n \geq 1$, and define $h = (b - a)/n$, $x_j = a + jh$ for $j = 0, 1, \ldots, n$. Further assume ψ is $2m + 2$ times differentiable on $[a, b]$ with $\psi^{(2m+2)} \in L^1(a, b)$. Then*

$$\int_a^b \psi(x)\, dx - h \sum_{j=0}^n{}' \psi(x_j) = \sum_{i=1}^m \frac{B_{2i}}{(2i)!} h^{2i} \left[\psi^{(2i-1)}(b) - \psi^{(2i-1)}(a)\right]$$

$$+ \frac{h^{2m+2}}{(2m+2)!} \int_a^b \overline{B}_{2m+2}\left(\frac{x-a}{h}\right) \psi^{(2m+2)}(x)\, dx. \tag{5.64}$$

In this formula, $\{B_k\}$ are the Bernoulli constants, $B_k(x)$ is the Bernoulli polynomial of degree k, and $\overline{B}_k(x)$ is the periodic extension of $B_k(x)$ on $[0,1]$. For definitions of $\{B_k\}$ and $\{B_k(x)\}$, see [9, Sect. 5.4].

Theorem 5.16. *Apply the trapezoidal rule*

$$I \equiv \int_0^\pi \gamma(\theta)\, d\theta \approx T_n \equiv h \sum_{j=0}^{n}{}'' \gamma(jh), \quad h = \frac{\pi}{n}$$

to the integral

$$\int_0^\pi t(\theta) \sin^m \theta\, d\theta$$

with $t(\theta)$ a sufficiently differentiable function and $m \geq 0$ an integer. More precisely, assume $t^{(p)} \in L^1(0,\pi)$ with

$$p = \begin{cases} m+2, & m \text{ even,} \\ m+1, & m \text{ odd.} \end{cases}$$

Then

$$I - T_n = \mathcal{O}(h^p). \tag{5.65}$$

The proof is an immediate corollary of the Euler–MacLaurin expansion (5.64) The theorem generalizes to non-integer values p; see Sidi [104, Appendix D], [105].

Theorem 5.17. *In the integral (5.62), assume $q \geq 1$ and $2q$ is a positive integer. Introduce*

$$p = \begin{cases} 2q, & 2q \text{ even,} \\ 2q+1, & 2q \text{ odd.} \end{cases}$$

Assume f is p-times differentiable with $f^{(p)} \in L^1(\mathbb{S}^2)$. Then the error in approximating (5.62) by (5.63) satisfies

$$I - I_n = \mathcal{O}(h^p). \tag{5.66}$$

A proof of this is given in [11, Theorem 2.1], and extensions to non-integer values of $2q$ are given in [105]. The values of q with $2q$ an odd integer are particularly good. In such cases, the rate of convergence improves to $I - I_n = \mathcal{O}(h^{4q})$; see [16, 105].

5.5.1 *Singular Integrands*

We examine the application of the above schema to the integral

$$I = \int_{\partial\Omega} \frac{\rho(\boldsymbol{y})}{|\boldsymbol{y} - \boldsymbol{x}|}\, d\sigma(\boldsymbol{y}), \tag{5.67}$$

where Ω is an open, bounded simply connected region in $\mathbb{R}^3$, its boundary $\partial\Omega$ is a smooth surface, and $\boldsymbol{x} \in \partial\Omega$. Other integral singularities can be treated similarly. As earlier in (5.2)–(5.3), we apply a transformation of variables

$$\mathcal{M} : \mathbb{S}^2 \xrightarrow[\text{onto}]{1\text{--}1} \partial\Omega,$$

obtaining

$$I = \int_{\mathbb{S}^2} \frac{\rho(\mathcal{M}(\boldsymbol{\eta}))\, J_{\mathcal{M}}(\boldsymbol{\eta})}{|\boldsymbol{x} - \mathcal{M}(\boldsymbol{\eta})|}\, dS^2(\boldsymbol{\eta}). \tag{5.68}$$

Since the transformation L of (5.59) is based on smoothing the integrand at the north pole $(0, 0, 1)$ and south pole $(0, 0, -1)$ of $\mathbb{S}^2$, the original coordinate system of $\mathbb{R}^3$ needs to be rotated to have one of the poles of $\mathbb{S}^2$ in the rotated system be the location of the singularity in the integrand.

Let $\mathcal{M}(\widehat{\boldsymbol{\xi}}) = \boldsymbol{x}$, $\widehat{\boldsymbol{\xi}} \in \mathbb{S}^2$. We introduce an orthogonal Householder transformation of $\mathbb{R}^3$,

$$\boldsymbol{\eta} = \mathcal{H}\boldsymbol{\eta}^*, \quad \boldsymbol{\eta}^* \in \mathbb{S}^2 \tag{5.69}$$

before we apply the final mapping L. Choose a Householder matrix $\mathcal{H} = I - 2\,\boldsymbol{w}\boldsymbol{w}^{\mathrm{T}}$, $\boldsymbol{w}^{\mathrm{T}}\boldsymbol{w} = 1$, such that

$$\mathcal{H}\begin{bmatrix} 0 \\ 0 \\ \pm 1 \end{bmatrix} = \widehat{\boldsymbol{\xi}}, \quad \text{or equivalently,} \quad \mathcal{H}\widehat{\boldsymbol{\xi}} = \begin{bmatrix} 0 \\ 0 \\ \pm 1 \end{bmatrix} \tag{5.70}$$

with the sign chosen later to minimize any loss of significance error. The requirement (5.70) means that $\widehat{\boldsymbol{\xi}}$ will be mapped to either the north or south pole of $\mathbb{S}^2$, or conversely, a pole of $\mathbb{S}^2$ is mapped to $\widehat{\boldsymbol{\xi}}$.

Finding $\boldsymbol{w}$ is straightforward and inexpensive, requiring only a small number of arithmetic operations; see [9, Sect. 9.3]. In evaluating (5.69), use

$$\begin{aligned} \boldsymbol{\eta} &= \mathcal{H}\boldsymbol{\eta}^* \\ &= \boldsymbol{\eta}^* - \boldsymbol{w}\left(2\,\boldsymbol{w}^{\mathrm{T}}\boldsymbol{\eta}^*\right) \\ &= \boldsymbol{\eta}^* - c\,\boldsymbol{w}, \quad c = 2\,\boldsymbol{w}^{\mathrm{T}}\boldsymbol{\eta}^* \end{aligned}$$

Computing $2\,\boldsymbol{w}^{\mathrm{T}}\boldsymbol{\eta}^*$ requires four multiplications and two additions; and computing $\boldsymbol{\eta}$ requires a further three multiplications and three subtractions (a total of 12 arithmetic operations for $\boldsymbol{\eta}^* \mapsto \boldsymbol{\eta}$).

In the integral (5.68), the transformation $\boldsymbol{\eta} = \mathcal{H}\boldsymbol{\eta}^*$ yields

$$I = \int_{\mathbb{S}^2} \frac{\rho(\mathcal{M}(\mathcal{H}\boldsymbol{\eta}^*))\, J_{\mathcal{M}}(\mathcal{H}\boldsymbol{\eta}^*)}{|\boldsymbol{x} - \mathcal{M}(\mathcal{H}\boldsymbol{\eta}^*)|}\, dS^2(\boldsymbol{\eta}^*)$$

since the absolute value of the determinant of the Jacobian of the mapping is 1. Now use the mapping

$$\boldsymbol{\eta}^* = L(\widetilde{\boldsymbol{\eta}})\,, \quad \widetilde{\boldsymbol{\eta}} \in \mathbb{S}^2$$

as before in (5.61)–(5.62), yielding

$$I = \int_{\mathbb{S}^2} \frac{\rho(\mathcal{M}(\mathcal{H}L(\widetilde{\boldsymbol{\eta}})))\, J_{\mathcal{M}}(\mathcal{H}L(\widetilde{\boldsymbol{\eta}}))}{|P - \mathcal{M}(\mathcal{H}L(\widetilde{\boldsymbol{\eta}}))|}\, J_L(\widetilde{\boldsymbol{\eta}})\, dS^2(\widetilde{\boldsymbol{\eta}})\,. \tag{5.71}$$

Now apply the scheme of (5.63) to I, obtaining a numerical approximation I_n of I. For the convergence of I_n to I, we have the following.

Theorem 5.18. *Let $q \geq 1$ be an integer, and introduce*

$$p = \begin{cases} q, & q \text{ even,} \\ q+1, & q \text{ odd.} \end{cases}$$

In the integral (5.71), assume that $\rho \in C^p(\partial\Omega)$. Assume that the surface $\partial\Omega$ is similarly differentiable, which is equivalent to assuming that the mapping $\mathcal{M}$ is suitably differentiable. Let I_n be the approximation of (5.71) based on the schema of (5.63). Then

$$I - I_n = \mathcal{O}(h^p)\,. \tag{5.72}$$

A proof is given in [11, Theorem 4.2]. It is also conjectured there that

$$I - I_n = \mathcal{O}(h^q) \tag{5.73}$$

holds for an arbitrary real $q \geq 1$; there are possibly improved rates for special choices of q, as with q an odd integer in (5.72). These conjectures are illustrated in the following example.

Example 5.19. We give numerical results for

$$I = \int_{\partial\Omega} \frac{e^{0.1(y_1+2y_2+3y_3)}}{|\boldsymbol{x} - \boldsymbol{y}|}\, d\sigma(\boldsymbol{y})\,. \tag{5.74}$$

Table 5.7 Numerical integrals for (5.74) with $q = 2.5$ and $q = 3$

n	N	$q = 2.5$			$q = 3$	
		$I_n - I_{\frac{1}{2}n}$	Ratio	EOC	$I_n - I_{\frac{1}{2}n}$	Ratio
4	24	1.99E + 1			2.21E + 01	
8	112	−5.93E − 1	−33.65		−2.46E + 00	−8.96
16	480	1.58E − 1	−3.76		−7.66E − 02	32.2
32	1,984	2.56E − 2	6.16	2.62	−1.00E − 03	76.5
64	8,064	4.53E − 3	5.64	2.50	7.07E − 08	−14,168
128	32,512	8.01E − 4	5.66	2.50	3.22E − 10	219
256	130,560	1.42E − 4	5.66	2.50	5.04E − 12	63.9
512	523,264	2.50E − 5	5.66	2.50	5.68E − 14	88.8
1,024	2,095,104	4.43E − 6	5.66	2.50	0	

We use the ellipsoidal surface $\partial\Omega$ of (5.4) with the surface parameters $(a, b, c) = (1, 2, 3)$. We choose the point $\boldsymbol{x}$ to correspond to the spherical coordinates $(\theta, \phi) = (\pi/4, \pi/4)$ under the mapping $\mathcal{M}$ of (5.5). In this case, the true value of I is given by

$$I \doteq 38.254918969803924.$$

In Table 5.7, we give numerical results for the choices $q = 2.5$ and $q = 3$. The column N gives the number of node points on $\mathbb{S}^2$ and the column EOC gives the estimated order of convergence for the case $q = 2.5$, as then the error has a regular behaviour. The rate of convergence for the $q = 2.5$ case agrees with the conjecture in (5.73). The case $q = 3$ is much better than predicted by (5.72), since an order $\mathcal{O}(h^4)$ would have the values of *Ratio* approaching 16 with increasing n. Clearly the convergence is much more rapid than $\mathcal{O}(h^4)$.

$\square$

5.6 Quadrature over the Unit Disk

Related to quadrature over the unit sphere $\mathbb{S}^2$ is quadrature over the unit disk $\mathbb{D}$ in the plane $\mathbb{R}^2$,

$$\mathbb{D} = \left\{ (x, y) : x^2 + y^2 \leq 1 \right\}.$$

We begin with a connection between these two domains, and then we discuss some numerical methods for quadrature over the unit disk.

Begin by looking at the upper hemisphere $\mathbb{S}^2_+$ as the image of the mapping

$$z = \sqrt{1 - x^2 - y^2}, \qquad (x, y) \in \mathbb{D}.$$

Then the surface integral over this hemisphere,

$$\int_{\mathbb{S}^2_+} f(\boldsymbol{\eta}) \, dS^2(\boldsymbol{\eta})$$

can be transformed to

$$\int_D f\left(x, y, \sqrt{1 - x^2 - y^2}\right) \sqrt{1 + \left(\frac{\partial z}{\partial x}\right)^2 + \left(\frac{\partial z}{\partial y}\right)^2} \, dx \, dy$$

$$= \int_D f\left(x, y, \sqrt{1 - x^2 - y^2}\right) \frac{dx \, dy}{\sqrt{1 - x^2 - y^2}}.$$

Consequently, we can write

$$\int_{\mathbb{S}^2} f(\boldsymbol{\eta}) \, dS^2(\boldsymbol{\eta}) = \int_D \left[f\left(x, y, \sqrt{1 - x^2 - y^2}\right) \right.$$

$$\left. + f\left(x, y, -\sqrt{1 - x^2 - y^2}\right) \right] \frac{dx \, dy}{\sqrt{1 - x^2 - y^2}}. \qquad (5.75)$$

Integration over $\mathbb{S}^2$ is equivalent to a weighted integration over the unit disk $\mathbb{D}$.

5.6.1   A Product Gauss Formula

For the integral

$$I(f) = \int_{\mathbb{D}} f(x, y) \, dx \, dy,$$

we develop a numerical method in a fashion analogous to that of the product Gauss formula (5.12). Using polar coordinates, we write

$$I(f) = \int_0^{2\pi} \int_0^1 r \, f(r \cos \theta, r \sin \theta) \, dr \, d\theta. \qquad (5.76)$$

For the integration in θ, we use the trapezoidal rule (5.7), as in (5.12). For the integration in r, we apply Gauss–Legendre quadrature to the integrand $r \, f(r \cos \theta, r \sin \theta)$.

Using Gauss–Legendre quadrature, we introduce the formula

$$I_n(f) = h \sum_{j=0}^{2n} \sum_{k=0}^{n} w_k r_k f(r_k \cos \theta_j, r_k \sin \theta_j). \qquad (5.77)$$

The nodes $\theta_j = jh$, $j = 0, 1, \ldots, 2n$, $h = 2\pi/(2n+1)$; and $\{w_k\}$ and $\{r_k\}$ are the weights and nodes, respectively, for the $(n+1)$-point Gauss–Legendre quadrature formula on $[0, 1]$.

Theorem 5.20. *If $f(x, y)$ is a polynomial of degree $\leq 2n$, then $I(f) = I_n(f)$. The quadrature formula (5.77) has degree of precision $2n$.*

Proof. Let $f(x, y) = x^\alpha y^\beta$, with α, β non-negative integers and $\alpha + \beta \leq m$. Then

$$I(f) = \left(\int_0^{2\pi} (\cos\theta)^\alpha (\sin\theta)^\beta \, d\theta \right) \left(\int_0^1 r^{\alpha+\beta+1} \, dr \right) \equiv J^{\alpha,\beta} K^{\alpha+\beta+1},$$

$$I_n(f) = \left(h \sum_{j=0}^{2n} (\cos\theta_j)^\alpha (\sin\theta_j)^\beta \right) \left(\sum_{k=0}^n w_k r_k^{\alpha+\beta+1} \right) \equiv J_n^{\alpha,\beta} K_n^{\alpha+\beta+1}$$

with

$$K^\ell = \int_0^1 r^\ell \, dr, \quad K_n^\ell = \sum_{k=0}^n w_k r_k^\ell.$$

If β is odd, the integrals $J^{\alpha,\beta} = J_n^{\alpha,\beta} = 0$, much as was done in the proof of Theorem 5.4. If β is even, we can convert the integrand $(\cos\theta)^\alpha (\sin\theta)^\beta$ into a polynomial of powers of $\cos\theta$ with degree $\alpha + \beta$. Using Lemma 5.2, we can then show

$$J_n^{\alpha,\beta} = J^{\alpha,\beta}, \quad \alpha + \beta \leq 2n.$$

The $(n+1)$-point Gauss–Legendre quadrature formula has a degree of precision of $2n + 1$, from which

$$K_n^{\alpha+\beta+1} = K^{\alpha+\beta+1}, \quad \alpha + \beta \leq 2n.$$

This proves $I_n(f) = I(f)$ for all $\alpha, \beta \geq 0$, $0 \leq \alpha + \beta \leq 2n$.

To show that (5.77) has degree of precision exactly $2n$, consider the polynomial $f(x, y) = r (r \cos\theta)^{2n+1}$. Then $J^{2n+1,0} = 0$ and both $J_n^{2n+1,0}$ and K_n^{2n+2} are nonzero. $\qquad\square$

5.7 Discrete Orthogonal Expansions

Recall the orthogonal expansions described in Chap. 4. In particular, recall the Laplace expansion (4.55) on $\mathbb{S}^2$ and the orthogonal polynomial expansion (4.80) over the unit disk D. In both cases the coefficients in the expansion are inner products, and generally they must be evaluated numerically. In this section we use product Gauss formulas to evaluate these integrals, obtaining approximations to the orthogonal projections $\mathcal{Q}_n$ defined earlier in (4.56) and (4.81).

5.7.1 *Hyperinterpolation over* $\mathbb{S}^2$

Recall the truncated Laplace expansion: for $f \in C(\mathbb{S}^2)$,

$$\mathcal{Q}_n f(\boldsymbol{\eta}) = \sum_{m=0}^{n} \sum_{k=1}^{2m+1} (f, Y_{m,k}) \, Y_{m,k}(\boldsymbol{\eta}), \qquad \boldsymbol{\eta} \in \mathbb{S}^2$$

with $\mathcal{Q}_n$ the associated orthogonal projection operator from $L^2(\mathbb{S}^2)$ onto Π_n; see (4.55). We introduce a "discrete inner product" using the product Gaussian quadrature formula (5.12), but with n replaced with $n + 1$,

$$I_{n+1}(F) = h \sum_{j=0}^{2n+1} \sum_{k=0}^{n} w_k F(\cos \phi_j \sin \theta_k, \sin \phi_j \sin \theta_k, \cos \theta_k). \qquad (5.78)$$

The nodes $\{z_\ell = \cos \theta_\ell\}$ and weights $\{w_\ell\}$ are for the $(n + 1)$-point Gauss–Legendre formula on $[0, 1]$; and for the azimuthal angle stepsize, $h = \pi / (n + 1)$. This formula has degree of precision $2n + 1$.
Define

$$(f, g)_n = I_{n+1}(fg), \qquad (5.79)$$

From the degree of precision of (5.78), this discrete inner product has the important property that

$$(f, g)_n = (f, g) \qquad \forall f, g \in \Pi_n(\mathbb{S}^2). \qquad (5.80)$$

Using (5.79), define

$$\widetilde{\mathcal{Q}}_n f(\boldsymbol{\eta}) = \sum_{m=0}^{n} \sum_{k=1}^{2m+1} (f, Y_{m,k})_n \, Y_{m,k}(\boldsymbol{\eta}), \quad \boldsymbol{\eta} \in \mathbb{S}^2. \qquad (5.81)$$

Applying (5.80),

$$\widetilde{\mathcal{Q}}_n f = f \quad \forall f \in \Pi_n(\mathbb{S}^2),$$

thus making $\widetilde{\mathcal{Q}}_n$ a projection operator from $C(\mathbb{S}^2)$ onto $\Pi_n(\mathbb{S}^2)$. The calculation of $\widetilde{\mathcal{Q}}_n f$ using (5.81) is commonly referred to as "hyperinterpolation". The operator $\widetilde{\mathcal{Q}}_n$ is also called a "discrete orthogonal projection operator".

For analyzing the error in $\widetilde{\mathcal{Q}}_n f$, we can follow exactly the derivation of (4.61), leading to

$$\| f - \widetilde{\mathcal{Q}}_n f \|_\infty \leq \left(1 + \| \widetilde{\mathcal{Q}}_n \|_{C \to C} \right) E_{n,\infty}(f). \tag{5.82}$$

Recalling (4.64), the orthogonal projection operator $\mathcal{Q}_n$ satisfies

$$\| \mathcal{Q}_n \|_{C \to C} = \mathcal{O}(\sqrt{n}).$$

Sloan and Womersley [106, Theorem 5.5.4] extended this as follows.

Theorem 5.21. *For the hyperinterpolation operator of* (5.81),

$$\| \widetilde{\mathcal{Q}}_n \|_{C \to C} = \mathcal{O}(\sqrt{n}). \tag{5.83}$$

The proof of this is quite complicated, and therefore, we prove only the weaker result

$$\| \widetilde{\mathcal{Q}}_n \|_{C \to C} = \mathcal{O}(n). \tag{5.84}$$

Before giving the proof of this result, we introduce some additional framework that is also useful in examining interpolation and other approximation theory problems over $\mathbb{S}^2$.

Begin by recalling the reproducing kernel

$$G_n(\boldsymbol{\xi}, \boldsymbol{\eta}) = \sum_{m=0}^{n} \sum_{k=1}^{2m+1} Y_{m,k}(\boldsymbol{\xi}) Y_{m,k}(\boldsymbol{\eta}), \quad \boldsymbol{\xi}, \boldsymbol{\eta} \in \mathbb{S}^2;$$

see (2.34) in Chap. 2. This satisfies

$$(G_n(\cdot, \boldsymbol{\eta}), p) = p(\boldsymbol{\eta}), \quad \boldsymbol{\eta} \in \mathbb{S}^2, \ \forall p \in \Pi_n, \tag{5.85}$$

using the standard L^2-inner product $(\cdot, \cdot)$ over $\mathbb{S}^2$. This property implies that G_n is unique; and thus G_n is independent of the particular basis of spherical harmonics being used. This is true in general for reproducing kernels. Recalling (2.30), the above formula for G_n can be simplified greatly, to

$$G_n(\boldsymbol{\xi}, \boldsymbol{\eta}) = \widetilde{G}_n(\boldsymbol{\xi} \cdot \boldsymbol{\eta}),$$

$$\widetilde{G}_n(t) = P_n^{(0,1)}(t); \tag{5.86}$$

also see [50, p. 1399]. This is the basis of formula (4.57) for $\mathcal{Q}_n f$. To simplify notation in the remainder of this section, let $\{Y_1, \dots, Y_N\}$ denote the basis $\{Y_{m,k} : 1 \le k \le 2m + 1, \, 0 \le m \le n\}$ for Π_n, $N = (n + 1)^2$.

Recall the discussion in Sect. 5.3.3 of interpolation using spherical polynomials. Let $\left\{\boldsymbol{\eta}_j : j = 1, \dots, N = (n+1)^2\right\}$ be a fundamental system, for example, one that maximizes the determinant function in (5.51), thus guaranteeing that the interpolation matrix

$$
\Psi_n = \begin{bmatrix}
Y_1(\boldsymbol{\eta}_1) & \cdots & Y_k(\boldsymbol{\eta}_1) & \cdots & Y_N(\boldsymbol{\eta}_1) \\
\vdots & \ddots & \vdots & \ddots & \vdots \\
Y_1(\boldsymbol{\eta}_\ell) & \cdots & Y_k(\boldsymbol{\eta}_\ell) & \cdots & Y_N(\boldsymbol{\eta}_\ell) \\
\vdots & \ddots & \vdots & \ddots & \vdots \\
Y_1(\boldsymbol{\eta}_N) & \cdots & Y_k(\boldsymbol{\eta}_N) & \cdots & Y_N(\boldsymbol{\eta}_N)
\end{bmatrix}
$$

is nonsingular. Introduce the functions

$$
g_j(\boldsymbol{\xi}) = G_n(\boldsymbol{\xi}, \boldsymbol{\eta}_j), \quad j = 1, \dots, N. \tag{5.87}
$$

These are another basis for Π_n. To see this, introduce the interpolation matrix

$$
K_n = \begin{bmatrix}
g_1(\boldsymbol{\eta}_1) & \cdots & g_k(\boldsymbol{\eta}_1) & \cdots & g_N(\boldsymbol{\eta}_1) \\
\vdots & \ddots & \vdots & \ddots & \vdots \\
g_1(\boldsymbol{\eta}_\ell) & \cdots & g_k(\boldsymbol{\eta}_\ell) & \cdots & g_N(\boldsymbol{\eta}_\ell) \\
\vdots & \ddots & \vdots & \ddots & \vdots \\
g_1(\boldsymbol{\eta}_N) & \cdots & g_k(\boldsymbol{\eta}_N) & \cdots & g_N(\boldsymbol{\eta}_N)
\end{bmatrix}
$$

whose elements are calculated easily from (5.86). Then

$$
K_n = \Psi_n^{\mathrm{T}} \Psi_n,
$$

implying that K_n is nonsingular. The matrix K_n occurs when interpolating data using linear combinations of $\{g_1, \dots, g_N\}$: given data values $\{f_1, \dots, f_N\}$, find

$$
(\mathcal{I}_n f)(\boldsymbol{\eta}) = \sum_{j=1}^{N} \alpha_j g_j(\boldsymbol{\eta})
$$

such that

$$
(\mathcal{I}_n f)(\boldsymbol{\eta}_i) = f_i, \quad i = 1, \dots, N.
$$

The coefficients $\{\alpha_j\}$ are obtained by solving the linear system

$$K_n \left[\alpha_1, \ldots, \alpha_N\right]^{\mathrm{T}} = \left[f_1, \ldots, f_N\right]^{\mathrm{T}}.$$

Returning to (5.81) for $\widetilde{\mathcal{Q}}_n f$, write it as

$$\widetilde{\mathcal{Q}}_n f(\boldsymbol{\eta}) = \sum_{k=1}^{N} (f, Y_k)_n \, Y_k(\boldsymbol{\eta}), \quad \boldsymbol{\eta} \in \mathbb{S}^2. \tag{5.88}$$

Write the quadrature formula (5.78) as

$$I_{n+1}(F) = \sum_{j=1}^{M} w_j F(\boldsymbol{\tau}_j).$$

with $M = 2(n+1)^2$, and recall the definition (5.79) for $(\cdot, \cdot)_n$. If we expand the formula (5.88) and re-arrange terms, we obtain

$$\widetilde{\mathcal{Q}}_n f(\boldsymbol{\eta}) = \sum_{j=1}^{M} w_j f(\boldsymbol{\tau}_j) \gamma_j(\boldsymbol{\eta}), \tag{5.89}$$

$$\gamma_j(\boldsymbol{\eta}) = G_n(\boldsymbol{\eta}, \boldsymbol{\tau}_j) = \widetilde{G}_n(\boldsymbol{\eta}{\cdot}\boldsymbol{\tau}_j), \quad j = 1, \ldots, M.$$

Lemma 5.22.

$$\|\widetilde{\mathcal{Q}}_n\|_{C \to C} = \max_{\boldsymbol{\eta} \in \mathbb{S}^2} \sum_{j=1}^{M} w_j \, |\gamma_j(\boldsymbol{\eta})|. \tag{5.90}$$

Proof. From (5.89),

$$\left|\widetilde{\mathcal{Q}}_n f(\boldsymbol{\eta})\right| \leq \|f\|_\infty \sum_{j=1}^{M} w_j \, |\gamma_j(\boldsymbol{\eta})|,$$

$$\|\widetilde{\mathcal{Q}}_n f\|_\infty \leq \|f\|_\infty \max_{\boldsymbol{\eta} \in \mathbb{S}^2} \sum_{j=1}^{M} w_j \, |\gamma_j(\boldsymbol{\eta})|,$$

$$\|\widetilde{\mathcal{Q}}_n\|_\infty \leq \max_{\boldsymbol{\eta} \in \mathbb{S}^2} \sum_{j=1}^{M} w_j \, |\gamma_j(\boldsymbol{\eta})|. \tag{5.91}$$

To prove equality, choose a point $\boldsymbol{\eta}_0 \in \mathbb{S}^2$ for which the maximum in the sum on the right side of (5.90) is attained. Then construct a continuous function $\widehat{f}$ for which

$$\widehat{f}(\boldsymbol{\tau}_j) = \operatorname{sign} \gamma_j(\boldsymbol{\eta}_0), \quad j = 1, \ldots, M.$$

and $\|\widehat{f}\|_\infty = 1$. For this function, we have

$$\widetilde{\mathcal{Q}}_n \widehat{f}(\boldsymbol{\eta}_0) = \sum_{j=1}^{M} w_j \, |\gamma_j(\boldsymbol{\eta}_0)|,$$

$$\|\widetilde{\mathcal{Q}}_n \widehat{f}\|_\infty \geq \max_{\boldsymbol{\eta} \in \mathbb{S}^2} \sum_{j=1}^{M} w_j \, |\gamma_j(\boldsymbol{\eta})|,$$

$$\|\widetilde{\mathcal{Q}}_n\|_\infty \geq \max_{\boldsymbol{\eta} \in \mathbb{S}^2} \sum_{j=1}^{M} w_j \, |\gamma_j(\boldsymbol{\eta})|.$$

When combined with (5.91), this completes the proof of (5.90).   □

Proof of (5.84). Using (5.90), let $\boldsymbol{\eta}_0$ be a point for which the maximum in the sum on the right side of (5.90) is attained,

$$\|\widetilde{\mathcal{Q}}_n\|_{C \to C} = \sum_{j=1}^{M} w_j \, |\gamma_j(\boldsymbol{\eta}_0)| = \sum_{j=1}^{M} w_j \, |G_n(\boldsymbol{\eta}_0, \boldsymbol{\tau}_j)|$$

$$= \sum_{j=1}^{M} \sqrt{w_j} \left(\sqrt{w_j} \, |G_n(\boldsymbol{\eta}_0, \boldsymbol{\tau}_j)| \right)$$

$$\leq \left(\sum_{j=1}^{M} w_j \right)^{\frac{1}{2}} \left(\sum_{j=1}^{M} w_j \, [G_n(\boldsymbol{\eta}_0, \boldsymbol{\tau}_j)]^2 \right)^{\frac{1}{2}}.$$

The last step uses the Cauchy–Schwarz inequality. Note that $[G_n(\boldsymbol{\eta}_0, \boldsymbol{\tau})]^2$ is a spherical polynomial of degree $2n$. The integration formula (5.78) has degree of precision $2n + 1$, and therefore

$$\sum_{j=1}^{M} w_j \, [G_n(\boldsymbol{\eta}_0, \boldsymbol{\tau}_j)]^2 = \int_{\mathbb{S}^2} [G_n(\boldsymbol{\eta}_0, \boldsymbol{\tau})]^2 \, dS^2(\boldsymbol{\tau}).$$

Apply the reproducing kernel property (5.85) with $p = G_n(\boldsymbol{\eta}_0, \cdot)$, obtaining

$$\int_{\mathbb{S}^2} [G_n(\boldsymbol{\eta}_0, \boldsymbol{\tau})]^2 \, dS^2(\boldsymbol{\tau}) = G_n(\boldsymbol{\eta}_0, \boldsymbol{\eta}_0).$$

From (5.86),

$$G_n(\boldsymbol{\eta}_0, \boldsymbol{\eta}_0) = \widetilde{G}_n(1) = P_n^{(0,1)}(1).$$

From Reimer [95],

$$P_n^{(0,1)}(1) = \frac{(n+1)^2}{4\pi}.$$

Combining these results, we have

$$\|\widetilde{\mathcal{Q}}_n\|_{C \to C} \leq \sqrt{4\pi}\sqrt{P_n^{(0,1)}(1)} \leq n+1,$$

thus proving (5.84).   □

The proof of the much better result (5.83) is more complicated, requiring a more careful analysis of the properties of the quadrature method; see [106, pp. 102–114]. The paper [106] examines other quadrature schemes for the discrete inner product $(\cdot, \cdot)_n$, although (5.79) seems the most widely used quadrature method.

Returning to the convergence analysis of $\widetilde{\mathcal{Q}}_n f$, use (5.82) to show that the rate of uniform convergence of $\widetilde{\mathcal{Q}}_n f$ to f should be the same as that of $\mathcal{Q}_n f$ to f. As with the derivation of Corollary 4.14, we have the following result.

Corollary 5.23. *Assume that $f \in C(\mathbb{S}^2)$ is Hölder continuous with exponent $\alpha \in (\frac{1}{2}, 1]$. Then*

$$\|f - \widetilde{\mathcal{Q}}_n f\|_\infty \leq \frac{c}{n^{\alpha - 1/2}}$$

for a suitable constant $c > 0$. The discrete orthogonal projection $\widetilde{\mathcal{Q}}_n f$ is uniformly convergent to f in $C(\mathbb{S}^2)$ as $n \to \infty$.

Computational cost. When compared with the Laplace expansion using spherical harmonics, methods based on the numerical evaluation of Fourier series have had the advantage of efficient evaluation by means of the fast Fourier transform; e.g. see [60], [61, Chap. 13], and [92, Sect. 10.9.2]. Improving the efficiency of evaluating the Laplace expansion while maintaining its accuracy has been a long term goal of many researchers. In chronological order, we note in particular the papers of Orszag [90], Swarztrauber [112,113], Driscoll and Healy [42], and Swarztrauber and Spotz [114]. Recent research, however, has led to methods of evaluating spherical polynomials and numerically approximating the discrete orthogonal projection $\mathcal{Q}_n f$ that are competitive with the fast Fourier transform; in particular, see Keiner and Potts [68], Mohlenkamp [83], and Rokhlin and Tyger [98]. The number of arithmetic operations to evaluate the coefficients in $\widetilde{\mathcal{Q}}_n f(\boldsymbol{\eta})$ appears at first hand to be $\mathcal{O}(n^4)$. An algorithm is given in [83] that reduces this to $\mathcal{O}(n^2 \log^2 n)$; and this is the same as the cost of evaluating a discrete double Fourier series using trigonometric polynomials of degree n in each variable.

5.7.2   Hyperinterpolation over the Unit Disk

Recall the truncated orthogonal polynomial expansion (4.80) over the unit disk $\mathbb{D}$:

$$\mathcal{Q}_n f(\boldsymbol{x}) = \sum_{m=0}^{n} \sum_{k=0}^{m} (f, \varphi_k^m) \, \varphi_k^m(\boldsymbol{x}), \quad \boldsymbol{x} \in \mathbb{D}$$

with $\{\varphi_k^m : 0 \le k \le m, \ m \ge 0\}$ a complete family of orthonormal polynomials over $\mathbb{D}$. Proceed as in Sect. 5.6. We apply the product Gauss formula (5.77) to define

$$(f, g)_n = I_n(fg). \tag{5.92}$$

Define

$$\widetilde{\mathcal{Q}}_n f(\boldsymbol{x}) = \sum_{m=0}^{n} \sum_{k=0}^{m} (f, \varphi_k^m)_n \, \varphi_k^m(\boldsymbol{x}), \quad \boldsymbol{x} \in \mathbb{D}. \tag{5.93}$$

Using Theorem 5.20, we have

$$(f, g)_n = (f, g) \quad \forall f, g \in \Pi_n(\mathbb{S}^2).$$

This implies that $\widetilde{\mathcal{Q}}_n$ is a projection from $C(\mathbb{D})$ onto $\Pi_n(\mathbb{R}^2)$, and as before

$$\| f - \widetilde{\mathcal{Q}}_n f \|_\infty \le \left(1 + \| \widetilde{\mathcal{Q}}_n \|_{C \to C} \right) E_{n,\infty}(f). \tag{5.94}$$

It is shown in [58] that

$$\| \widetilde{\mathcal{Q}}_n \|_{C \to C} = \mathcal{O}(n \log n).$$

This is only slightly worse than the result $\| \mathcal{Q}_n \|_{C \to C} = \mathcal{O}(n)$ of (4.85). As in (4.86), we have the following.

Corollary 5.24. *Assume $f \in C^r(\mathbb{D})$ with $r \ge 1$; and further assume that the rth-derivatives satisfy a Hölder condition with exponent $\alpha \in (0, 1]$. Then $\widetilde{\mathcal{Q}}_n f$ converges uniformly to f on $\mathbb{S}^2$, and*

$$\| f - \widetilde{\mathcal{Q}}_n f \|_\infty \le \frac{c \log n}{n^{r-1+\alpha}}, \quad n \ge 2 \tag{5.95}$$

for some constant $c \ge 0$.

Chapter 6
Applications: Spectral Methods

This chapter begins with two illustrations of the application of the material from the preceding chapters. The first, given in Sect. 6.1, is to solve the Dirichlet problem

$$-\Delta u(\boldsymbol{x}) = 0, \qquad \boldsymbol{x} \in \Omega \subset \mathbb{R}^3, \tag{6.1}$$

$$u(\boldsymbol{x}) = f(\boldsymbol{x}), \qquad \boldsymbol{x} \in \partial\Omega, \tag{6.2}$$

with Ω an open simply-connected region having a smooth boundary. This is converted to an integral equation over $\mathbb{S}^2$ and then a Galerkin method is used to solve it numerically, obtaining an approximation related to spherical polynomials.

Our second illustration, given in Sect. 6.2, is to solve the Neumann problem

$$-\Delta u(\boldsymbol{x}) + \gamma(\boldsymbol{x})\, u(\boldsymbol{x}) = f(\boldsymbol{x}), \qquad \boldsymbol{x} \in \Omega \subset \mathbb{R}^2,$$

$$\frac{\partial u(\boldsymbol{x})}{\partial \boldsymbol{\nu}_{\boldsymbol{x}}} = g(\boldsymbol{x}), \qquad \boldsymbol{x} \in \partial\Omega,$$

with Ω an open simply-connected region having a smooth boundary. This is converted to an equivalent elliptic problem over the unit disk $\mathbb{D}$ and it is solved numerically with Galerkin's method, obtaining an approximation based on polynomials over $\mathbb{R}^2$.

In the final section of the chapter, we discuss a Galerkin method for the following Beltrami-type equation

$$-\Delta^* u + c_0 u = f \quad \text{in } \mathbb{S}^{d-1},$$

where $c_0 > 0$ and f are given. Here the dimension $d \geq 3$ is arbitrary.

K. Atkinson and W. Han, *Spherical Harmonics and Approximations on the Unit* 211
Sphere: An Introduction, Lecture Notes in Mathematics 2044,
DOI 10.1007/978-3-642-25983-8_6, © Springer-Verlag Berlin Heidelberg 2012

6.1   A Boundary Integral Equation

One method of solving (6.1)–(6.2) is to use the classical boundary integral equation that arises from representing the solution u as a double layer potential,

$$u(\boldsymbol{x}) = \int_{\partial\Omega} \rho(\boldsymbol{y}) \frac{\partial}{\partial\boldsymbol{\nu_y}} \left(\frac{1}{|\boldsymbol{x}-\boldsymbol{y}|} \right) d\sigma(\boldsymbol{y}), \quad \boldsymbol{x} \in \Omega. \tag{6.3}$$

In this representation, ρ is an unknown "double-layer density function". The notation $\partial/\partial\boldsymbol{\nu_y}$ denotes the normal derivative in the direction of $\boldsymbol{\nu_y}$, the outer normal at $\boldsymbol{y} \in \partial\Omega$. For all integrable functions ρ, this function u is harmonic, meaning it satisfies (6.1). By requiring it to also satisfy the Dirichlet boundary condition (6.2), one obtains the boundary integral equation

$$-2\pi \rho(\boldsymbol{x}) + \int_{\partial\Omega} \rho(\boldsymbol{y}) \frac{\partial}{\partial\boldsymbol{\nu_y}} \left(\frac{1}{|\boldsymbol{x}-\boldsymbol{y}|} \right) d\sigma(\boldsymbol{y}) = f(\boldsymbol{x}), \quad \boldsymbol{x} \in \partial\Omega. \tag{6.4}$$

As notation, let

$$L(\boldsymbol{x}, \boldsymbol{y}) = \frac{\partial}{\partial\boldsymbol{\nu_y}} \left(\frac{1}{|\boldsymbol{x}-\boldsymbol{y}|} \right)$$

and introduce the integral operator $\mathcal{L}$,

$$\mathcal{L}(\boldsymbol{x}) = \int_{\partial\Omega} L(\boldsymbol{x}, \boldsymbol{y}) g(\boldsymbol{y}) d\sigma(\boldsymbol{y}), \quad \boldsymbol{x} \in \partial\Omega$$

for a generic function g. The integral equation (6.4) is represented abstractly in operator notation as

$$(-2\pi + \mathcal{L})\rho = f.$$

There is a well-developed theory for this equation. In particular, the integral operator $\mathcal{L}$ is compact from $C(\partial\Omega)$ to $C(\partial\Omega)$ and from $L^2(\partial\Omega)$ to $L^2(\partial\Omega)$. Let $\mathcal{Y}$ denote either of these spaces. Then it is well-known that

$$-2\pi + \mathcal{L} : \mathcal{Y} \xrightarrow[onto]{1-1} \mathcal{Y} \tag{6.5}$$

and the inverse $(-2\pi + \mathcal{L})^{-1}$ is a bounded linear operator from $\mathcal{Y}$ onto $\mathcal{Y}$. For a complete development of these ideas, see Kress [70, Chap. 6].

We convert (6.4) to another integral equation, but defined on $\mathbb{S}^2$. To accomplish this, assume that a smooth mapping

$$M : \mathbb{S}^2 \xrightarrow[onto]{1-1} \partial\Omega \tag{6.6}$$

is known. Recalling the discussion in Sect. 4.2.2 about the differentiation of functions defined on $\mathbb{S}^2$, we assume that the components of the mapping M are continuously differentiable. Using M, change the variable of integration in (6.4). For a generic function G,

$$\int_{\partial\Omega} G(\boldsymbol{x}, \boldsymbol{y})\, d\sigma(\boldsymbol{y}) = \int_{\mathbb{S}^2} G(\boldsymbol{x}, M(\boldsymbol{\eta}))\, J(\boldsymbol{\eta})\, dS^2(\boldsymbol{\eta})$$

with $J(\boldsymbol{\eta})$ the Jacobian of the mapping. If

$$M(\boldsymbol{\eta}) = M(\cos\phi\sin\theta, \sin\phi\sin\theta, \cos\theta) \equiv \widehat{M}(\theta, \phi),$$

then

$$J(\boldsymbol{\eta}) = \frac{1}{\sin\theta} \left| \frac{\partial\widehat{M}}{\partial\theta} \times \frac{\partial\widehat{M}}{\partial\phi} \right|.$$

An example of such a mapping M for an ellipsoidal boundary is given in Example 5.1.

Applying this to (6.4), obtain

$$-2\,\pi\, R(\boldsymbol{\xi}) + \int_{\mathbb{S}^2} R(\boldsymbol{\eta})\, L(M(\boldsymbol{\xi}), M(\boldsymbol{\eta}))\, J(\boldsymbol{\eta})\, dS^2(\boldsymbol{\eta}) = F(\boldsymbol{\xi}), \quad \boldsymbol{\xi} \in \mathbb{S}^2, \quad (6.7)$$

with $R(\boldsymbol{\xi}) = \rho(M(\boldsymbol{\xi}))$ and $F(\boldsymbol{\xi}) = f(M(\boldsymbol{\xi}))$. As additional notation, let

$$K(\boldsymbol{\xi}, \boldsymbol{\eta}) = L(M(\boldsymbol{\xi}), M(\boldsymbol{\eta}))\, J(\boldsymbol{\eta}),$$

and let $\mathcal{K}$ denote the associated integral operator in (6.7). Let $\mathcal{X}$ denote either $C(\mathbb{S}^2)$ or $L^2(\mathbb{S}^2)$. Then from the properties of the double-layer integral operator $\mathcal{L}$, cited earlier, and from the properties of the boundary mapping M, it is straightforward to show that $\mathcal{K}$ is a compact mapping from $\mathcal{X}$ to $\mathcal{X}$ and

$$-2\pi + \mathcal{K} : \mathcal{X} \xrightarrow[\text{onto}]{1-1} \mathcal{X}. \tag{6.8}$$

The inverse $(-2\pi + \mathcal{K})^{-1}$ is a bounded linear operator on $\mathcal{X}$ to $\mathcal{X}$. The kernel function $K(\boldsymbol{\xi}, \boldsymbol{\eta})$ is singular when $\boldsymbol{\xi} = \boldsymbol{\eta}$, and this affects the implementation of numerical methods for the solution of

$$(-2\pi + \mathcal{K})\, R = F. \tag{6.9}$$

We now describe such a numerical method.

Recall from (4.55) the orthogonal projection $\mathcal{Q}_n : L^2(\mathbb{S}^2) \to \Pi_n(\mathbb{S}^2)$. To approximate (6.9), find $R_n \in L^2(\mathbb{S}^2)$ such that

$$(-2\pi + \mathcal{Q}_n\mathcal{K})\, R_n = \mathcal{Q}_n F. \tag{6.10}$$

Equivalently, find $R_n \in \Pi_n(\mathbb{S}^2)$ such that

$$\mathcal{Q}_n \left[(-2\pi + \mathcal{K})\, R_n - F \right] = 0. \tag{6.11}$$

This is Galerkin's method, written in the form used with integral equations of the second kind. For the analysis of the existence and convergence of R_n, use (6.10), and for actually computing R_n, use (6.11).

To implement the method, begin by expressing the projection $\mathcal{Q}_n$ in the form

$$\mathcal{Q}_n g = \sum_{j=1}^{N_n} (g, \varphi_j)\, \varphi_j \tag{6.12}$$

with $\{\varphi_1, \ldots, \varphi_N\}$ an orthonormal basis of $\Pi_n(\mathbb{S}^2)$. For example, see the basis (4.7). Represent the solution R_n of (6.11) as

$$R_n(\boldsymbol{\eta}) = \sum_{j=1}^{N_n} \alpha_j^{(n)} \varphi_j(\boldsymbol{\eta}). \tag{6.13}$$

Then (6.10) is equivalent to solving for $\left[\alpha_1^{(n)}, \ldots, \alpha_N^{(n)} \right]^{\mathrm{T}}$ in the linear system

$$-2\pi \alpha_i^{(n)} + \sum_{j=1}^{N} \alpha_j^{(n)} (\varphi_i, \mathcal{K}\varphi_j) = (F, \varphi_i), \qquad i = 1, \ldots, N_n. \tag{6.14}$$

The coefficient matrix $[(\varphi_i, \mathcal{K}\varphi_j)]$ is $N_n \times N_n$. All of the coefficients of this linear system are twofold surface integrals over $\mathbb{S}^2$, and thus they are fourfold single integrals. The size of the linear system is usually quite small for a partial differential equation in $\mathbb{R}^3$, but the numerical integration of the coefficients can be quite expensive if not done carefully.

Note that the numerical method (6.10)–(6.14) also makes sense within the space $C(\mathbb{S}^2)$, although $\mathcal{Q}_n$ is no longer an orthogonal projection. As earlier, let $\mathcal{X}$ denote either $L^2(\mathbb{S}^2)$ or $C(\mathbb{S}^2)$.

6.1.1 Convergence Theory

For a review of the analysis of Galerkin methods for the solution of integral equations of the second kind, see [10, Chap. 3]. A crucial step is showing that

$$\|\mathcal{K} - \mathcal{Q}_n\mathcal{K}\|_{\mathcal{X}\to\mathcal{X}} \to 0 \quad \text{as } n \to \infty \tag{6.15}$$

for the operator norm on $\mathcal{X}$. In the case that $\mathcal{X} = L^2(\mathbb{S}^2)$, this is relatively straightforward. The projections $\{\mathcal{Q}_n\}$ are pointwise convergent on $L^2(\mathbb{S}^2)$, i.e. $\mathcal{Q}_n g \to g$ as $n \to \infty$, for every $g \in L^2(\mathbb{S}^2)$; see Theorem 2.34. In addition, $\mathcal{K}$ is compact on $L^2(\mathbb{S}^2)$. The result (6.15) then follows from [10, Lemma 3.1.2].

For the case $\mathcal{X} = C(\mathbb{S}^2)$, showing (6.15) requires a closer examination of the properties of $\mathcal{K}$, which are derived from those of $\mathcal{L}$ and of the surface mapping M. Recall the discussion in Sect. 4.2.2 about the differentiation of functions defined on $\mathbb{S}^2$. We assume that the components of the mapping M are differentiable, at least once, and further that the first-order derivatives are Hölder continuous with exponent $\lambda \in (\frac{1}{2}, 1]$. It then follows from the argument in [8, p. 268] that

$$\|\mathcal{K} - \mathcal{Q}_n\mathcal{K}\|_{C\to C} \le \frac{c}{n^{\lambda - 1/2}}$$

for all sufficiently large n, for some constant $c > 0$. This shows (6.15) for $\mathcal{X} = C(\mathbb{S}^2)$.

To show the invertibility of $-2\pi + \mathcal{Q}_n\mathcal{K}$, use the identity

$$-2\pi + \mathcal{Q}_n\mathcal{K} = (-2\pi + \mathcal{K}) - (\mathcal{K} - \mathcal{Q}_n\mathcal{K})$$

$$= (-2\pi + \mathcal{K})\left[I - (-2\pi + \mathcal{K})^{-1}(\mathcal{K} - \mathcal{Q}_n\mathcal{K})\right],$$

which makes use of the existence and boundedness of $(-2\pi + \mathcal{K})^{-1}$, which follows from (6.8). From (6.15) it follows that

$$\left\|(-2\pi + \mathcal{K})^{-1}(\mathcal{K} - \mathcal{Q}_n\mathcal{K})\right\|_{\mathcal{X}\to\mathcal{X}} \le \frac{1}{2}$$

for all sufficiently large n, say $n \ge n_0$. The geometric series theorem [13, Theorem 2.3.1] then implies the existence and uniform boundedness of $(-2\pi + \mathcal{Q}_n\mathcal{K})^{-1}$ for $n \ge n_0$, thus also showing the solvability of (6.10):

$$(-2\pi + \mathcal{Q}_n\mathcal{K})^{-1} = \left[I - (-2\pi + \mathcal{K})^{-1}(\mathcal{K} - \mathcal{Q}_n\mathcal{K})\right]^{-1}(-2\pi + \mathcal{K})^{-1},$$

$$\left\|(-2\pi + \mathcal{Q}_n\mathcal{K})^{-1}\right\|_{\mathcal{X}\to\mathcal{X}} \le \frac{\left\|(-2\pi + \mathcal{K})^{-1}\right\|_{\mathcal{X}\to\mathcal{X}}}{1 - \left\|(-2\pi + \mathcal{K})^{-1}(\mathcal{K} - \mathcal{Q}_n\mathcal{K})\right\|_{\mathcal{X}\to\mathcal{X}}}$$

$$\le 2\left\|(-2\pi + \mathcal{K})^{-1}\right\|_{\mathcal{X}\to\mathcal{X}}, \qquad n \ge n_0. \tag{6.16}$$

For convergence, use the identity

$$R - R_n = -2\pi \left(-2\pi + \mathcal{Q}_n \mathcal{K}\right)^{-1} \left(R - \mathcal{Q}_n R\right). \tag{6.17}$$

Then,

$$\|R - R_n\|_{\mathcal{X}} \leq 2\pi \left\| \left(-2\pi + \mathcal{Q}_n \mathcal{K}\right)^{-1} \right\|_{\mathcal{X} \to \mathcal{X}} \|R - \mathcal{Q}_n R\|_{\mathcal{X}}. \tag{6.18}$$

From (6.16) and (6.17), we have $R_n \to R$ if and only if $\mathcal{Q}_n R \to R$.
When $\mathcal{X} = L^2(\mathbb{S}^2)$,

$$\|R - R_n\|_{L^2(\mathbb{S}^2)} = \mathcal{O}\left(\|R - \mathcal{Q}_n R\|_{L^2(\mathbb{S}^2)}\right). \tag{6.19}$$

The error $R - \mathcal{Q}_n R$ was investigated in Sect. 4.2.4, and $\mathcal{Q}_n R \to R$ for any $R \in L^2(\mathbb{S}^2)$. For rates of convergence as a function of the differentiability of R, see (4.60). In particular, if $R \in C^{k,\alpha}(\mathbb{S}^2)$ with $k \geq 0$, $\alpha \in (0,1]$, then

$$\|R - R_n\|_{L^2(\mathbb{S}^2)} = \mathcal{O}\left(n^{-(k+\alpha)}\right). \tag{6.20}$$

With $\mathcal{X} = C(\mathbb{S}^2)$, the sequence $\{\|\mathcal{Q}_n\|_{C \to C} : n \geq 0\}$ is unbounded (see (4.64)), and then the principle of uniform boundedness (see [13, Sect. 2.4.3]) implies that there is a function $R \in C(\mathbb{S}^2)$ for which $\mathcal{Q}_n R$ does not converge uniformly to R. Using Corollary 4.14, we have convergence of $\mathcal{Q}_n R$ to R provided $R \in C^{k,\alpha}(\mathbb{S}^2)$ with $k \geq 0$, $\alpha \in (0,1]$, and $k + \alpha > \frac{1}{2}$, namely

$$\|R - \mathcal{Q}_n R\|_{\infty} \leq \frac{c}{n^{k+\alpha-1/2}}, \tag{6.21}$$

for a suitable $c > 0$. When combined with (6.18), we have uniform convergence of R_n to R whenever $R \in C^{k,\alpha}(\mathbb{S}^2)$ with $k + \alpha > \frac{1}{2}$, and

$$\|R - R_n\|_{\infty} \leq \mathcal{O}\left(n^{-(k+\alpha-1/2)}\right). \tag{6.22}$$

In addition to the convergence of R_n to R, it is necessary to also consider the convergence of the resulting approximation of the double layer potential (6.3). As notation, let $\rho_n(\boldsymbol{y}) = R_n(\boldsymbol{\eta})$ for $\boldsymbol{y} = M(\boldsymbol{\eta})$, $\boldsymbol{\eta} \in \mathbb{S}^2$. For $\boldsymbol{x} \in \Omega$, define

$$u_n(\boldsymbol{x}) = \int_{\partial \Omega} \rho_n(\boldsymbol{y}) \frac{\partial}{\partial \boldsymbol{\nu_y}} \left(\frac{1}{|\boldsymbol{x} - \boldsymbol{y}|}\right) d\sigma(\boldsymbol{y}) \tag{6.23}$$

$$= \int_{\mathbb{S}^2} \rho_n(M(\boldsymbol{\eta})) \frac{\partial}{\partial \boldsymbol{\nu_y}} \left(\frac{1}{|\boldsymbol{x} - \boldsymbol{y}|}\right) \bigg|_{\boldsymbol{y} = M(\boldsymbol{\eta})} J(\boldsymbol{\eta}) \, dS^2(\boldsymbol{\eta}). \tag{6.24}$$

This will need to be evaluated numerically, but we first consider the error $u - u_n$. The function $u - u_n$ is harmonic on Ω, and consequently we can apply the maximum principle (Corollary 3.19) for harmonic functions to obtain

$$\max_{x \in \overline{\Omega}} |u(x) - u_n(x)| = \max_{x \in \partial\Omega} |u(x) - u_n(x)|. \tag{6.25}$$

Subtracting (6.23) from (6.3),

$$u(x) - u_n(x) = \int_{\partial\Omega} e_n(y) \frac{\partial}{\partial \nu_y} \left(\frac{1}{|x - y|} \right) d\sigma(y), \quad x \in \Omega$$

with $e_n = \rho - \rho_n$. Form the limit as x approaches a point on the boundary $\partial\Omega$, as was done in obtaining (6.4). This yields

$$u(x) - u_n(x) = -2\pi\, e_n(x) + \int_{\partial\Omega} e_n(y) \frac{\partial}{\partial \nu_y} \left(\frac{1}{|x - y|} \right) d\sigma(y), \quad x \in \partial\Omega.$$

So

$$u - u_n = (-2\pi + \mathcal{L})\, e_n$$

for $u - u_n$ restricted to $\partial\Omega$. Forming bounds, the error $u - u_n$ on $\partial\Omega$ satisfies

$$\|u - u_n\|_\infty \le (2\pi + \|\mathcal{L}\|_{C \to C})\, \|\rho - \rho_n\|_\infty. \tag{6.26}$$

Error bounds for ρ_n from (6.22) then yield the same rates of convergence for u_n to u over $\partial\Omega$. If $\|\rho - \rho_n\|_\infty \to 0$ on $\partial\Omega$, then (6.25) implies that the convergence $u_n(x)$ to $u(x)$ is uniform over Ω.

6.1.2 Quadrature

Referring to the approximating linear system (6.14), the matrix coefficients $(\varphi_i, \mathcal{K}\varphi_j)$ and the right sides (F, φ_i) must be evaluated numerically in almost all cases. There are two integrations to consider: (i) approximating the inner product $(\cdot, \cdot)$, and (ii) evaluating the integral operator term $\mathcal{K}\varphi_i$. The inner product can be dealt with exactly as described in Sect. 5.7.1. Define a discrete inner product $(\cdot, \cdot)_n$ as in (5.80). Then replace (6.14) with

$$-2\pi\tilde{\alpha}_i^{(n)} + \sum_{j=1}^N \tilde{\alpha}_j^{(n)} (\varphi_i, \mathcal{K}\varphi_j)_n = (F, \varphi_i)_n, \quad i = 1, \ldots, N_n. \tag{6.27}$$

The numerical solution is then

$$\tilde{R}_n(\eta) = \sum_{j=1}^{N_n} \tilde{\alpha}_j^{(n)} \varphi_j(\eta).$$

Note that $\mathcal{K}\varphi_j$ is not being approximated at this point. This "semi-discrete Galerkin method" is analyzed in [53, Sect. 3], and it is shown to have the same convergence analysis and results as given in (6.15)–(6.26) when analyzed as an approximation within $C(\mathbb{S}^2)$.

The integrand in

$$\mathcal{K}g(\boldsymbol{\xi}) = \int_{\mathbb{S}^2} g(\boldsymbol{\eta})\, L(M(\boldsymbol{\xi}), M(\boldsymbol{\eta}))\, J(\boldsymbol{\eta})\, dS^2(\boldsymbol{\eta}), \qquad \boldsymbol{\xi} \in \mathbb{S}^2$$

is singular at $\boldsymbol{\eta} = \boldsymbol{\xi}$. This can be reduced to a bounded discontinuity by means of the identity

$$\int_{\partial\Omega} \frac{\partial}{\partial\boldsymbol{\nu}_{\boldsymbol{y}}} \left(\frac{1}{|\boldsymbol{x} - \boldsymbol{y}|} \right) d\sigma(\boldsymbol{y}) \equiv -2\pi, \qquad \boldsymbol{x} \in \Omega;$$

see [70, Example 6.16]. Then we can write

$$\int_{\mathbb{S}^2} g(\boldsymbol{\eta})\, L(M(\boldsymbol{\xi}), M(\boldsymbol{\eta}))\, J(\boldsymbol{\eta})\, dS^2(\boldsymbol{\eta})$$

$$= -2\pi\, g(\boldsymbol{\xi}) + \int_{\mathbb{S}^2} [g(\boldsymbol{\eta}) - g(\boldsymbol{\xi})]\, L(M(\boldsymbol{\xi}), M(\boldsymbol{\eta}))\, J(\boldsymbol{\eta})\, dS^2(\boldsymbol{\eta}).$$

If the function g is Lipschitz continuous on $\mathbb{S}^2$, then the new integrand has a bounded discontinuity at $\boldsymbol{\eta} = \boldsymbol{\xi}$. However, this is still a difficult integral to evaluate numerically, and a number of different approaches have been used. We use something similar to the ideas in Sect. 5.5.

Assume we are approximating an integral

$$\int_{\mathbb{S}^2} g(\boldsymbol{\eta})\, dS^2(\boldsymbol{\eta})$$

in which the integrand g is singular at $\boldsymbol{\eta} = \boldsymbol{\xi}$. Perform a spherical rotation (see (5.69)) to change the integrand to one in which the singularity is located at the north or south poles of $\mathbb{S}^2$ (at $(0, 0, \pm 1)$), resulting in a new integral

$$\int_{\mathbb{S}^2} \widehat{g}(\boldsymbol{\zeta})\, dS^2(\boldsymbol{\zeta}).$$

Use spherical coordinates to write this as the iterated integral

$$I(g) = \int_0^{2\pi} \int_{-1}^1 \widehat{g}\Big(\sqrt{1 - z^2}\cos\phi,\, \sqrt{1 - z^2}\sin\phi,\, z \Big)\, dz\, d\phi.$$

For the integration in ϕ, use the standard trapezoidal method, as was done with the singular integration method (5.63). For the integration in z, we use a method analogous to that of (5.63) to handle the singularity at one of the poles. It is called the "IMT method", is due to Iri, Moriguti, and Takasawa, and is discussed in [9, pp. 306–307]. It is especially suitable for integrands with endpoint singularities. Combining these rules, we have the quadrature method

$$I(\widehat{g}) \approx I_{m_1,m_2}(\widehat{g}) \tag{6.28}$$

where $I_{m_1,m_2}(\widehat{g})$ equals

$$\frac{\pi}{m_1} \sum_{j=1}^{2m_1} \sum_{k=1}^{m_2-1} w_{k,m_2} \widehat{g}\left(\sqrt{1 - z_{k,m_2}^2} \cos \phi_{j,m_1}, \sqrt{1 - z_{k,m_2}^2} \sin \phi_{j,m_1}, z_{k,m_2}\right).$$

The nodes $\{z_{k,m_2}\}$ and weights $\{w_{k,m_2}\}$ are those of the IMT method on $[-1,1]$. For the trapezoidal rule, $\phi_{j,m_1} = j\pi/m_1$, $j = 1,\ldots,2m_1$. Note that the periodicity of the integrand in ϕ has been used to simplify the trapezoidal rule.

The evaluation of the coefficients $(\varphi_i, \mathcal{K}\varphi_j)$, $i,j = 1,\ldots,N$, using the discrete inner product $(\cdot,\cdot)_n$ and the IMT method for approximating $\mathcal{K}\varphi_j$ is computationally expensive. The total number of arithmetic operations is $\mathcal{O}(n^2 m_1 m_2)$. We evaluate the coefficients simultaneously, doing so as to avoid unnecessary re-calculation of quantities shared between the various coefficients. It is also convenient to use parallel computation, greatly increasing the speed. To lessen the cost when solving the Dirichlet problems with multiple boundary functions $f(\boldsymbol{x})$, evaluate these coefficients and save them for later use with the various boundary functions.

An alternative approach. Graham and Sloan [53, Sect. 4] use an alternative approach to approximating $\mathcal{K}\varphi_j$. Begin again by rotating the coordinate system to place the singularity at a pole, $\boldsymbol{\eta} \to \mathcal{T}\boldsymbol{\eta} \equiv \boldsymbol{\zeta}$, and denote the new kernel function by $\widetilde{K}(\boldsymbol{\xi},\boldsymbol{\zeta}) = K(\boldsymbol{\xi},\boldsymbol{\eta})$. Then write the kernel function $\widetilde{K}$ as

$$\widetilde{K}(\boldsymbol{\xi},\boldsymbol{\zeta}) = \frac{\widetilde{K}_1(\boldsymbol{\xi},\boldsymbol{\zeta})}{|\boldsymbol{\xi} - \boldsymbol{\zeta}|},$$

where

$$\widetilde{K}_1(\boldsymbol{\xi},\boldsymbol{\zeta}) = |\boldsymbol{\xi} - \boldsymbol{\zeta}| \, \widetilde{K}(\boldsymbol{\xi},\boldsymbol{\zeta}).$$

The function $\widetilde{K}_1$ has a bounded discontinuity at $\boldsymbol{\zeta} = \mathcal{T}\boldsymbol{\xi}$. Expand $\widetilde{K}_1(\boldsymbol{\xi},\cdot)$ as a truncated Laplace series,

$$\widetilde{K}_1(\boldsymbol{\xi},\boldsymbol{\zeta}) \approx \sum_{j=1}^{N_\ell} \left(\widetilde{K}_1(\boldsymbol{\xi},\cdot), \varphi_j\right)_\ell \varphi_j(\boldsymbol{\zeta})$$

with integration parameter ℓ and $N_\ell = (\ell + 1)^2$. To complete the method, use the identity (3.58) with dimension $d = 3$:

$$\int_{\mathbb{S}^2} \frac{\varphi_j(\boldsymbol{\eta})}{|\boldsymbol{\xi} - \boldsymbol{\eta}|} \, dS^2(\boldsymbol{\eta}) = \frac{4\pi}{2d_j + 1} \varphi_j(\boldsymbol{\xi}),$$

where d_j is the degree of the spherical harmonic φ_j.

It is shown in [53, Theorem 4.1] that if $\ell = an + 1$ for some $a > 1$, then the overall rate of convergence of this discretization for implementing Galerkin's method is the same as that given in (6.22) for the original Galerkin method.

As another approach to a spectral method for boundary integral equations on $\mathbf{S}^2$, see [51].

Evaluating the solution. Returning to the evaluation of the approximate solution $u_n(\boldsymbol{x})$ in (6.23), use the product Gauss formula (5.12). Let n_q denote the integration parameter used in (5.12). For any fixed n_q, the integration error in evaluating $u_n(\boldsymbol{x})$ will increase as $\boldsymbol{x}$ approaches $\partial\Omega$ because the integrand becomes increasingly peaked. To check the accuracy of our numerical method, we choose a rectangular grid covering the unit ball $\mathbb{B}^3$, and then we map it onto another grid in Ω using

$$\boldsymbol{x} \mapsto |\boldsymbol{x}| \, M(\boldsymbol{x}/|\boldsymbol{x}|), \quad \boldsymbol{x} \in \mathbb{B}^3 \backslash \{\mathbf{0}\}, \tag{6.29}$$

with the origin in $\mathbb{B}^3$ mapping to the origin in Ω. We note that our following numerical example is for a region Ω that is "starlike" with respect to the origin. In the following tables and graphs, we look at the error at only those points $\boldsymbol{x} \in \mathbb{B}^3$ for which $|\boldsymbol{x}| \leq r < 1$, for some r.

6.1.3   A Numerical Example

As an interesting non-convex surface, consider the surface defined by the mapping

$$M(x, y, z) = \beta(x, y, z) \, (ax, by, cz), \quad (x, y, z) \in \mathbb{S}^2,$$

$$\beta(x, y, z) = 2 - \frac{\alpha}{1 + 5 \left(-z + x^2 + 1\right)^2}$$

with constants $a, b, c > 0$ and $\alpha \in (0, 2)$. This is called a "bean-shaped region". We use the particular surface $\partial\Omega$ obtained with $(a, b, c) = (2, 1, 1)$ and $\alpha = 1.5$. Figure 6.1 shows the surface $\partial\Omega$, and Fig. 6.2 shows the cross-sections with the xz and yz coordinate planes.

To study empirically our Galerkin method for solving (6.1)–(6.2), we choose two illustrative true solutions u, and we attempt to retrieve them from their boundary values.

Fig. 6.1 The bean-shaped
region Ω

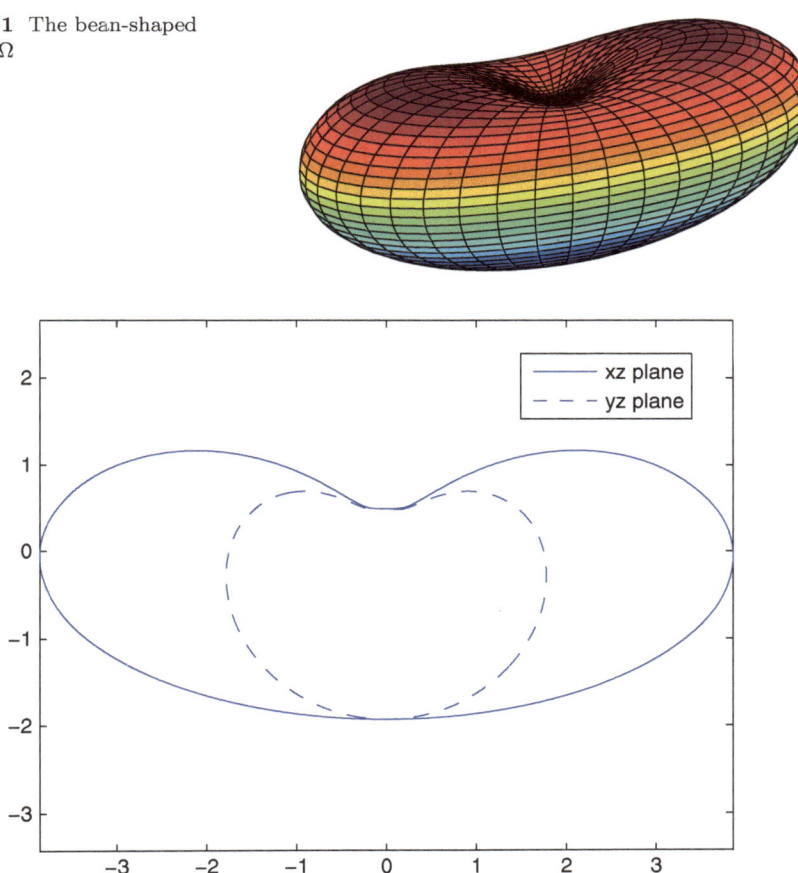

Fig. 6.2 Cross-sections of Ω with the xz and yz coordinate planes

$$u^{(1)}(x, y, z) = \frac{1}{|(x, y, z) - (5, 4, 3)|}, \tag{6.30}$$

$$u^{(2)}(x, y, z) = e^{\gamma x} \cos(\gamma y) + e^{\gamma z} \sin(\gamma x), \tag{6.31}$$

with $\gamma = 1$. The solution $u^{(1)}$ is very well-behaved, whereas the solution $u^{(2)}$ contains greater variation in size and also some oscillatory behaviour.

We use the numerical quadrature described in Sect. 6.1.2, using very large integration parameters to ensure that our integrations are exact to machine accuracy, which is much more accuracy than is needed ordinarily. The resulting Galerkin coefficients are saved; and then they are retrieved for use when solving for the coefficients $\left\{\alpha_j^{(n)}\right\}$ in (6.14); we do so with various choices of the boundary function f. The accuracy of the solution ρ_n must

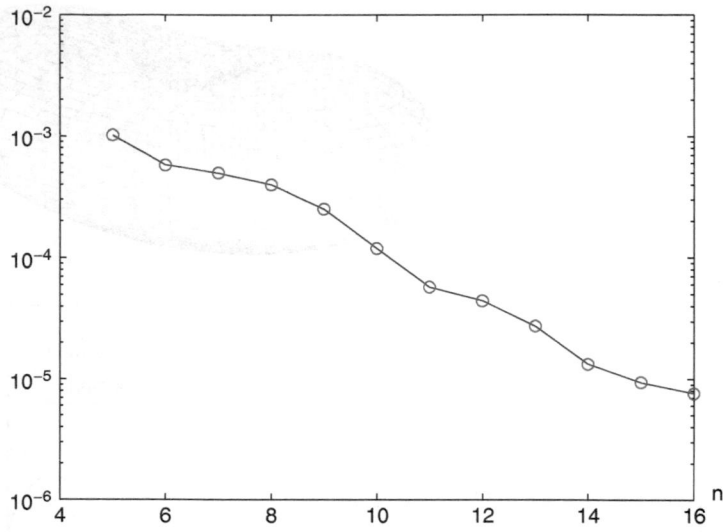

Fig. 6.3 Maximum errors for solution $u_n^{(1)}$ (see (6.30)) at points $\boldsymbol{\xi} = M(\boldsymbol{x}) \in \Omega$, $\boldsymbol{x} \in \mathbb{B}^3$, $|\boldsymbol{x}| \leq 0.8$

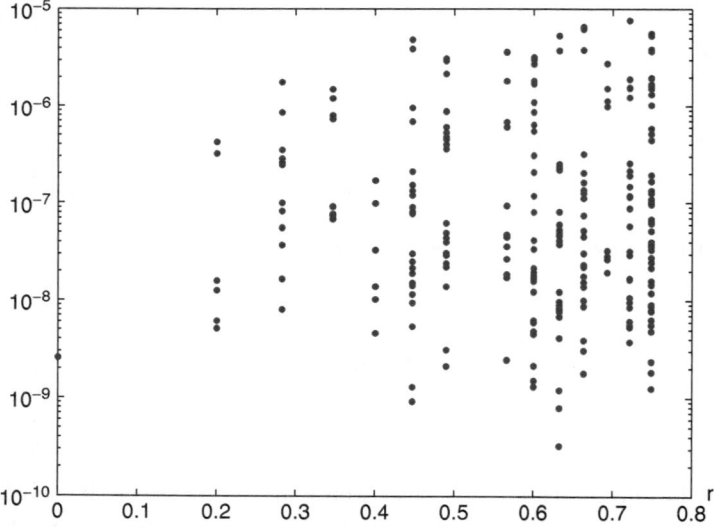

Fig. 6.4 Errors for $u_{16}^{(1)}$ at points $\boldsymbol{\xi} = M(\boldsymbol{x}) \in \Omega$, $\boldsymbol{x} \in \mathbb{B}^3$, $|\boldsymbol{x}| \leq 0.8$

be checked indirectly. We evaluate $u_n(\boldsymbol{x})$ using (6.23) and compare it to the true solution $u(\boldsymbol{x})$ at a variety of points within Ω. The approximation $u_n(\boldsymbol{x})$ must be evaluated numerically, as described preceding (6.29).

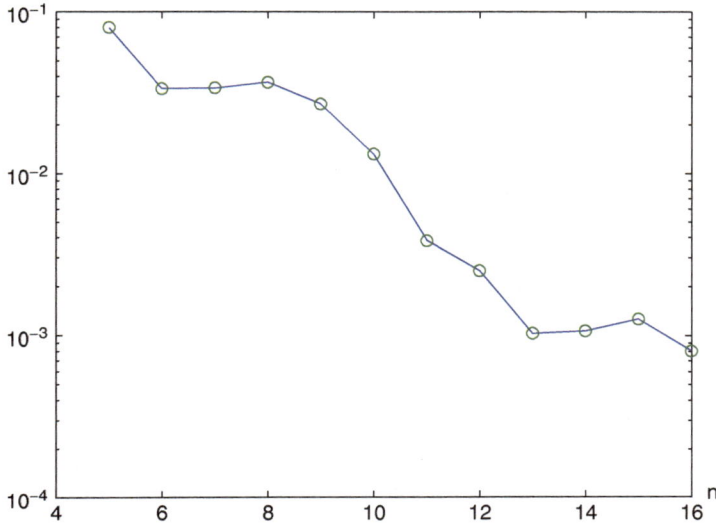

Fig. 6.5 Maximum errors for solution $u_n^{(2)}$ (see (6.31)) at points $\boldsymbol{\xi} = M(\boldsymbol{x}) \in \Omega$, $\boldsymbol{x} \in \mathbb{B}^3$, $|\boldsymbol{x}| \leq 0.8$

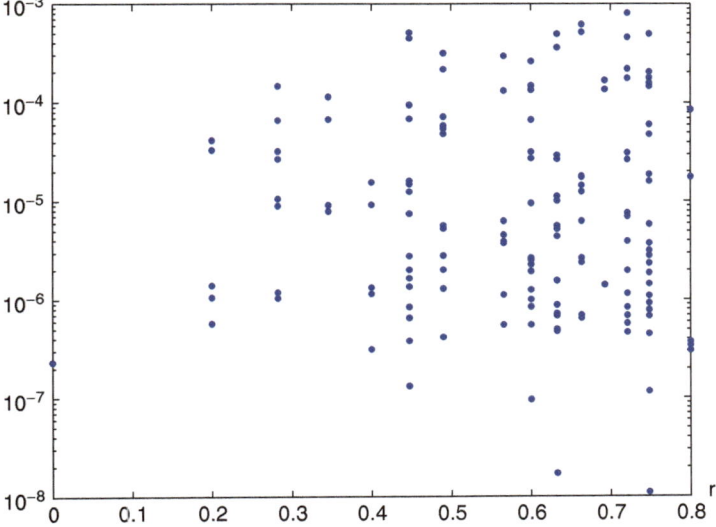

Fig. 6.6 Errors for $u_{16}^{(2)}$ at points $\boldsymbol{\xi} = M(\boldsymbol{x}) \in \Omega$, $\boldsymbol{x} \in \mathbb{B}^3$, $|\boldsymbol{x}| \leq 0.8$

To check the accuracy, we proceed as described preceding and following (6.29). In the following graphs, we use $r = 0.8$. We choose a large value for the integration parameter n_q to ensure maximal accuracy in the evaluation of $u_n(\boldsymbol{x})$ from (6.23). Figures 6.3 and 6.4 show errors for the case of the true solution $u^{(1)}$, see (6.30); and Figs. 6.5 and 6.6 show errors for the case of the true solution $u^{(2)}$, see (6.31). The horizontal axis in Figs. 6.3 and 6.5 is the degree n, varying from 5 to 16. An exponential rate of convergence as a function of n would result in a straight line, and that is consistent with these two graphs. The horizontal axis in Figs. 6.4 and 6.6 is $r = |\boldsymbol{x}|$. They show the large variation in the error as $\boldsymbol{x}$ varies, even for constant $|\boldsymbol{x}|$.

6.2 A Spectral Method for a Partial Differential Equation

There is a rich literature on spectral methods for solving partial differential equations. For the more recent literature, see the books [22, 27, 28, 41, 64, 102, 103]. Their bibliographies contain references to many earlier papers on spectral methods. Almost all of the spectral methods have been based on one-variable approximations, considering all multi-variable problems as some combination of one-variable problems. In the following we give a spectral method that directly applies multivariable approximation, using the material presented earlier in Sect. 4.3.

Consider the Neumann problem

$$-\Delta u(\boldsymbol{s}) + \gamma(\boldsymbol{s})\, u(\boldsymbol{s}) = f(\boldsymbol{s}), \qquad \boldsymbol{s} \in \Omega \subset \mathbb{R}^2, \tag{6.32}$$

$$\frac{\partial u(\boldsymbol{s})}{\partial \boldsymbol{\nu_s}} = g(\boldsymbol{s}), \qquad \boldsymbol{s} \in \partial\Omega \tag{6.33}$$

with Ω an open simply-connected region having a smooth boundary $\partial\Omega$ and $\boldsymbol{\nu_s}$ denoting the outer unit normal at $\boldsymbol{s} \in \partial\Omega$. For simplicity in our presentation, assume that the function γ is positive and continuous on $\overline{\Omega}$, and thus

$$\gamma(\boldsymbol{s}) \geq c_\gamma > 0, \qquad \boldsymbol{s} \in \overline{\Omega}, \tag{6.34}$$

for some c_γ. Further assume $f \in C(\overline{\Omega})$, $g \in C(\partial\Omega)$. We seek a solution $u \in C^2(\Omega) \cap C^1(\overline{\Omega})$.

This problem has the following variational reformulation: Find $u \in H^1(\Omega)$ for which

$$\mathcal{A}(u, v) = \ell_1(v) + \ell_2(v) \quad \forall v \in H^1(\Omega). \tag{6.35}$$

In this equation,

$$\mathcal{A}(v_1, v_2) \equiv \int_\Omega \left[\nabla v_1(\boldsymbol{s}) \cdot \nabla v_2(\boldsymbol{s}) + \gamma(\boldsymbol{s})\, v_1(\boldsymbol{s})\, v_2(\boldsymbol{s}) \right] d\boldsymbol{s}, \tag{6.36}$$

for $v_1, v_2 \in H^1(\Omega)$, and

$$\ell_1(v) \equiv \int_\Omega v(s)\, f(s)\, ds,$$

$$\ell_2(v) \equiv \int_{\partial\Omega} v(s)\, g(s)\, d\sigma(s)$$

for $v \in H^1(\Omega)$. The reformulation is obtained by multiplying (6.32) by $v(s)$, integrating over Ω, and then applying the divergence theorem to integrate by parts.

The theory for this variational formulation is very well-developed; see [13, Chap. 8]. We give a very minimal outline of this theory, briefly reviewing notation and results that are needed in discussing the numerical solution of (6.35) by Galerkin's method. First recall the Lax–Milgram Lemma.

Theorem 6.1. (LAX–MILGRAM LEMMA) *Assume V is a Hilbert space, $a(\cdot, \cdot)$ is a bounded, V-elliptic bilinear form on V, $\ell \in V'$. Then the problem*

$$u \in V, \quad a(u, v) = \ell(v) \quad \forall\, v \in V \tag{6.37}$$

has a unique solution. Moreover, for some constant $c > 0$ independent of ℓ,

$$\|u\|_V \leq c\, \|\ell\|_{V'}.$$

The norm in $H^1(\Omega)$ is $\|v\|_1 = \sqrt{(v, v)_1}$,

$$(v_1, v_2)_1 = \int_\Omega [\nabla v_1(s) \cdot \nabla v_2(s) + v_1(s)\, v_2(s)]\, ds, \quad v_1, v_2 \in H^1(\Omega).$$

The bilinear functional $\mathcal{A}$ is bounded,

$$|\mathcal{A}(v_1, v_2)| \leq c_\mathcal{A} \|v_1\|_1 \|v_2\|_1, \quad v_1, v_2 \in H^1(\Omega),$$

$$c_\mathcal{A} = \max\{1, \|\gamma\|_\infty\}.$$

With the assumption (6.34), $\mathcal{A}$ is "strongly elliptic",

$$|\mathcal{A}(v, v)| \geq \min\{1, c_\gamma\} \|v\|_1^2, \quad v \in H^1(\Omega). \tag{6.38}$$

The linear functionals ℓ_1 and ℓ_2 are bounded linear functionals on $H^1(\Omega)$. By the Cauchy–Schwarz inequality,

$$|\ell_1(v)| \leq \|f\|_{L^2} \|v\|_{L^2} \leq \|f\|_{L^2} \|v\|_1. \tag{6.39}$$

Showing the boundedness of ℓ_2 is more complicated, requiring looking at the trace on $\partial\Omega$ of functions $v \in H^1(\Omega)$. Begin by noting that the restriction mapping $\rho : H^1(\Omega) \to H^{1/2}(\partial\Omega)$ is continuous [80, Theorem 3.37] and the embedding $\iota : H^{1/2}(\partial\Omega) \hookrightarrow L^2(\partial\Omega)$ is compact [80, Theorem 3.27]. If we further denote by l_g the continuous mapping

$$l_g : w \mapsto \int_{\partial\Omega} w(\boldsymbol{s})\, g(\boldsymbol{s})\, d\sigma(\boldsymbol{s}), \qquad w \in L^2(\partial\Omega),$$

then we see $\ell_2 = l_g \circ \iota \circ \rho$, and therefore ℓ_2 is bounded on $H^1(\Omega)$.

Under our assumptions on $\mathcal{A}$, including the strong ellipticity in (6.38), Theorem 6.1 implies the existence of a unique solution u to (6.35) with

$$\|u\|_1 \le \frac{1}{c_e} \left(\|\ell_1\| + \|\ell_2\| \right). \tag{6.40}$$

In order to define a numerical method for solving (6.35), we introduce a change of variables to transform the Neumann problem (6.32)–(6.33) on Ω to an equivalent problem on the unit disk $\mathbb{B}^2$. Introduce a change of variables,

$$\boldsymbol{s} = \Phi(\boldsymbol{x}), \quad |\boldsymbol{x}| \le 1$$

with Φ a twice-differentiable mapping that is one-to-one from $\overline{\mathbb{B}}^2$ onto $\overline{\Omega}$. Let $\Psi = \Phi^{-1} : \overline{\Omega} \xrightarrow[onto]{1-1} \overline{\mathbb{B}}^2$. Finding such a mapping can be difficult, but there are also simple examples. For Ω the ellipse given by

$$\left(\frac{s_1}{a}\right)^2 + \left(\frac{s_2}{b}\right)^2 \le 1,$$

define

$$\Phi(\boldsymbol{x}) = (ax_1, bx_2), \quad |\boldsymbol{x}| \le 1.$$

This Φ is easily seen to be a twice-differentiable mapping.

As notation to make it easier to see the relation between a function defined on Ω and the corresponding function defined on $\mathbb{B}^2$, we introduce the following. For $v \in L^2(\Omega)$, let

$$\widetilde{v}(\boldsymbol{x}) = v(\Phi(\boldsymbol{x})), \quad \boldsymbol{x} \in \overline{\mathbb{B}}^2,$$

and conversely,

$$v(\boldsymbol{s}) = \widetilde{v}(\Psi(\boldsymbol{s})), \quad \boldsymbol{s} \in \overline{\Omega}.$$

Assuming $v \in H^1(\Omega)$, it follows that

$$\nabla_x \widetilde{v}(\boldsymbol{x}) = J(\boldsymbol{x})^{\mathrm{T}} \nabla_s v(\boldsymbol{s}), \quad \boldsymbol{s} = \Phi(\boldsymbol{x})$$

with $J(\boldsymbol{x})$ the Jacobian matrix for Φ over the unit ball $\mathbb{B}^2$,

$$J(\boldsymbol{x}) \equiv (D\Phi)(\boldsymbol{x}) = \left[\frac{\partial \Phi_i(\boldsymbol{x})}{\partial x_j} \right]_{i,j=1}^2, \quad \boldsymbol{x} \in \overline{\mathbb{B}}^2.$$

Similarly,

$$\nabla_s v(\boldsymbol{s}) = K(\boldsymbol{s})^{\mathrm{T}} \nabla_x \widetilde{v}(\boldsymbol{x}), \quad \boldsymbol{x} = \Psi(\boldsymbol{s})$$

with $K(\boldsymbol{s})$ the Jacobian matrix for Ψ over Ω. Also,

$$K(\Phi(\boldsymbol{x})) = J(\boldsymbol{x})^{-1}.$$

The differentiability of a function $\widetilde{v}(\boldsymbol{x})$ depends on that of $v(\boldsymbol{s})$ and the mapping $\Phi(\boldsymbol{x})$.

Using the change of variables $\boldsymbol{s} = \Phi(\boldsymbol{x})$, the formula (6.36) converts to

$$\begin{aligned}
\mathcal{A}(v_1, v_2) &= \int_{\mathbb{B}^2} \{ [K(\Phi(\boldsymbol{x}))^{\mathrm{T}} \nabla_x \widetilde{v}_1(\boldsymbol{x})]^{\mathrm{T}} [K(\Phi(\boldsymbol{x}))^{\mathrm{T}} \nabla_x \widetilde{v}_2(\boldsymbol{x})] \\
&\qquad + \gamma(\Phi(\boldsymbol{x})) v_1(\Phi(\boldsymbol{x})) v_2(\Phi(\boldsymbol{x})) \} \, |\det[J(\boldsymbol{x})]| \, d\boldsymbol{x} \\
&= \int_{\mathbb{B}^2} \{ [J(\boldsymbol{x})^{-\mathrm{T}} \nabla_x \widetilde{v}_1(\boldsymbol{x})]^{\mathrm{T}} [J(\boldsymbol{x})^{-\mathrm{T}} \nabla_x \widetilde{v}_2(\boldsymbol{x})] \\
&\qquad + \widetilde{\gamma}(\boldsymbol{x}) \widetilde{v}_1(\boldsymbol{x}) \widetilde{v}_2(\boldsymbol{x}) \} \, |\det[J(\boldsymbol{x})]| \, d\boldsymbol{x} \\
&= \int_{\mathbb{B}^2} \{ \nabla_x \widetilde{v}_1(\boldsymbol{x})^{\mathrm{T}} A(\boldsymbol{x}) \nabla_x \widetilde{v}_2(\boldsymbol{x}) + \widetilde{\gamma}(\boldsymbol{x}) \widetilde{v}_1(\boldsymbol{x}) \widetilde{v}_2(\boldsymbol{x}) \} \, |\det[J(\boldsymbol{x})]| \, d\boldsymbol{x} \\
&\equiv \widetilde{\mathcal{A}}(\widetilde{v}_1, \widetilde{v}_2)
\end{aligned} \tag{6.41}$$

with

$$A(\boldsymbol{x}) = J(\boldsymbol{x})^{-1} J(\boldsymbol{x})^{-\mathrm{T}}.$$

We can also introduce analogues to ℓ_1 and ℓ_2 following a change of variables, calling them $\widetilde{\ell}_1$ and $\widetilde{\ell}_2$ and defined on $H^1(\mathbb{B}^2)$. For example,

$$\widetilde{\ell}_1(\widetilde{v}) = \int_{\mathbb{B}^2} \widetilde{v}(\boldsymbol{x}) f(\Phi(\boldsymbol{x})) \, |\det[J(\boldsymbol{x})]| \, d\boldsymbol{x}.$$

We can then convert (6.35) to an equivalent problem over $H^1(\mathbb{B}^2)$. The variational problem becomes the following: Find $\widetilde{u} \in H^1(\mathbb{B}^2)$ for which

$$\widetilde{\mathcal{A}}(\widetilde{u}, \widetilde{v}) = \widetilde{\ell}_1(\widetilde{v}) + \widetilde{\ell}_2(\widetilde{v}) \quad \forall \widetilde{v} \in H^1(\mathbb{B}^2). \tag{6.42}$$

The assumptions and results in (6.36)–(6.39) extend to this new problem on $H^1(\mathbb{B}^2)$. The strong ellipticity condition (6.38) becomes

$$\widetilde{\mathcal{A}}(\widetilde{v}, \widetilde{v}) \geq \widetilde{c}_e \|\widetilde{v}\|_1^2, \quad \widetilde{v} \in H^1(\mathbb{B}^2),$$

$$\widetilde{c}_e = c_e \frac{\min_{\boldsymbol{x} \in \overline{\mathbb{B}}^2} |\det J(\boldsymbol{x})|}{\max \left[1, \max_{\boldsymbol{x} \in \overline{\mathbb{B}}^2} \|J(\boldsymbol{x})\|_2^2\right]},$$

where $\|J(\boldsymbol{x})\|_2$ denotes the operator matrix 2-norm of $J(\boldsymbol{x})$ for $\mathbb{R}^2$,

$$\|J(\boldsymbol{x})\|_2 = \sqrt{\max\{\lambda_1(\boldsymbol{x}), \lambda_2(\boldsymbol{x})\}}$$

with $\lambda_1(\boldsymbol{x})$, $\lambda_2(\boldsymbol{x})$ the eigenvalues of $J(\boldsymbol{x})^{\mathrm{T}} J(\boldsymbol{x})$. Also,

$$\left|\widetilde{\mathcal{A}}(\widetilde{v}, \widetilde{w})\right| \leq \widetilde{c}_{\mathcal{A}} \|\widetilde{v}\|_1 \|\widetilde{w}\|_1,$$

$$\widetilde{c}_{\mathcal{A}} = \left\{\max_{\boldsymbol{x} \in \overline{\mathbb{B}}^2} |\det [J(\boldsymbol{x})]|\right\} \cdot \max\left\{\max_{\boldsymbol{x} \in \overline{\mathbb{B}}^2} \|A(\boldsymbol{x})\|_2, \|\gamma\|_\infty\right\}.$$

The Lax–Milgram Theorem can be applied to (6.42), just as was done earlier for (6.35) to obtain (6.40). Thus we have the existence of a unique solution $\widetilde{u} \in H^1(\mathbb{B}^2)$ with

$$\|\widetilde{u}\|_1 \leq \frac{1}{\widetilde{c}_e} \left(\|\widetilde{\ell}_1\| + \|\widetilde{\ell}_2\|\right). \tag{6.43}$$

To obtain a numerical solution to (6.42), use Galerkin's method with the approximating subspace $\Pi_n \equiv \Pi_n(\mathbb{B}^2)$. We want to find $\widetilde{u}_n \in \Pi_n$ such that

$$\widetilde{\mathcal{A}}(\widetilde{u}_n, \widetilde{v}) = \widetilde{\ell}_1(\widetilde{v}) + \widetilde{\ell}_2(\widetilde{v}) \quad \forall \widetilde{v} \in \Pi_n. \tag{6.44}$$

The Lax–Milgram theorem implies the existence of a unique $\widetilde{u}_n$ for all n. For the error in this Galerkin method, apply Cea's Lemma [13, Proposition 9.1.3],

$$\|\widetilde{u} - \widetilde{u}_n\|_1 \leq \frac{\widetilde{c}_{\mathcal{A}}}{\widetilde{c}_e} \inf_{\widetilde{v} \in \Pi_n} \|\widetilde{u} - \widetilde{v}\|_1. \tag{6.45}$$

Since the set of all polynomials is dense in $H^1(\mathbb{B}^2)$, it follows that the Galerkin method of (6.44) is convergent for all solutions $\widetilde{u} \in H^1(\mathbb{B}^2)$.

The best approximation error on the right side of this inequality can be bounded by using Theorem 4.16. In particular, if $\widetilde{u} \in C^{m+1}(\overline{\mathbb{B}}^2)$, then it can be shown that

$$\|\widetilde{u} - \widetilde{u}_n\|_1 \leq \frac{c}{n^m}\, \omega_{m+1}(\widetilde{u}, 1/n) \tag{6.46}$$

for some $c > 0$, where

$$\omega_{m+1}(\widetilde{u}, \delta) = \max_{|\alpha|=m+1} \sup_{|\boldsymbol{x}-\boldsymbol{y}| \leq \delta} |D^\alpha \widetilde{u}(\boldsymbol{x}) - D^\alpha \widetilde{u}(\boldsymbol{y})|.$$

6.2.1 Implementation

We seek

$$\widetilde{u}_n(\boldsymbol{x}) = \sum_{k=1}^{N} \alpha_k \varphi_k(\boldsymbol{x}) \tag{6.47}$$

with $\{\varphi_1, \ldots, \varphi_N\}$ a basis of Π_n. Then (6.42) is equivalent to

$$\sum_{k=1}^{N} \alpha_k \int_{\mathbb{B}^2} \left[\sum_{i,j=1}^{2} a_{i,j}(\boldsymbol{x}) \frac{\partial \varphi_k(\boldsymbol{x})}{\partial x_j} \frac{\partial \varphi_\ell(\boldsymbol{x})}{\partial x_i} + \gamma(\boldsymbol{x})\varphi_k(\boldsymbol{x})\varphi_\ell(\boldsymbol{x}) \right] |\det[J(\boldsymbol{x})]|\, d\boldsymbol{x}$$

$$= \int_{\mathbb{B}^2} f(\boldsymbol{x})\, \varphi_\ell(\boldsymbol{x}) \, |\det[J(\boldsymbol{x})]|\, d\boldsymbol{x}$$

$$+ \int_{\mathbb{S}^1} g(\boldsymbol{x})\, \varphi_\ell(\boldsymbol{x}) \, |J_{bdy}(\boldsymbol{x})|\, dS^1(\boldsymbol{x}), \qquad \ell = 1, \ldots, N. \tag{6.48}$$

The function $|J_{bdy}(\boldsymbol{x})|$ arises from the transformation of an integral over $\partial\Omega$ to one over $\mathbb{S}^1 = \partial\mathbb{B}^2$, associated with the change from ℓ_2 to $\widetilde{\ell}_2$ as discussed preceding (6.42). For example, $\partial\Omega$ is often represented as a mapping

$$\chi(\theta) = (\chi_1(\theta), \chi_2(\theta)), \qquad 0 \leq \theta \leq 2\pi.$$

In that case, $|J_{bdy}(\boldsymbol{x})|$ is simply $|\chi'(\theta)|$ and the associated integral is

$$\int_0^{2\pi} g(\chi(\theta))\, \varphi_\ell(\chi(\theta)) \, |\chi'(\theta)|\, d\theta. \tag{6.49}$$

The basis $\{\varphi_1, \ldots, \varphi_N\}$ is often taken to be orthonormal using the standard inner product of $L^2(\mathbb{B}^2)$. The reason for doing so is the empirical observation that the resulting linear system in (6.48) is then fairly well-conditioned. The setup of (6.48) requires calculating the orthonormal polynomials and their first partial derivatives; and following, the integrals in the linear system need to be evaluated, usually numerically. For the integrals over $\mathbb{B}^2$, we use the product Gauss method (5.77). For the one-variable integral in (6.49), we use the trapezoidal rule. In the following numerical illustration, the ridge polynomials (4.77) are used for the basis functions.

6.2.2   A Numerical Example

As an illustrative example, consider the mapping $s = \Phi(x)$,

$$s_1 = x_1 - x_2 + ax_1^2,$$
$$s_2 = x_1 + x_2.$$
(6.50)

Figure 6.7 shows the images in $\overline{\Omega}$ of the circles $r = j/10$, $j = 1, \ldots, 10$ and the azimuthal lines $\theta = j\pi/10$, $j = 1, \ldots, 20$.

   The following information is needed when implementing the transformation from $-\Delta u + \gamma u = f$ on Ω to a new equation on $\mathbb{B}^2$:

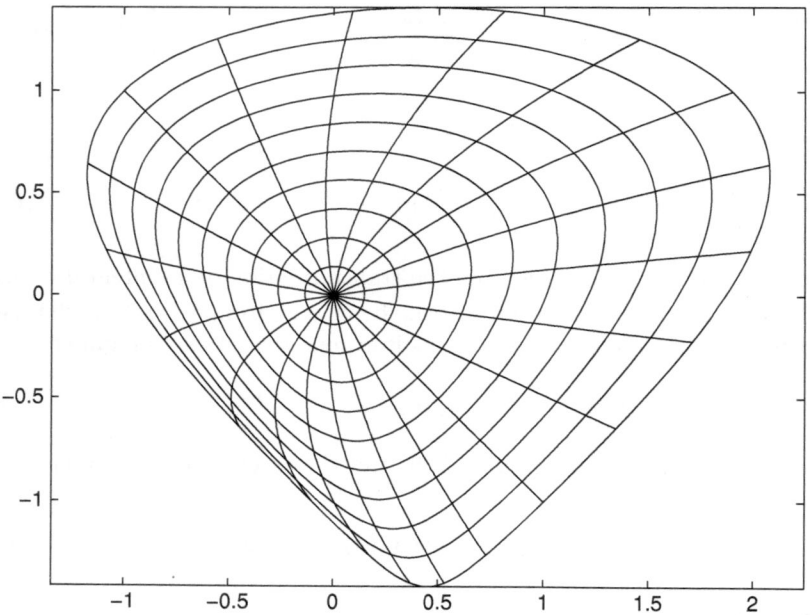

Fig. 6.7 Images of (6.50), with $a = 0.5$, for lines of constant radius and constant azimuth on the unit disk

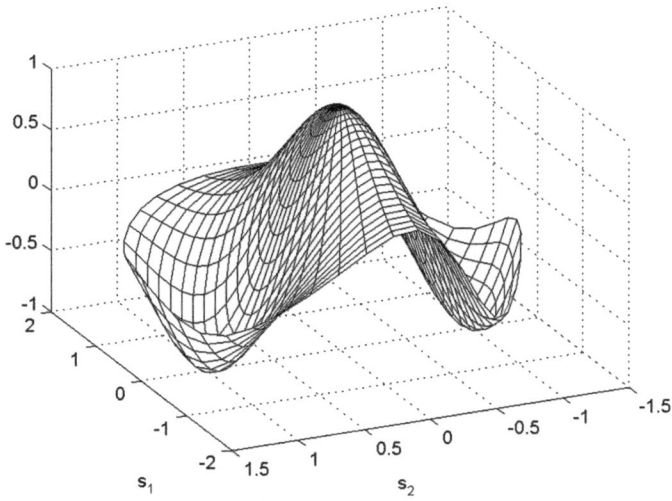

Fig. 6.8 The function $u(s, t)$ of (6.52)

$$D\Phi(\boldsymbol{x}) = J(x_1, x_2) = \begin{pmatrix} 1 + 2ax_1 & -1 \\ 1 & 1 \end{pmatrix},$$

$$\det(J) = 2\,(1 + ax_1),$$

$$J(\boldsymbol{x})^{-1} = \frac{1}{2\,(1 + ax_1)} \begin{pmatrix} 1 & 1 \\ -1 & 1 + 2ax_1 \end{pmatrix},$$

$$A(\boldsymbol{x}) = J(\boldsymbol{x})^{-1} J(\boldsymbol{x})^{-\mathrm{T}} = \frac{1}{2\,(1 + ax_1)^2} \begin{pmatrix} 1 & ax_1 \\ ax_1 & 2a^2 x_1^2 + 2ax_1 + 1 \end{pmatrix}.$$

The latter are the coefficients needed to define $\widetilde{\mathcal{A}}$ in (6.41).

   We give numerical results for solving the equation

$$-\Delta u(\boldsymbol{s}) + e^{s_1 - s_2} u(\boldsymbol{s}) = f(\boldsymbol{s}), \qquad \boldsymbol{s} \in \Omega. \qquad (6.51)$$

As a test case, we choose

$$u(\boldsymbol{s}) = e^{-s_1^2} \cos(\pi s_2), \qquad \boldsymbol{s} \in \Omega. \qquad (6.52)$$

The solution is pictured in Fig. 6.8. To find $f(\boldsymbol{s})$, we substitute this u into (6.51). We use the domain parameter $a = 0.5$, with Ω pictured in Fig. 6.7. The numerical integrals in the system (6.48) were evaluated using the product Gauss method (5.77) with a sufficiently large integration parameter.

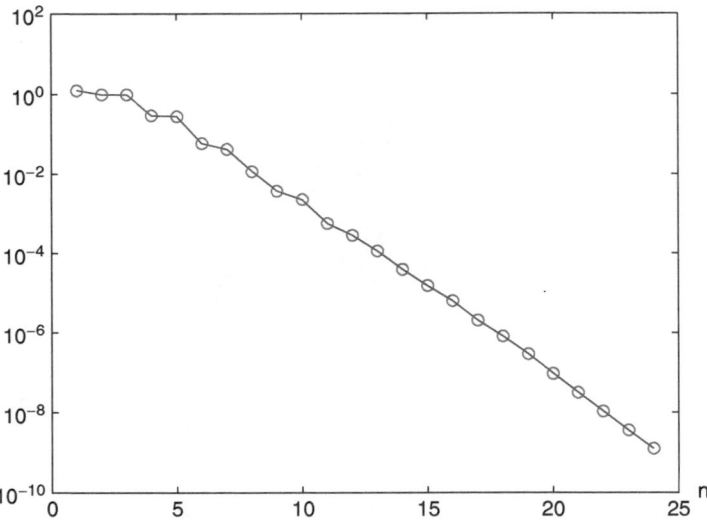

Fig. 6.9 The errors $\|u - u_n\|_\infty$ vs. the degree n

To calculate the error, we evaluate the numerical solution $\widetilde{u}_n$ and the analytical solution $\widetilde{u}$ on the grid

$$\Phi(x_{i,j}, y_{i,j}) = \Phi(r_i \cos\theta_j, r_i \sin\theta_j)$$

$$(r_i, \theta_j) = \left(\frac{i}{10}, \frac{j\pi}{10}\right), \qquad i = 0, 1, \ldots, 10; \quad j = 1, \ldots, 20.$$

The results are shown graphically in Fig. 6.9. The use of a semi-log scale demonstrates the exponential convergence of the method as the degree increases. For simplicity, we have chosen to use the uniform norm $\|u - u_n\|_\infty$ rather than the Sobolev norm $\|u - u_n\|_1$ of $H^1(\Omega)$ that arises in the theoretical error bounds of (6.45)–(6.46).

This spectral method is developed at length in the series of papers [12, 14, 15], for both Dirichlet and Neumann boundary conditions.

6.3 A Galerkin Method for a Beltrami-Type Equation

We take the following Beltrami-type equation

$$-\Delta^* u + c_0 u = f \quad \text{in } \mathbb{S}^{d-1} \tag{6.53}$$

as an example to develop the Galerkin method using spherical harmonics to solve differential equations posed over the unit sphere. It is possible to extend the discussion of the Galerkin method for solving equations on a portion of the sphere, but we limit ourselves to (6.53). For the given data, we assume $c_0 > 0$ and $f \in L^2(\mathbb{S}^{d-1})$. Note that in this section, the dimension parameter $d \geq 3$ is arbitrary.

Through a standard procedure, we arrive at the following weak formulation of (6.53): Find $u \in H^1(\mathbb{S}^{d-1})$ such that

$$(\nabla^* u, \nabla^* v)_{L^2(\mathbb{S}^{d-1})^{d-1}} + c_0 (u, v)_{L^2(\mathbb{S}^{d-1})} = (f, v)_{L^2(\mathbb{S}^{d-1})} \quad \forall\, v \in H^1(\mathbb{S}^{d-1}). \tag{6.54}$$

We apply the Lax–Milgram Lemma, Theorem 6.1, to show the solution existence and uniqueness of the problem (6.54). For this purpose, it is convenient to note that

$$\|v\|_{H^1(\mathbb{S}^{d-1})} := \left[(\nabla^* v, \nabla^* v)_{L^2(\mathbb{S}^{d-1})^{d-1}} + (v, v)_{L^2(\mathbb{S}^{d-1})} \right]^{1/2} \tag{6.55}$$

defines an equivalent norm over the Sobolev space $H^1(\mathbb{S}^{d-1})$. Let us demonstrate this equivalence. Expand the function v in terms of an orthonormal basis $\{Y_{n,j} : 1 \leq j \leq N_{n,d}\}$ of $\mathbb{Y}_n^d$ for each $n \in \mathbb{N}_0$:

$$v(\boldsymbol{\xi}) = \sum_{n=0}^{\infty} \sum_{j=1}^{N_{n,d}} v_{n,j} Y_{n,j}(\boldsymbol{\xi}), \quad v_{n,j} = (v, Y_{n,j})_{L^2(\mathbb{S}^{d-1})}, \quad \boldsymbol{\xi} \in \mathbb{S}^{d-1}. \tag{6.56}$$

From Definition 3.23, the norm on the space $H^1(\mathbb{S}^{d-1})$ is

$$\left[\sum_{n=0}^{\infty} \sum_{j=1}^{N_{n,d}} (n + \delta_d)^2 |v_{n,j}|^2 \right]^{1/2}. \tag{6.57}$$

Apply (3.19) to $Y_{n,j}$,

$$-\Delta_{(d-1)}^* Y_{n,j}(\boldsymbol{\xi}) = n (n + d - 2) Y_{n,j}(\boldsymbol{\xi}).$$

Multiply both sides by $Y_{m,k}(\boldsymbol{\xi})$, integrate over $\mathbb{S}^{d-1}$ and then perform an integration by parts to obtain

$$\int_{\mathbb{S}^{d-1}} \nabla^* Y_{n,j} \cdot \nabla^* Y_{m,k} \, dS^{d-1} = n (n + d - 2) \int_{\mathbb{S}^{d-1}} Y_{n,j} Y_{m,k} \, dS^{d-1}.$$

Therefore, we have the orthogonality of the gradient of the spherical harmonic basis functions:

$$\int_{\mathbb{S}^{d-1}} \nabla^* Y_{n,j} \cdot \nabla^* Y_{m,k} \, dS^{d-1} = n (n + d - 2) \delta_{nm} \delta_{jk}. \tag{6.58}$$

Take the gradient of the equality (6.56),

$$\nabla^* v(\boldsymbol{\xi}) = \sum_{n=0}^{\infty} \sum_{j=1}^{N_{n,d}} v_{n,j} \nabla^* Y_{n,j}(\boldsymbol{\xi}).$$

Then use the orthogonality relation (6.58) to find

$$\int_{\mathbb{S}^{d-1}} |\nabla^* v|^2 \, dS^{d-1} = \sum_{n=0}^{\infty} \sum_{j=1}^{N_{n,d}} n\,(n+d-2) \, |v_{n,j}|^2. \qquad (6.59)$$

Thus, we have the following expression for the norm $\|v\|_{H^1(\mathbb{S}^{d-1})}$ defined by (6.55):

$$\|v\|_{H^1(\mathbb{S}^{d-1})} = \left\{ \sum_{n=0}^{\infty} \sum_{j=1}^{N_{n,d}} [n\,(n+d-2) + 1] \, |v_{n,j}|^2 \right\}^{1/2}.$$

From this expression, the equivalence of the two norms defined in (6.55) and (6.57) is evident.

Take $V = H^1(\mathbb{S}^{d-1})$ with the norm $\|\cdot\|_{H^1(\mathbb{S}^{d-1})}$ and

$$a(u,v) = (\nabla^* u, \nabla^* v)_{L^2(\mathbb{S}^{d-1})^{d-1}} + c_0\,(u,v)_{L^2(\mathbb{S}^{d-1})},$$

$$\ell(v) = (f,v)_{L^2(\mathbb{S}^{d-1})}.$$

It is then a routine matter to verify all the conditions stated in Theorem 6.1 are valid. Therefore, the problem (6.54) has a unique solution $u \in V$. Moreover, it can be shown that actually $u \in H^2(\mathbb{S}^{d-1})$ and the equality (6.53) holds a.e. in $\mathbb{S}^{d-1}$.

To develop a Galerkin method with spherical harmonic basis functions, let

$$V_N = \text{span}\,\{Y_{n,j} : 1 \le j \le N_{n,d},\ 0 \le n \le N\}, \qquad N \in \mathbb{N}$$

and consider the problem of finding $u_N \in V_N$ such that

$$(\nabla^* u_N, \nabla^* v)_{L^2(\mathbb{S}^{d-1})^{d-1}} + c_0\,(u_N, v)_{L^2(\mathbb{S}^{d-1})} = (f,v)_{L^2(\mathbb{S}^{d-1})} \qquad \forall\, v \in V_N. \tag{6.60}$$

Theorem 6.1 can be applied again to conclude that the Galerkin scheme (6.60) admits a unique solution $u_N \in V_N$.

For convergence and error estimation of the Galerkin solution u_N, we apply the Cea's Lemma [13, Proposition 9.1.3],

$$\|u - u_N\|_{H^1(\mathbb{S}^{d-1})} \le c \inf \left\{ \|u - v\|_{H^1(\mathbb{S}^{d-1})} : v \in V_N \right\} \tag{6.61}$$

for some constant $c > 0$ independent of N and u.

To proceed further, we explore the solution smoothness for (6.53). For the right side of the equation, assume

$$f \in H^s(\mathbb{S}^{d-1}), \quad s \geq 0. \tag{6.62}$$

Write

$$f(\boldsymbol{\xi}) = \sum_{n=0}^{\infty} \sum_{j=1}^{N_{n,d}} f_{n,j} Y_{n,j}(\boldsymbol{\xi}), \quad f_{n,j} = (f, Y_{n,j})_{L^2(\mathbb{S}^{d-1})}, \ \boldsymbol{\xi} \in \mathbb{S}^{d-1}$$

and similarly for the solution u:

$$u(\boldsymbol{\xi}) = \sum_{n=0}^{\infty} \sum_{j=1}^{N_{n,d}} u_{n,j} Y_{n,j}(\boldsymbol{\xi}), \quad u_{n,j} = (u, Y_{n,j})_{L^2(\mathbb{S}^{d-1})}, \ \boldsymbol{\xi} \in \mathbb{S}^{d-1}.$$

The assumption (6.62) implies

$$\|f\|_{H^s(\mathbb{S}^{d-1})} = \left[\sum_{n=0}^{\infty} \sum_{j=1}^{N_{n,d}} (n + \delta_d)^{2s} |f_{n,j}|^2 \right]^{1/2} < \infty. \tag{6.63}$$

Since

$$-\Delta^* u(\boldsymbol{\xi}) = \sum_{n=0}^{\infty} \sum_{j=1}^{N_{n,d}} n \, (n + d - 2) \, u_{n,j} Y_{n,j}(\boldsymbol{\xi}),$$

from (6.53) we get the equality

$$\sum_{n=0}^{\infty} \sum_{j=1}^{N_{n,d}} [n \, (n + d - 2) + 1] \, u_{n,j} Y_{n,j}(\boldsymbol{\xi}) = \sum_{n=0}^{\infty} \sum_{j=1}^{N_{n,d}} f_{n,j} Y_{n,j}(\boldsymbol{\xi}).$$

Therefore,

$$u_{n,j} = \frac{1}{n \, (n + d - 2) + 1} f_{n,j}, \quad 1 \leq j \leq N_{n,d}, \ n = 0, 1, \ldots.$$

The condition (6.63) then implies

$$\|u\|_{H^{s+2}(\mathbb{S}^{d-1})} = \left[\sum_{n=0}^{\infty} \sum_{j=1}^{N_{n,d}} (n + \delta_d)^{2\,(s+2)} |u_{n,j}|^2 \right]^{1/2} < \infty$$

and we have the solution regularity

$$u \in H^{s+2}(\mathbb{S}^{d-1}), \quad \|u\|_{H^{s+2}(\mathbb{S}^{d-1})} \leq c \, \|f\|_{H^s(\mathbb{S}^{d-1})} \tag{6.64}$$

for some constant $c > 0$ independent of f.

Thus, under the assumption (6.62), we have (6.64). Combining (6.61), (6.64), and (3.105), we deduce the following error bound for the Galerkin solution

$$\|u - u_N\|_{H^1(\mathbb{S}^{d-1})} \leq \frac{c}{N^{s+1}} \, \|f\|_{H^s(\mathbb{S}^{d-1})}, \tag{6.65}$$

where the constant $c > 0$ is independent of f and N. It can be shown that the convergence order $(s + 1)$ is optimal under the smoothness assumption (6.62) on the right side f of (6.53).

For a further discussion of solving elliptic partial differential equations on a sphere, see [72].

References

1. M. Abramowitz, I. Stegun (eds.), *Handbook of Mathematical Functions* (Dover Publications, New York, 1965)
2. L.V. Ahlfors, *Complex Analysis*, 2nd edn. (McGraw-Hill, New York, 1966)
3. C. Ahrens, G. Beylkin, Rotationally invariant quadratures for the sphere. Proc. Roy. Soc. A **465**, 3103–3125 (2009)
4. C. An, X. Chen, I. Sloan, R. Womersley, Well conditioned spherical designs for integration and interpolation on the two-sphere. SIAM J. Numer. Anal. **48**, 2135–2157 (2010)
5. G.E. Andrews, R. Askey, R. Roy, *Special Functions* (Cambridge University Press, Cambridge, 1999)
6. S.R. Arridge, Optical tomography in medical imaging. Inverse Probl. **15**, R41–R93 (1999)
7. S.R. Arridge, J.C. Schotland, Optical tomography: forward and inverse problems. Inverse Probl. **25**, 123010-1-59 (2009)
8. K. Atkinson, The numerical solution of Laplace's equation in three dimensions. SIAM J. Numer. Anal. **19**, 263–274 (1982)
9. K. Atkinson, *An Introduction to Numerical Analysis*, 2nd edn. (Wiley, New York, 1989)
10. K. Atkinson, *The Numerical Solution of Integral Equations of the Second Kind* (Cambridge University Press, Cambridge, 1997)
11. K. Atkinson, Quadrature of singular integrands over surfaces. Electr. Trans. Numer. Anal. **17**, 133–150 (2004)
12. K. Atkinson, D. Chien, O. Hansen, A spectral method for elliptic equations: the Dirichlet problem. Adv. Comput. Math. **33**, 169–189 (2010)
13. K. Atkinson, W. Han, *Theoretical Numerical Analysis*, 3rd edn. (Springer, Berlin, 2009)
14. K. Atkinson, O. Hansen, A spectral method for the eigenvalue problem for elliptic equations. Electron. Trans. Numer. Anal. **37**, 386–412 (2010)
15. K. Atkinson, O. Hansen, D. Chien, A spectral method for elliptic equations: the Neumann problem. Adv. Comput. Math. **34**, 295–317 (2011)
16. K. Atkinson, A. Sommariva, Quadrature over the sphere. Electr. Trans. Numer. Anal. **20**, 104–118 (2005)
17. J. Avery, A formula for angular and hyperangular integration. J. Math. Chem. **24**, 169–174 (1998)
18. T. Bagby, L. Bos, N. Levenberg, Multivariate simultaneous approximation. Constr. Approx. **18**, 569–577 (2002)

K. Atkinson and W. Han, *Spherical Harmonics and Approximations on the Unit Sphere: An Introduction*, Lecture Notes in Mathematics 2044, DOI 10.1007/978-3-642-25983-8, © Springer-Verlag Berlin Heidelberg 2012

19. G. Bal, Inverse transport theory and applications. Inverse Probl. **25**, 053001, 48 (2009)
20. Eiichi Bannai and Etsuko Bannai, A survey on spherical designs and algebraic combinatorics on spheres. Eur. J. Combinator. **30**, 1392–1425 (2009)
21. M. de Berg, M. van Kreveld, M. Overmars, O. Schwarzkopf, *Computational Geometry: Algorithms and Applications* (Springer, Berlin, 1997)
22. J. Boyd, *Chebyshev and Fourier Spectral Methods*, 2nd edn. (Dover Publiations, New York, 2000)
23. G. Brown, F. Dai, Approximation of smooth functions on compact two-point homogeneous spaces. J. Funct. Anal. **220**, 401–423 (2005)
24. M. Buhmann, in *Acta Numerica*. Radial Basis Functions (Cambridge University Press, Cambridge, 2000), pp. 1–38
25. M. Buhmann, *Radial Basis Functions* (Cambridge University Press, Cambridge, 2003)
26. M. Burkardt, Delaunay triangulation of points on the unit sphere. http://people.sc. fsu.edu/~burkardt/m_src/sphere_delaunay/sphere_delaunay.html, 2011
27. C. Canuto, A. Quarteroni, My. Hussaini, T. Zang, *Spectral Methods in Fluid Mechanics* (Springer, New York, 1988)
28. C. Canuto, A. Quarteroni, My. Hussaini, T. Zang, *Spectral Methods: Fundamentals in Single Domains* (Springer, New York, 2006)
29. K.M. Case, P.F. Zweifel, *Linear Transport Theory* (Addison-Wesley, Reading, 1967)
30. S. Chandrasekhar, *Radiative Transfer* (Dover Publications, New York, 1960)
31. Y. Chen, *Galerkin Methods for Solving Single Layer Integral Equations in Three Dimensions*, Ph.D. thesis, University of Iowa, 1994
32. R. Cools, An encyclopaedia of cubature formulas. J. Complex. **19**, 445–453 (2003)
33. R. Cools, I. Mysovskikh, H. Schmid, Cubature formulae and orthogonal polynomials. J. Comp. Appl. Math. **127**, 121–152 (2001)
34. J. Cui, W. Freeden, Equidistribution on the sphere. SIAM J. Sci. Comp. **18**, 595–609 (1997)
35. F. Dai, Jackson-type inequality for doubling weights on the sphere. Constr. Approx. **24**, 91–112 (2006)
36. F. Dai, Y. Xu, Moduli of smoothness and approximation on the unit sphere and the unit ball. Adv. Math. **224**, 1233–1310 (2010)
37. F. Dai, Y. Xu, Polynomial approximation in Sobolev spaces on the unit sphere and the unit ball, J. Approx. Thy. **163**, 1400–1418 (2011)
38. B. Davison, *Neutron Transport Theory* (Oxford University Press, Oxford, 1957)
39. Z. Ditzian, A modulus of smoothness on the unit sphere. J. Anal. Math. **79**, 189–200 (1999)
40. Z. Ditzian, Jackson-type inequality on the sphere. Acta Math. Hungar. **102**, 1–35 (2004)
41. E. Doha, W. Abd-Elhameed, Efficient spectral-Galerkin algorithms for direct solution of second-order equations using ultraspherical polynomials. SIAM J. Sci. Comput. **24**, 548–571 (2002)
42. J. Driscoll, D. Healy, Computing Fourier transforms and convolutions on the 2-sphere. Adv. Appl. Math. **15**, 202–250 (1994)
43. J.J. Duderstadt, W.R. Martin, *Transport Theory* (Wiley, New York, 1979)
44. C. Dunkl, Y. Xu, *Orthogonal Polynomials of Several Variables* (Cambridge University Press, Cambridge, 2001)
45. G. Fasshauer, *Meshfree Approximation Methods with MATLAB* (World Scientific Publishing, Singapore, 2007)
46. G. Fasshauer, L. Schumaker, in *Mathematical Methods for Curves and Surface II*, ed. by M. Dæhlen, T. Lyche, L. Schumaker. Scattered data fitting on the sphere (Vanderbilt University Press, Nashville, 1998)
47. W. Freeden, T. Gervens, M. Schreiner, *Constructive Approximation on the Sphere: With Applications to Geomathematics* (Oxford Science Publications, Oxford, 1998)
48. W. Freeden, V. Michel, *Multiscale Potential Theory: With Applications to Geoscience* (Birkhäuser, Boston, 2004)

49. W. Freeden, M. Schreiner, *Special Functions of Mathematical Geosciences: A Scalar, Vectorial, and Tensorial Setup* (Springer, Berlin, 2009)

50. M. Ganesh, I. Graham, J. Sivaloganathan, A pseudospectral three-dimensional boundary integral method applied to a nonlinear model problem from finite elasticity. SIAM J. Numer. Anal. **31**, 1378–1414 (1994)

51. M. Ganesh, I. Graham, J. Sivaloganathan, A new spectral boundary integral collocation method for three-dimensional potential problems. SIAM J. Numer. Anal. **35**, 778–805 (1998)

52. W. Gautschi, *Orthogonal Polynomials: Computation and Approximation* (Oxford University Press, Oxford, 2004)

53. I. Graham, I. Sloan, Fully discrete spectral boundary integral methods for Helmholtz problems on smooth closed surfaces in $\mathbb{R}^3$. Numer. Math. **92**, 289–323 (2002)

54. T. Gronwall, On the degree of convergence of Laplace's series. Trans. Amer. Math. Soc. **15**, 1–30 (1914)

55. W. Han, K. Atkinson, H. Zheng, Some integral identities for spherical harmonics in an arbitrary dimension, to appear in J. of Math. Chem.

56. W. Han, J. Eichholz, J. Huang, J. Lu, RTE based bioluminescence tomography: a theoretical study, Inverse Probl. Sci. Eng. **19**, 435–459 (2011)

57. W. Han, J. Huang, J. Eichholz, Discrete-ordinate discontinuous Galerkin methods for solving the radiative transfer equation. SIAM J. Sci. Comput. **32**, 477–497 (2010)

58. O. Hansen, K. Atkinson, D. Chien, On the norm of the hyperinterpolation operator on the unit disk. IMA J. Numer. Anal. **29**, 257–283 (2009)

59. D. Hardin, E. Saff, Discretizing manifolds via minimum energy points. Not. AMS **51**, 1186–1193 (2004)

60. P. Henrici, Fast Fourier methods in computational complex analysis. SIAM Rev. **21**, 481–527 (1979)

61. P. Henrici, *Applied and Computational Complex Analysis*, vol. III (Wiley, New York, 1986)

62. L. Henyey, J. Greenstein, Diffuse radiation in the galaxy. Astrophys. J. **93**, 70–83 (1941)

63. K. Hesse, I.H. Sloan, R. Womersley, in *Handbook of Geomathematics*. Numerical Integration on the Sphere, Chapter 40 (Springer, Berlin, 2010)

64. J. Hesthaven, S. Gottlieb, D. Gottlieb, *Spectral Methods for Time-Dependent Problems* (Cambridge University Press, Cambridge, 2007)

65. H. Hochstadt, *Integral Equations* (Wiley, New York, 1973)

66. M. Jaswon, G. Symm, *Integral Equation Methods in Potential Theory and Elastostatics* (Academic, London, 1977)

67. P. Keast, J. Diaz, Fully symmetric integration formulas for the surface of the sphere in *s* dimensions. SIAM J. Num. Anal. **20**, 406–419 (1983)

68. J. Keiner, D. Potts, Fast evaluation of quadrature formulae on the sphere. Math. Comput. **77**, 397–419 (2008)

69. V.A. Kozlov, V.G. Maz'ya, J. Rossmann, *Spectral Problems Associated with Corner Singularities of Solutions to Elliptic Equations* (AMS, Providence, 2001)

70. R. Kress, *Linear Integral Equations*, 2nd edn. (Springer, Berlin, 1999)

71. M.-J. Lai, L.L. Schumaker, *Spline Functions on Triangulations* (Cambridge University Press, Cambridge, 2007)

72. Q. Le Gia, Galerkin approximation for elliptic PDEs on spheres. J. Approx. Thy. **130**, 123–147 (2004)

73. V. Lebedev, Quadratures on a sphere. USSR Comp. Math. Math. Physics **16**, 10–24 (1976)

74. E.E. Lewis, W.F. Miller, *Computational Methods of Neutron Transport* (Wiley, New York, 1984)

75. J. Lyness, D. Jespersen, Moderate degree symmetric quadrature rules for the triangle. J. Inst. Math. Appl. **15**, 19–32 (1975)

76. B. Logan, L. Shepp, Optimal reconstruction of a function from its projections. Duke Math. J. **42**, 645–659 (1975)
77. G. Lorentz, *Approximation of Functions*, 2nd edn. (Chelsea Publishing Co., New York, 1986)
78. T. MacRobert, *Spherical Harmonics*, 3rd edn. (Pergamon Press, Oxford, 1967)
79. A. McLaren, Optimal integration on a sphere. Math. Comput. **17**, 361–383 (1963)
80. W. McLean, *Strongly Elliptic Systems and Boundary Integral Equations* (Cambridge University Press, Cambridge, 2000)
81. S.G. Mikhlin, *Mathematical Physics: An Advanced Course* (North-Holland, Amsterdam, 1970)
82. M.F. Modest, *Radiative Heat Transfer*, 2nd edn. (Academic, London, 2003)
83. M. Mohlenkamp, A fast transform for spherical harmonics. J. Fourier Anal. Appl. **5**, 159–184 (1999)
84. C. Müller, *Spherical Harmonics*. Lecture Notes in Mathematics, vol. 17 (Springer, New York, 1966)
85. C. Müller, *Analysis of Spherical Symmetries in Euclidean Spaces* (Springer, New York, 1998)
86. F. Natterer, F. Wübbeling, *Mathematical Methods in Image Reconstruction* (SIAM, Philadelphia, 2001)
87. M. Neamtu, L. Schumaker, On the approximation order of splines on spherical triangulations. Adv. Comp. Math. **21**, 3–20 (2004)
88. D. Newman, H. Shapiro, in *Proceedings Conference on Approximation Theory* (Oberwolfach, 1963). Jackson's Theorem in Higher Dimensions (Birkhäuser, Basel), pp. 208–219 (1964)
89. F. Olver, D. Lozier, R. Boisvert, C. Clark (eds.), *NIST Handbook of Mathematical Functions* (Cambridge University Press, Cambridge, 2010)
90. S. Orszag, Fourier series on spheres. Mon. Weather Rev. **102**, 56–75 (1974)
91. M. Powell, *Approximation Theory and Methods* (Cambridge University Press, Cambridge, 1981)
92. A. Quarteroni, R. Sacco, F. Saleri, *Numerical Mathematics* (Springer, Berlin, 2000)
93. D. Ragozin, Constructive polynomial approximation on spheres and projective spaces. Trans. Amer. Math. Soc. **162**, 157–170 (1971)
94. D. Ragozin, Uniform convergence of spherical harmonic expansions. Math. Annalen **195**, 87–94 (1972)
95. M. Reimer, Interpolation on the sphere and bounds for the Lagrangian square sums. Results Math. **11**, 144–166 (1987)
96. R. Renka, Algorithm 772: STRIPACK: Delaunay triangulation and Voronoi diagram on the surface of a sphere. ACM Trans. Math. Soft. **23**, 416–434 (1997)
97. T. Rivlin, *An Introduction to the Approximation of Functions* (Dover Publications, New York, 1981)
98. V. Rokhlin, M. Tygert, Fast algorithms for spherical harmonic expansions. SIAM J. Sci. Comput. **27**, 1903–1928 (2006)
99. K. Rustamov, On approximation of functions on the sphere. Russ. Acad. Sci. Izv. Math. **43**, 311–329 (1994)
100. E. Saff, A. Kuijlaars, Distributing many points on a sphere. Math. Intel. **19**, 5–11 (1997)
101. I. Schoenberg, Positive definite functions on the sphere. Duke Math. J. **9**, 96–108 (1942)
102. J. Shen, T. Tang, *Spectral and High-Order Methods with Applications* (Science Press, Beijing, 2006)
103. J. Shen, L. Wang, Analysis of a spectral-Galerkin approximation to the Helmholtz equation in exterior domains. SIAM J. Numer. Anal. **45**, 1954–1978 (2007)
104. A. Sidi, *Practical Extrapolation Methods: Theory and Applications* (Cambridge University Press, Cambridge, 2003)

105. A. Sidi, Analysis of Atkinson's variable transformation for numerical integration over smooth surfaces in $\mathbb{R}^3$. Numer. Math. **100**, 519–536 (2005)

106. I.H. Sloan, R. Womersley, Constructive polynomial approximation on the sphere. J. Approx. Thy. **103**, 91–118 (2000)

107. I.H. Sloan, R. Womersley, Good approximation on the sphere, with application to geodesy and the scattering of sound. J. Comp. Appl. Math. **149**, 227–237 (2002)

108. I.H. Sloan, R. Womersley, Extremal systems of points and numerical integration on the sphere. Adv. Comp. Math. **21**, 107–125 (2004)

109. S. Sobolev, Cubature formulas on the sphere invariant under finite groups of rotations. Sov. Math. **3**, 1307–1310 (1962)

110. A. Sommariva, R. Womersley, Integration by RBF over the sphere, Applied Mathematics report AMR 05/17, University of NSW, Sydney, 2005

111. A. Stroud, *Approximate Calculation of Multiple Integrals* (Prentice-Hall, Englewood Cliffs, 1971)

112. P. Swarztrauber, On the spectral approximation of discrete scalar and vector functions on the sphere. SIAM J. Numer. Anal. **16**, 934–949 (1979)

113. P. Swarztrauber, The approximation of vector functions and their derivatives on the sphere. SIAM J. Numer. Anal. **18**, 191–210 (1981)

114. P. Swarztrauber, W. Spotz, Generalized discrete spherical harmonic transforms. J. Comput. Phys. **159**, 213–230 (2000)

115. G. Szegö, *Orthogonal Polynomials* (AMS, Providence, 1939)

116. G.E. Thomas, K. Stamnes, *Radiative Transfer in the Atmosphere and Ocean* (Cambridge University Press, Cambridge, 1999)

117. E. Weisstein, *CRC Concise Encyclopedia of Mathematics*, 2nd edn. (Chapman & Hall/CRC, Boca Raton, 2003)

118. H. Wendland, *Scattered Data Approximation* (Cambridge University Press, Cambridge, 2010)

119. R. Womersley, Extremal (Maximum Determinant) points on the sphere $\mathbb{S}^2$. http:// web.maths.unsw.edu.au/~rsw/Sphere/Extremal/New/extremal1.html, 2011

120. Y. Xu, Representation of reproducing kernels and the Lebesgue constants on the ball. J. Approx. Thy. **112**, 295–310 (2001)

121. Y. Xu, in *Advances in the Theory of Special Functions and Orthogonal Polynomials*. Lecture Notes on Orthogonal Polynomials of Several Variables (Nova Science Publishers, New York, 2004), pp. 135–188

122. W. Zdunkowski, T. Trautmann, A. Bott, *Radiation in the Atmosphere: A Course in Theoretical Meteorology* (Cambridge University Press, Cambridge, 2007)

123. A. Zygmund, *Trigonometric Series*, vols. I and II (Cambridge University Press, Cambridge, 1959)

Index

K. Atkinson and W. Han, *Spherical Harmonics and Approximations on the Unit Sphere: An Introduction*, Lecture Notes in Mathematics 2044, DOI 10.1007/978-3-642-25983-8, © Springer-Verlag Berlin Heidelberg 2012

LECTURE NOTES IN MATHEMATICS  Springer

Edited by J.-M. Morel, B. Teissier; P.K. Maini

Editorial Policy (for the publication of monographs)

1. Lecture Notes aim to report new developments in all areas of mathematics and their applications - quickly, informally and at a high level. Mathematical texts analysing new developments in modelling and numerical simulation are welcome.

   Monograph manuscripts should be reasonably self-contained and rounded off. Thus they may, and often will, present not only results of the author but also related work by other people. They may be based on specialised lecture courses. Furthermore, the manuscripts should provide sufficient motivation, examples and applications. This clearly distinguishes Lecture Notes from journal articles or technical reports which normally are very concise. Articles intended for a journal but too long to be accepted by most journals, usually do not have this "lecture notes" character. For similar reasons it is unusual for doctoral theses to be accepted for the Lecture Notes series, though habilitation theses may be appropriate.

2. Manuscripts should be submitted either online at www.editorialmanager.com/lnm to Springer's mathematics editorial in Heidelberg, or to one of the series editors. In general, manuscripts will be sent out to 2 external referees for evaluation. If a decision cannot yet be reached on the basis of the first 2 reports, further referees may be contacted: The author will be informed of this. A final decision to publish can be made only on the basis of the complete manuscript, however a refereeing process leading to a preliminary decision can be based on a pre-final or incomplete manuscript. The strict minimum amount of material that will be considered should include a detailed outline describing the planned contents of each chapter, a bibliography and several sample chapters.

   Authors should be aware that incomplete or insufficiently close to final manuscripts almost always result in longer refereeing times and nevertheless unclear referees' recommendations, making further refereeing of a final draft necessary.

   Authors should also be aware that parallel submission of their manuscript to another publisher while under consideration for LNM will in general lead to immediate rejection.

3. Manuscripts should in general be submitted in English. Final manuscripts should contain at least 100 pages of mathematical text and should always include

   - a table of contents;
   - an informative introduction, with adequate motivation and perhaps some historical remarks: it should be accessible to a reader not intimately familiar with the topic treated;
   - a subject index: as a rule this is genuinely helpful for the reader.

   For evaluation purposes, manuscripts may be submitted in print or electronic form (print form is still preferred by most referees), in the latter case preferably as pdf- or zipped psfiles. Lecture Notes volumes are, as a rule, printed digitally from the authors' files. To ensure best results, authors are asked to use the LaTeX2e style files available from Springer's web-server at:

   ftp://ftp.springer.de/pub/tex/latex/svmonot1/ (for monographs) and
   ftp://ftp.springer.de/pub/tex/latex/svmultt1/ (for summer schools/tutorials).

Additional technical instructions, if necessary, are available on request from lnm@springer.com.

4. Careful preparation of the manuscripts will help keep production time short besides ensuring satisfactory appearance of the finished book in print and online. After acceptance of the manuscript authors will be asked to prepare the final LaTeX source files and also the corresponding dvi-, pdf- or zipped ps-file. The LaTeX source files are essential for producing the full-text online version of the book (see http://www.springerlink.com/openurl.asp?genre=journal&issn=0075-8434 for the existing online volumes of LNM). The actual production of a Lecture Notes volume takes approximately 12 weeks.

5. Authors receive a total of 50 free copies of their volume, but no royalties. They are entitled to a discount of 33.3 % on the price of Springer books purchased for their personal use, if ordering directly from Springer.

6. Commitment to publish is made by letter of intent rather than by signing a formal contract. Springer-Verlag secures the copyright for each volume. Authors are free to reuse material contained in their LNM volumes in later publications: a brief written (or e-mail) request for formal permission is sufficient.

Addresses:
Professor J.-M. Morel, CMLA,
École Normale Supérieure de Cachan,
61 Avenue du Président Wilson, 94235 Cachan Cedex, France
E-mail: morel@cmla.ens-cachan.fr

Professor B. Teissier, Institut Mathématique de Jussieu,
UMR 7586 du CNRS, Équipe "Géométrie et Dynamique",
175 rue du Chevaleret
75013 Paris, France
E-mail: teissier@math.jussieu.fr

For the "Mathematical Biosciences Subseries" of LNM:

Professor P. K. Maini, Center for Mathematical Biology,
Mathematical Institute, 24-29 St Giles,
Oxford OX1 3LP, UK
E-mail : maini@maths.ox.ac.uk

Springer, Mathematics Editorial, Tiergartenstr. 17,
69121 Heidelberg, Germany,
Tel.: +49 (6221) 4876-8259

Fax: +49 (6221) 4876-8259
E-mail: lnm@springer.com